advanced

LEVEL study aids

chemistry

Derek McMonagle

053121

JOHN MURRAY

Titles in this series:
Advanced Level Study Aids Biology   0 7195 7630 X
Advanced Level Study Aids Chemistry   0 7195 7631 8
Advanced Level Study Aids Physics   0 7195 7629 6

First published in 2002
by John Murray (Publishers) Ltd
50 Albemarle Street
London W1S 4BD

Layouts by Amanda Easter
Artwork by Oxford Designers & Illustrators
Cover design by John Townson/Creation

Typeset in 10/12pt Garamond by Wearset Ltd, Boldon, Tyne and Wear
Printed and bound in Great Britain by St Edmundsbury Press, Bury St Edmunds

A catalogue entry for this title is available from the British Library

ISBN 0 7195 7631 8

Contents

Acknowledgements

Especially for Jess, Ben, Dom and Becky.

Exam questions, answers and mark schemes have been reproduced with kind permission from the following examination boards:

AQA: The Association Examining Board (AEB)
       Northern Examining and Assessment Board
       (NEAB)
Edexcel, including London Examinations
OCR (Oxford, Cambridge and RSA Examinations)
WJEC (Welsh Joint Education Committee).

AS and A level specification matrix

Board	AQA					Edexcel				OCR						WJEC			
Module	1	2	3	4	5	1	2	4	5	2811	2812	2813	2814	2815	2816	CH1	CH2	CH4	CH5
1	✓					✓				✓						✓			
2	✓					✓				✓						✓			
3	✓					✓				✓						✓			
4	✓					✓				✓						✓			
5	✓					✓				✓						✓			
6	✓					✓				✓						✓			
7	✓					✓				✓						✓			
8	✓					✓				✓							✓		
9			✓				✓				✓						✓		
10			✓				✓				✓						✓		
11			✓				✓				✓						✓		
12			✓				✓				✓							✓	
13	✓					✓				✓						✓			
14		✓				✓				✓						✓			
15		✓					✓					✓					✓		
16		✓					✓					✓					✓		
17					✓				✓					✓					✓
18					✓				✓					✓					✓
19				✓				✓							✓	✓			
20		✓		✓			✓		✓			✓			✓	✓			✓
21			✓	✓			✓		✓		✓		✓					✓	
22				✓				✓					✓					✓	
23				✓					✓				✓					✓	
24				✓					✓				✓					✓	
25					✓			✓						✓					✓
26				✓				✓							✓		✓		
27					✓				✓										
28					✓				✓					✓					✓
29				✓					✓		✓		✓					✓	

*Modules in tinted columns make up the A2 part of the course

Introduction

This book is a Study Aid and not simply a revision guide to be used at the end of an A level Chemistry course.

The content of the A level Chemistry specification is divided into a series of short topics. After this there are some simple multiple choice questions, followed by questions requiring both short and extended answers. Full answers are provided to all the questions. This book can be used to complement the lessons and course notes of the A level Chemistry course you may be following. Each topic contains all the fundamental knowledge you will need to know with key words and terms in **bold**. Equally importantly, you are provided with an opportunity to test what you have learnt.

This Study Aid will also be useful to you at the conclusion of your course. There is a section containing synoptic questions, some written in the same style as those that follow each topic, and some taken from recent examination papers of all the awarding bodies. Guidance is given on which topics to consult in answering the examination questions and full answers to all questions are given.

The index contains words and terms that you will come across within your course and indicates on which page(s) each is to be found.

1 Atomic structure

Atom

An atom is the smallest particle of an element that can take part in a chemical reaction.

Sub-atomic particles

Atoms are composed of three sub-atomic particles: **protons**, **neutrons** and **electrons** (Table 1.1).

Table 1.1 Mass and charge of sub-atomic particles

Particle	Relative atomic mass	Relative charge
electron	$\frac{1}{1836}$	-1
neutron	1	0
proton	1	$+1$

Although the proton and electron are very different in mass, their charges are exactly equal but opposite.

Atomic structure

Almost all atoms have the same basic structure: a central **nucleus** of protons and neutrons that is surrounded by a number of electrons (Figure 1.1). The only exception to this is the commonest form of hydrogen, which has no neutrons in its nucleus.

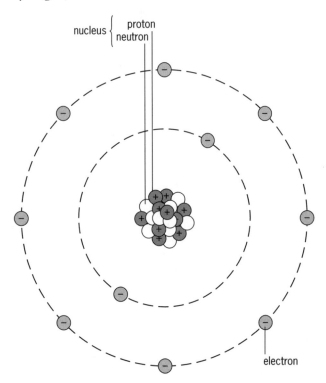

Figure 1.1 The structure of an atom

Most of the mass of an atom is contained in its nucleus.

The number of protons in the nucleus of an atom is always equal to the number of electrons around the nucleus. Atoms have no overall charge.

Atomic number and mass number

Atomic number, Z, and mass number, A, are used to describe an atom.

- The **atomic number** (or proton number) of an atom is the number of protons contained in its nucleus.
- The **mass number** of an atom is the total number of protons and neutrons contained in its nucleus. (As electrons have a much smaller mass than the other sub-atomic particles, they are ignored in the calculation.)

Atomic number and mass number do not have any units.

Figure 1.2 represents an atom of lithium. The atomic number $Z = 3$ and the mass number $A = 3 + 4 = 7$.

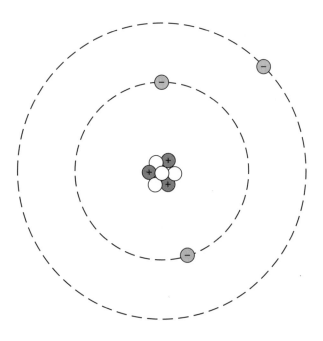

Figure 1.2 The structure of a lithium atom

Element

An **element** is a substance that cannot be decomposed into simpler substances.

Elements can be written using their mass number and atomic number and the atomic symbol for the element.

$$^{\text{mass number}}_{\text{atomic number}}\text{Symbol}$$

For example:

$$^{4}_{2}\text{He} \quad ^{14}_{7}\text{N} \quad ^{27}_{13}\text{Al} \quad ^{32}_{16}\text{S} \quad ^{127}_{53}\text{I}$$

Isotopes

All atoms of an element have the same atomic number. However, they may not have the same mass number since the number of neutrons in the nucleus may not always be the same.

Atoms of an element that have different mass numbers are called **isotopes**.

Isotopes can be written using their mass number and atomic number and the atomic symbol for the element. However, since all isotopes of an element have the same atomic number, isotopes are often represented using only their mass number.

They are frequently written as 'symbol-mass number' or 'element name-mass number'. For example, the three hydrogen isotopes are shown in Figure 1.3.

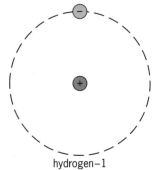

hydrogen–1

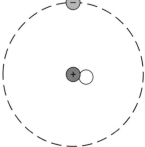

hydrogen–2

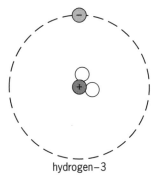

hydrogen–3

Figure 1.3 The three isotopes of hydrogen

Similarly the isotopes of uranium may be written as U-235 or uranium-235, and U-238 or uranium-238.

C-12 scale

In order to compare the masses of different atoms accurately, it is necessary to have a standard mass against which all other masses can be calculated. Masses are given *relative* to this standard, hence the term **relative atomic mass**. Relative atomic mass may be represented by the symbol A_r or RAM.

Before the discovery of isotopes, hydrogen was used as the standard against which the masses of other atoms were compared because chemists realised that it had the smallest atoms. Hydrogen was given an atomic mass of 1.

After the discovery of isotopes it became necessary to choose a single isotope, rather than an element, as the standard for relative atomic masses.

In 1961, the isotope carbon-12 was chosen as the new standard. On this scale, atoms of carbon-12 are given a mass of exactly 12. The relative atomic masses of all other atoms are given relative to this standard.

Relative atomic mass

If an element contained only one isotope, its relative atomic mass would simply be the relative mass of that isotope. However, samples of most elements contain a mixture of several isotopes in varying amounts. Natural **abundance** gives the proportion of each isotope in a sample of the element.

For example, there are two common isotopes of lithium (Table 1.2).

Table 1.2 Common isotopes of lithium

Isotope	Natural abundance
lithium-6	7.5%
lithium-7	92.5%

The relative atomic mass of lithium is given by:

$$\frac{(6 \times 7.5)}{100} + \frac{(7 \times 92.5)}{100} = 6.925$$

For most calculations a value of 7 is sufficiently accurate.

Similarly, there are two common isotopes of chlorine (Table 1.3).

Table 1.3 Common isotopes of chlorine

Isotope	Natural abundance
chlorine-35	75.77%
chlorine-37	24.23%

The relative atomic mass of chlorine is given by:

$$\frac{(35 \times 75.77)}{100} + \frac{(37 \times 24.23)}{100} = 35.4846$$

For most calculations a value of 35.5 is sufficiently accurate.

Rounding the relative atomic mass of chlorine to the nearest whole number would lead to significant errors in calculations so it is rounded to the nearest 0.5.

ATOMIC STRUCTURE

1 Which of the following is always true of an atom?
  A The number of protons is equal to the number of neutrons
  B There is no overall charge on it
  C The number of neutrons is equal to the number of electrons
  D There are no charged particles in the nucleus

2 One isotope of sodium can be represented by $^{24}_{11}$Na. The number of neutrons in one atom of this isotope is:
  A 11   **B** 13
  C 24   **D** 35

3 One isotope of bromine can be represented by $^{79}_{35}$Br.
  The number of electrons in one atom of this isotope is:
  A 35   **B** 44
  C 79   **D** 114

4 The following diagram represents an isotope of beryllium.

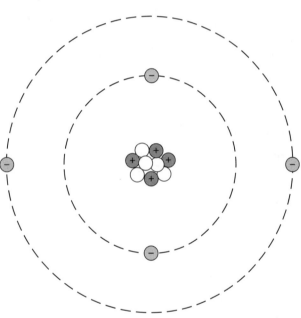

This isotope may be written as
  A beryllium-1
  B beryllium-4
  C beryllium-5
  D beryllium-9

5 Here is some information about four isotopes.

Isotope	Number of neutrons in nucleus	Number of protons in nucleus
A	22	18
B	18	20
C	22	16
D	16	18

Which two are isotopes of the same element?

6 Copy and complete the following table about the sub-atomic particles contained in an atom.

Particle	Mass in atomic mass units	Relative charge	Position in the atom
electron		−1	
neutron			in the nucleus
	1	+1	

7 Explain why:
  a most of the mass of an atom is contained in its nucleus
  b there is no net charge on an atom
  c it is not necessary to include the mass of electrons when calculating the mass number.

8 Copy and complete the following table. Use a Periodic Table to find the atomic symbols of the elements.

Number of protons	Number of neutrons	Atomic number	Mass number	Symbol
				$^{12}_{6}$C
9	10			
		8	18	
15			31	
	10	10		

9 There are two common isotopes of copper. The table contains some information about them.

Isotope	Natural abundance
copper-63	69.2%
copper-65	30.8%

  a Calculate the relative atomic mass of copper correct to 1 decimal place.
  b Explain why the relative atomic mass of copper is not rounded to the nearest whole number.

10 Boron has two common isotopes, boron-10 and boron-11. The relative atomic mass of boron is 10.8, correct to 1 decimal place.
  Use this information to estimate the natural abundance of the two isotopes of boron.

1 B

2 B

3 A

4 D

5 A and D

6

Particle	Mass in atomic mass units	Relative charge	Position in the atom
electron	$\frac{1}{1836}$	−1	around the nucleus
neutron	1	0	in the nucleus
proton	1	+1	in the nucleus

7 a The particles that form the nucleus, that is the protons and neutrons, have a much greater mass than the electrons outside the nucleus.
b The number of positively charged protons is equal to the number of negatively charged electrons so there is no net charge.
c Mass number is defined as the total number of protons and neutrons. The mass of an electron is insignificant compared to the mass of a proton or a neutron.

8

Number of protons	Number of neutrons	Atomic number	Mass number	Symbol
6	6	6	12	$^{12}_{6}C$
9	10	9	19	$^{19}_{9}F$
8	10	8	18	$^{18}_{8}O$
15	16	15	31	$^{31}_{15}P$
10	10	10	20	$^{20}_{10}Ne$

9 a Relative atomic mass $= \dfrac{63 \times 69.2}{100} + \dfrac{65 \times 30.8}{100}$

$= 63.616$

$= 63.6$

b Rounding the relative atomic mass to 64 would result in a significant error.

10 Let the natural abundance of boron-10 be $X\%$.

$\dfrac{10 \times X}{100} + \dfrac{11 \times (100 - X)}{100} = 10.8$

$10X + 1100 - 11X = 1080$

$X = 20$

The natural abundance of each isotope is:
boron-10 = 20%
boron-11 = 80%

2 Electron configuration

Levels, sub-levels and orbitals

The electrons that surround the nucleus of an atom are arranged in **energy levels**. The main, or **principal**, energy levels are identified as levels 1, 2, 3, 4, and so on, where level 1 is nearest to the nucleus.

Within each level there are **sub-levels** identified by the letters s, p, d and f (Table 2.1). The number of possible sub-levels increases with the main level. Each sub-level consists of a series of **orbitals**. The number of possible orbitals increases with sub-level.

Table 2.1 Electron levels and sub-levels

Main energy level	Sub-levels possible
1	1s
2	2s, 2p
3	3s, 3p, 3d
4	4s, 4p, 4d, 4f

In determining **electron configuration** it helps to explain the properties of the electrons if they are considered to **spin**. They may spin in one of two directions: either clockwise or anticlockwise. Each orbital can hold a maximum of two electrons, provided that they spin in opposite directions (Table 2.2).

Table 2.2 Orbitals in sub-levels

Sub-level	Number of orbitals	Maximum number of electrons
s	1	2
p	3	6
d	5	10
f	7	14

An s orbital is spherical in shape (Figure 2.1).

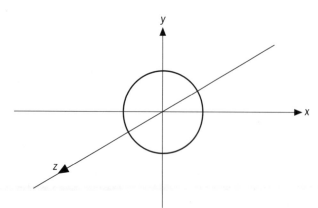

Figure 2.1 An s orbital

Each p orbital is a pair of lobes. The three p orbitals in a sub-level are perpendicular to each other and are shown as being on the x, y and z axes of an orthogonal set of axes (Figures 2.2, 2.3 and 2.4).

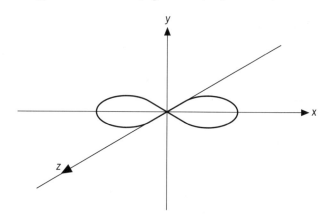

Figure 2.2 A p_x orbital

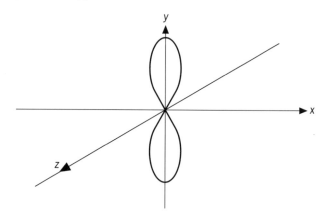

Figure 2.3 A p_y orbital

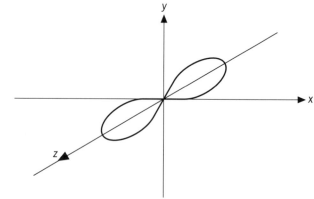

Figure 2.4 A p_z orbital

The d and f orbitals have more complicated shapes. As a result of differences in **shielding** from the nucleus, the different sub-levels within a level have slightly different energies. The energy diagram in Figure 2.5 shows the sequence of sub-levels in order of increasing energy.

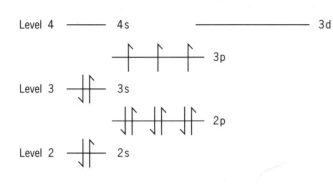

Figure 2.5 Energy level diagram for a phosphorus atom

The energy level diagram in Figure 2.5 shows the normal electron configuration of an atom of phosphorus, which contains 15 electrons. The electron configuration of phosphorus can be written as $1s^22s^22p^63s^23p^3$.

Notice the following points.

- The electrons are placed in the lowest energy orbitals possible.
- The symbols $\downarrow$ and $\uparrow$ are used to denote electrons. The opposite directions of the arrows denote that the electrons have opposite spin.
- The three electrons in the 3p orbitals are unpaired. If there are several orbitals of equal energy, electrons will arrange themselves, when possible, so that they remain unpaired, with the same spin, and thus occupy the maximum number of these orbitals. (This is **Hund's rule** of maximum multiplicity.)

The Periodic Table as s, p, d and f block elements

Within the Periodic Table, elements may be classified as s, p, d or f block (Figure 2.6).

Within each block, elements have their highest-energy electrons in either s, p, d or f electron sub-levels.

For example:

Na ($Z = 11$) $1s^22s^22p^63s^1$
Ca ($Z = 20$) $1s^22s^22p^63s^23p^64s^2$

Sodium and calcium are s block elements.

N ($Z = 7$) $1s^22s^22p^3$
Cl ($Z = 17$) $1s^22s^22p^63s^23p^5$

Nitrogen and chlorine are p block elements.

Fe ($Z = 26$) $1s^22s^22p^63s^23p^63d^64s^2$
Pd ($Z = 46$) $1s^22s^22p^63s^23p^63d^{10}4s^24p^64d^85s^2$

Iron and palladium are d block elements.

When writing an electron configuration remember that:

$$\begin{array}{c}\text{number of}\\\text{electrons}\end{array} = \begin{array}{c}\text{number of}\\\text{protons}\end{array} = \begin{array}{c}\text{atomic}\\\text{number}\end{array}$$

so the total number of electrons in all of the sub-levels should be equal to the atomic number of the element.

Group	1	2												3	4	5	6	7	0
Period 1	H																		He
2	Li	Be												B	C	N	O	F	Ne
3	Na	Mg												Al	Si	P	S	Cl	Ar
4	K	Ca	Sc	Ti	V	Cr	Mn	Fe	Co	Ni	Cu	Zn		Ga	Ge	As	Se	Br	Kr
5	Rb	Sr	Y	Zr	Nb	Mo	Tc	Ru	Rh	Pd	Ag	Cd		In	Sn	Sb	Te	I	Xe
6	Cs	Ba		Hf	Ta	W	Re	Os	Ir	Pt	Au	Hg		Tl	Pb	Bi	Po	At	Rn
7	Fr	Ra		Rf	Db	Sg	Bh	Hs	Mt	Uun	Uuu	Uub			Uuq		Uuh		Uno

s block

d block

p block

elements with atomic numbers 57–71
elements with atomic numbers 89–103 } f block

Figure 2.6 Periodic Table marked into s, p, d and f regions

Ionisation energy

The **first ionisation energy** of an element is the energy needed to remove a single electron from 1 mole of atoms of the element in the gaseous state, in order to form 1 mole of positively charged ions.

The first ionisation energy of an element is more correctly defined as the enthalpy change (see topic 15) for 1 mole for the reaction:

$$A(g) \rightarrow A^+(g) + e^-$$

Table 2.3 gives the first ionisation energy values for the elements of Group 1. There is a decrease in value from lithium to caesium. This can be explained by considering the electron configuration of the elements in the group.

Table 2.3 First ionisation energy of Group 1 elements

Group 1 element	First ionisation energy/kJ mol^{-1}
lithium	520
sodium	496
potassium	419
rubidium	403
caesium	376

The outer electron of sodium is in a 3s sub-level, whereas the outer electron of lithium is in a 2s sub-level. In comparison to the 2s sub-level, the 3s sub-level is higher in energy. The 3s electron is further from the nucleus and, to some extent, is shielded from the nucleus by the inner electrons. Since it is at higher energy and because of the shielding, less energy is needed to remove the electron from sodium so the ionisation energy is lower. A similar explanation applies to the sequence of elements in the group.

This trend in first ionisation energies provides evidence for the electrons of atoms being arranged in levels at different distances from the nucleus.

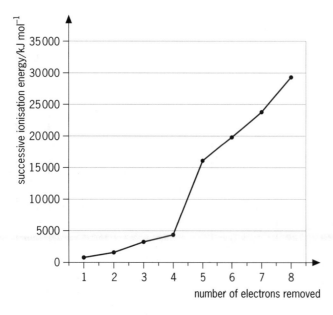

Figure 2.7 Ionisation energies for silicon

The second and third ionisation energies are the enthalpy changes for the reactions:

$$A^+(g) \rightarrow A^{2+}(g) + e^-$$
$$A^{2+}(g) \rightarrow A^{3+}(g) + e^-$$

Subsequent ionisation energies also support the electron arrangement model described in this topic. Figure 2.7 shows some of the ionisation energies for silicon ($1s^2 2s^2 2p^6 3s^2 3p^2$).

The first four electrons are removed from level 3. After this there is a substantial jump in ionisation energy because the next electron must be removed from level 2, which is nearer to the nucleus. Electrons at level 2 receive less shielding than those at level 3.

Ionisation energy across a period

There is a general increase in first ionisation energies across a period (see topic 6 for the structure of the Periodic Table). Table 2.4 gives the first ionisation energies of the elements in Period 3 (sodium to argon).

Table 2.4 First ionisation energy of Period 3 elements

Element	Na	Mg	Al	Si	P	S	Cl	Ar
First ionisation energy/ kJ mol^{-1}	496	738	577	787	1060	1000	1255	1520

The increase is due to electrons at the same main energy level being attracted by an increasing nuclear charge across the period. The increasing nuclear charge is caused by the increasing number of protons in the nucleus and makes it progressively more difficult to remove an electron, thus more energy is needed.

The fall in energy from magnesium to aluminium occurs because the outer electron in aluminium is in a 3p orbital ($1s^2 2s^2 2p^6 3s^2 3p^1$) whereas in magnesium the outer electron is in a 3s orbital ($1s^2 2s^2 2p^6 3s^2$). The 3p orbital is at a higher energy than the 3s orbital and thus less energy is needed to remove the electron.

There is also a fall in energy from phosphorus to sulphur. This is a consequence of the arrangement of electrons in the 3p orbitals of these elements.

phosphorus (3p) —↑— ↑— ↑—

sulphur (3p) —↑↓— ↑— ↑—

The 3p electrons in phosphorus are unpaired; however, in sulphur the additional electron is, by necessity, paired with an electron of opposite spin. Since electrons carry a charge, there is some repulsion between a pair of electrons as like charges repel. This repulsion increases the energy of the paired electrons and thus makes it easier to remove one of them.

The variation of first ionisation energy across a period provides evidence of the existence of energy sub-levels in electron arrangement.

1 Which one of the following is the electron arrangement of sulphur ($Z = 16$)?

 A $1s^22s^22p^63s^23p^5$

 B $1s^22s^22p^63s^23p^2$

 C $1s^22s^22p^63s^23p^4$

 D $1s^22s^22p^63s^23p^6$

2 What is the maximum number of electrons that can be held in a d sub-level?

 A 2

 B 5

 C 10

 D 14

3 Which of the following shows the normal configuration of the outer three electrons in an atom of nitrogen?

 A $\underline{\uparrow}\ \underline{\uparrow}\ \underline{\uparrow}$

 B $\underline{\uparrow\downarrow}\ \underline{\uparrow}\ \underline{}$

 C $\underline{\downarrow}\ \underline{\uparrow}\ \underline{\uparrow}$

 D $\underline{\uparrow\uparrow}\ \underline{\uparrow}\ \underline{}$

4 The third ionisation energy of an element, M, is the enthalpy change for the reaction

 A $M^+(g) \rightarrow M^{4+}(g) + 3e^-$

 B $M^{2+}(g) \rightarrow M^{3+}(g) + e^-$

 C $M(g) \rightarrow M^{3+}(g) + 3e^-$

 D $M^{3+}(g) \rightarrow M^{4+}(g) + e^-$

5 Which of the following elements of Period 3 has the highest first ionisation energy?

 A aluminium

 B chlorine

 C phosphorus

 D silicon

6 State the maximum number of orbitals in, and draw a diagram to show the shape of:

 a an s sub-level

 b a p sub-level.

7 The following table contains information about the element potassium. Give similar information for the other elements in the table.

Element	Atomic number	Electron configuration	Block
potassium	19	$1s^22s^22p^63s^23p^64s^1$	s
fluorine	9		
magnesium	12		
selenium	34		
titanium	22		

8 The following table gives the first ionisation energies of the elements lithium to oxygen.

Element	Li	Be	B	C	N	O
Ionisation energy/ kJmol^{-1}	520	900	801	1086	1402	1314

a Explain the term **first ionisation energy**.

b Explain why beryllium has a higher first ionisation energy than lithium.

c Explain why boron has a lower first ionisation energy than beryllium.

d Explain why oxygen has a lower first ionisation energy than nitrogen.

e Predict a likely value for the first ionisation energy of fluorine and explain how you determined this value.

9 The following diagram shows the electronic configuration of boron.

a Why does the electron in the 2p orbital have a higher energy than the electrons in the 2s orbital?

b In the diagram half-arrows represent electrons. What is the significance of the direction of these half-arrows?

c Draw a similar diagram to show the electronic structure of nitrogen.

10 The following graph shows successive ionisation energies for fluorine.

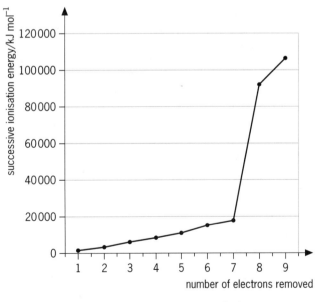

a Explain why more energy is needed to remove the second electron than the first electron.

b Write an equation to represent the enthalpy change for the fifth ionisation energy.

c Explain the large increase after the removal of the seventh electron.

ELECTRON CONFIGURATION

1 C

2 C

3 A

4 B

5 B

6 a Maximum of one orbital in an s sub-level.

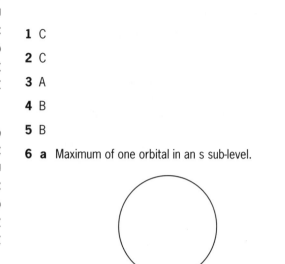

b Maximum of three orbitals in a p sub-level.

7

Element	Atomic number	Electron configuration	Block
potassium	19	$1s^22s^22p^63s^23p^64s^1$	s
fluorine	9	$1s^22s^22p^5$	p
magnesium	12	$1s^22s^22p^63s^2$	s
selenium	34	$1s^22s^22p^63s^23p^63d^{10}4s^24p^4$	p
titanium	22	$1s^22s^22p^63s^23p^63d^24s^2$	d

8 a The first ionisation energy of an element is the enthalpy change when a single electron is removed from each atom of one mole of the element to form ions, where both the atom and the ion are in the gaseous state.

b The outer electrons in a beryllium atom are attracted more strongly by the greater nuclear charge.

c There is a fall in energy from beryllium to boron because the outer electron in boron is in a p sub-level, which is higher in energy than the outer electrons in beryllium which are in an s sub-level. Also the outer electron in boron receives more shielding from the nucleus.

d The three outer electrons in a nitrogen atom are unpaired and exist in different 2p orbitals. The extra 2p electron in oxygen is paired. There is some repulsion between the pair of electrons thus it is easier to remove one of the paired 2p electrons from oxygen than to remove an unpaired 2p electron from nitrogen.

e There is a general increase across the period. The first ionisation energy of fluorine is likely to be greater than that of any other element in the period since the nuclear charge in fluorine is greatest. Expect $1500-2000\,kJ\,mol^{-1}$ (actually 1681).

9 a The electron in the 2p orbital is further from the nucleus than those in the 2s orbital.

b The half-arrow represents the direction of spin of the electron. Pairs of electrons in the same orbital must have opposite spins.

c

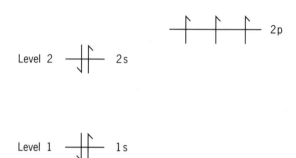

10 a After the removal of the first electron, the negative charge on the surrounding electrons is no longer equal to the positive charge on the nucleus. There is a net positive charge that makes it more difficult to remove subsequent electrons.

b $F^{4+}(g) \rightarrow F^{5+}(g) + e^-$

c The large increase after the removal of the seventh electron indicates that the eighth electron must be removed from an inner electron shell.

3 Bonding

Ionic bonding

Ionic bonds are formed between oppositely charged, positive and negative, ions.

Positive ions are formed when atoms lose one or more electrons.

$$\begin{array}{ll} \text{Li} & \rightarrow \text{Li}^+ + \text{e}^- \\ 1s^2 2s^1 & 1s^2 \end{array}$$

$$\begin{array}{ll} \text{Mg} & \rightarrow \text{Mg}^{2+} + 2\text{e}^- \\ 1s^2 2s^2 2p^6 3s^2 & 1s^2 2s^2 2p^6 \end{array}$$

Negative ions are formed when atoms gain one or more electrons.

$$\begin{array}{lll} \text{Cl} & + \text{e}^- \rightarrow & \text{Cl}^- \\ 1s^2 2s^2 2p^6 3s^2 3p^5 & & 1s^2 2s^2 2p^6 3s^2 3p^6 \end{array}$$

$$\begin{array}{lll} \text{O} & + 2\text{e}^- \rightarrow & \text{O}^{2-} \\ 1s^2 2s^2 2p^4 & & 1s^2 2s^2 2p^6 \end{array}$$

All ionic compounds are solids at room temperature. In these solids, ionic bonds do not exist in isolation, thus it makes no sense to talk of an ionic molecule. The ions form a **giant ionic lattice** in which each ion is surrounded by ions of opposite charge (Figure 3.1).

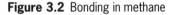

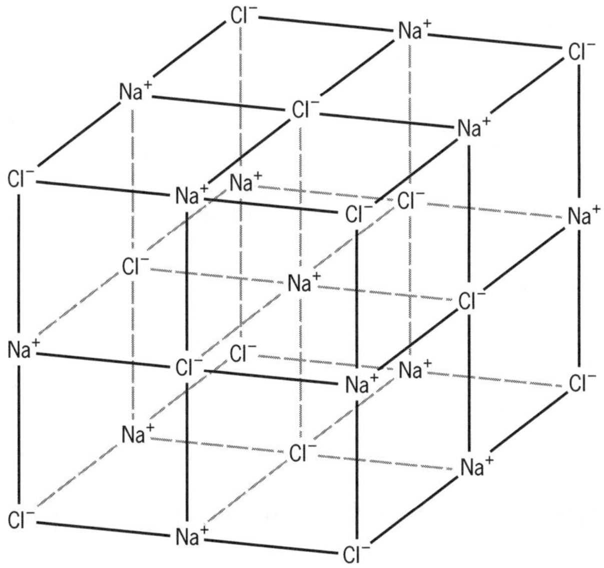

Figure 3.1 Ionic lattice for sodium chloride

In sodium chloride, a typical ionic compound, each sodium ion is surrounded by six chloride ions, and each chloride ion is surrounded by six sodium ions. The oppositely charged ions are electrostatically attracted to each other. The number of oppositely charged ions that surround each other depends on the sizes of the ions and how they pack together.

Ionic compounds are able to conduct electricity when in the molten state or in aqueous solution. Under these conditions the ions are able to move and carry charge.

Covalent bonding

Covalent bonds are formed when two atoms share a pair of electrons. In 'dot–cross' diagrams the electrons of the two atoms forming a covalent bond are shown by the symbols '•' and '×' respectively. These symbols do not imply that the electrons are different in any way but only that they are from different atoms. It is usual in such diagrams to show only the outer level of electrons. Figure 3.2 shows the dot–cross diagram for methane.

In a C—H bond in methane, the carbon atom and the hydrogen atom each provide one electron that is to be shared. Unlike ionic bonding, there is no transfer of electrons from one atom to another. The shared electron pair creates a bond between the atoms because it attracts the nuclei of both atoms. Covalent compounds do not conduct electricity in any state since there are no ions to carry charge.

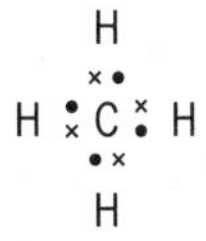

Figure 3.2 Bonding in methane

Coordinate bonding

Coordinate bonding is a particular form of covalent bonding in which one atom provides both of the electrons that form the covalent bond. It is sometimes referred to as **dative covalent bonding**.

Figure 3.3 shows the coordinate bond in the ammonium ion.

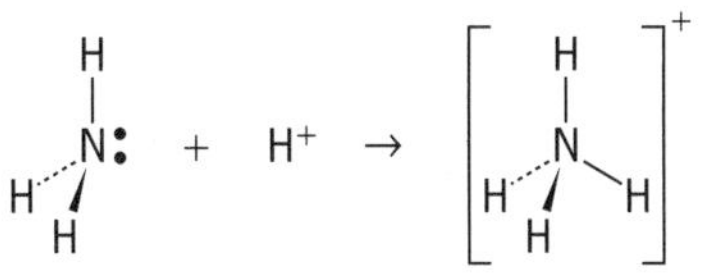

Figure 3.3 Bonding in the ammonium ion

On the nitrogen atom of an ammonia molecule there is a pair of electrons which is not used in the formation of the three N—H bonds. This pair of electrons is called a **non-bonding** or **lone pair** of electrons. They can be donated to make a coordinate bond with a hydrogen ion, in order to form the ammonium ion. The nitrogen atom provides both of the electrons for this bond.

Electronegativity

Electronegativity is a measure of the power of an atom to draw electron density from a covalent bond. There have been various attempts to define an electronegativity scale. The scale that is most frequently used was devised by the chemist Linus Pauling and is thus known as the **Pauling scale** (Figure 3.4). It is based on enthalpy values that have been determined experimentally.

H 2.1							He
Li 1.0	Be 1.5	B 2.0	C 2.5	N 3.0	O 3.5	F 4.0	Ne
Na 0.9	Mg 1.2	Al 1.5	Si 1.8	P 2.1	S 2.5	Cl 3.0	Ar
K 0.8	Ca 1.0					Br 2.8	Kr
Rb 0.8	Sr 1.0					I 2.5	Xe

Figure 3.4 Pauling electronegativity values

Small atoms with large numbers of protons in the nucleus attract electron density most strongly. As a result:

- electronegativity increases across a period, as the charge on the nucleus increases
- electronegativity decreases down a group, as the size of the atom increases.

No electronegativity values are given for the elements of Group 0 since they form relatively few molecules.

Covalent bonds with ionic character

When a covalent bond exists between two atoms of the same electronegativity, such as occurs in the diatomic molecule of an element, the pair of electrons is shared equally in the bond (Figure 3.5).

Figure 3.5 Electron cloud in hydrogen

However, where a covalent bond exists between two atoms of differing electronegativity this is not the case (Figure 3.6).

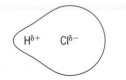

Figure 3.6 Electron cloud in hydrogen chloride

The chlorine atom has a higher electronegativity (3.0) than the hydrogen atom (2.1), so the pair of electrons that forms the covalent bond between the atoms lies closer to the chlorine atom than the hydrogen atom. The result is that the hydrogen atom has a slight deficiency in electron density, which is represented by $\delta+$ while the chlorine atom has a slight excess of electron density, which is represented by $\delta-$. This is not of the same magnitude as a positive or negative charge, so the symbol δ is used to imply 'very small' in this context.

This charge separation creates an electric dipole. Molecules that contain such bonds are described as **polar molecules**.

In an ionic compound, the difference in electronegativity between the atoms is such that an electron is completely transferred from one atom to the other, hence the formation of a pair of ions.

The percentage of ionic character between two atoms is related to the difference in their electronegativities (Figure 3.7).

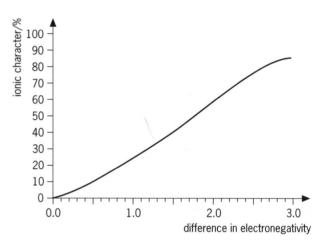

Figure 3.7 Electronegativity and bond ionic character

The greater the difference in electronegativity between the atoms forming a bond the greater the percentage of ionic character of the bond.

Ionic bonds with covalent character

The ions in a sodium chloride lattice are perfectly spherical, thus the bonds in this compound are said to be perfectly ionic (Figure 3.8).

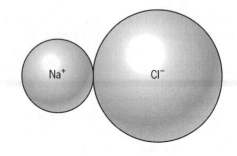

Figure 3.8 Sodium and chloride ions

However, this is not the case with all ionic compounds. In an ionic compound consisting of a small, highly charged, positive ion and a large negative ion, such as lithium iodide, the positive ion attracts electron charge away from the negative ion (Figure 3.9). The result is that the negative ion is distorted and electron density becomes concentrated between the ions similar to a covalent bond. Compounds like lithium iodide are said to be ionic with covalent character.

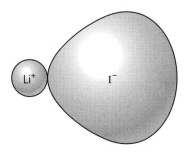

Figure 3.9 Lithium and iodide ions

The ability of a cation to polarise an anion is called its **polarising power**. Polarising power, and hence covalent character, increase if:

• the cation is small
• the anion is large
• the charges on the ions are high.

An approximate measure of polarising power of a cation is given by its **charge density**. This is the ratio of charge : ionic radius. The greater the charge density the greater the polarising power. Table 3.1 shows how the charge density and polarising power decrease down Group 1.

Table 3.1 Polarising power of Group 1 cations

Cation	Charge	Ionic radius /nm	Charge density	
Li^+	+1	0.074	13.5	decreasing polarising power
Na^+	+1	0.102	9.8	
K^+	+1	0.138	7.2	
Rb^+	+1	0.149	6.7	
Cs^+	+1	0.170	5.9	

The Al^{3+} ion is very small and carries a high charge. It has a charge density of 56.6 (+3/0.053). It attracts electrons so strongly that aluminium chloride is a covalent compound. It exists as Al_2Cl_6 molecules in which two $AlCl_3$ molecules are linked by coordinate bonds formed by the donation of lone pairs of electrons from two chlorine atoms (Figure 3.10).

In an H^+ ion there are no electrons surrounding the nucleus and the ionic radius is very small (of the order of 10^{-4} nm). The charge density is very large and thus the H^+ ion has a very high polarising power.

Figure 3.10 Coordinate bonding in $AlCl_3$

Intermolecular forces

Covalent molecules are attracted to each other by intermolecular forces of different strengths.

van der Waals	permanent dipoles	hydrogen bonding

——————————— increasing strength ———————→

Van der Waals forces are temporary induced dipole–dipole attractions. The electrons that form bonds in covalent compounds are not static. At any moment in time, even in a non-polar covalent molecule, movement of electrons may result in an asymmetrical electron distribution. This leads to a temporary dipole that, in turn, induces an opposite dipole in adjacent molecules (Figure 3.11). The result is that the first molecule is attracted to the adjacent molecules. The larger the molecules the greater the van der Waals forces of attraction between them.

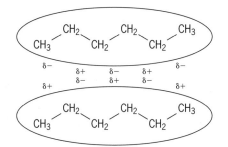

Figure 3.11 Van der Waals forces

Permanent dipole–dipole forces exist between polar molecules (Figure 3.12). Permanent dipoles are stronger than temporary dipoles, therefore this form of dipole–dipole intermolecular attraction is significantly stronger than van der Waals forces.

$$\overset{\delta+}{H}-\overset{\delta-}{Cl}----\overset{\delta+}{H}-\overset{\delta-}{Cl}$$

Figure 3.12 Dipole–dipole interactions

Hydrogen bonding is a special form of dipole–dipole force that exists between a lone pair of electrons on a fluorine, nitrogen or oxygen atom, and a hydrogen atom. Due to its relatively low electronegativity, a hydrogen atom in a covalent bond has a slight deficiency in electron density and is thus slightly positive ($\delta+$). Also, it has no inner electrons so lone pairs of electrons on other atoms are strongly attracted to the positively charged hydrogen nucleus.

A hydrogen-bonded hydrogen atom is always collinear with the atom to which it is attached and the atom to which it is attracted. The three atoms form a straight line (Figure 3.13).

Hydrogen bonds are the strongest intermolecular forces; however, they are only of the order of 5–10% of the strength of a covalent bond.

The forces between atoms in a covalent bond (intramolecular bonds) are much greater than those that exist between covalent molecules (intermolecular bonds).

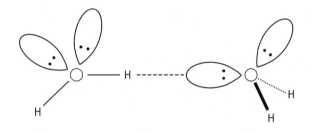

Figure 3.13 Hydrogen bonding between water molecules

1 Which of the following is a property of all ionic compounds?
  A They carry an electric current when molten
  B They carry an electric current when solid
  C They have a low melting point
  D They are very soluble in water

2 Which of the following is true of electronegativity?
  A Electronegativity decreases across a period
  B Electronegativity increases down a group
  C Metals generally have lower electronegativity values than non-metals
  D Noble gases have the highest electronegativity values

3 Which of the following shows intermolecular forces in order of increasing strength?
  A van der Waals → hydrogen bonding → permanent dipoles
  B permanent dipoles → van der Waals → hydrogen bonding
  C van der Waals → permanent dipoles → hydrogen bonding
  D permanent dipoles → hydrogen bonding → van der Waals

4 In which of the following covalent bonds is polarisation greatest?
  A Cl—Cl
  B $H_3C—NH_2$
  C H—Cl
  D $H_3C—H$

5 Which of the following ionic compounds is likely to show the greatest covalent character?
  A barium chloride
  B lithium bromide
  C magnesium fluoride
  D potassium iodide

6 a i Give the atomic number and electron configuration of potassium and bromine.
   ii Draw dot–cross diagrams to show the formation of potassium bromide (KBr) from its constituent atoms. You need only show outer electrons.
  b i Give the atomic number and electron configuration of carbon and chlorine.
   ii Draw dot–cross diagrams to show the formation of tetrachloromethane (CCl_4) from its constituent atoms. You need only show outer electrons.
  c In terms of their structure, explain why potassium bromide has a much higher melting point than tetrachloromethane.

7 The following table contains information about two organic compounds, butane and propan-1-ol.

Compound	Formula	M_r	Boiling point/°C
butane	$CH_3CH_2CH_2CH_3$	58	−0.5
propan-1-ol	$CH_3CH_2CH_2OH$	60	97.2

With the help of suitable diagrams, explain, in terms of intermolecular forces, why the boiling points of these compounds are significantly different.

8 The following table contains information about some of the halogens and the hydrogen halides.

	Electronegativity of the halogen	Boiling point of the hydrogen halide/°C
fluorine	4.0	20
chlorine	3.0	−85
bromine	2.8	−67
iodine		−35

  a Define electronegativity and suggest a likely value for iodine.
  b Explain why there is a general increase in the boiling points of the hydrogen halides between chlorine and iodine.
  c Explain why the boiling point of hydrogen fluoride is much higher than might be expected from considering the pattern shown by the remaining hydrogen halides.

9 a Indicate the polarity of the covalent bonds shown in the following compounds, using the symbols $\delta+$ and $\delta-$.
    i Br—F
    ii $H_3C—Cl$
    iii $(H_3C)_2N—H$
  b Using the electronegativity values and the graph given in Figures 3.4 and 3.7 on page 12, estimate the percentage of ionic character in the hydrogen halide bonds and tabulate your results. Comment on the trend down the group.

10 a Sketch a graph to show how electronegativity changes across Period 3.
  b Explain why electronegativity decreases down Group 1.
  c State the type of bonding between phosphorus and oxygen in the compound P_4O_{10} and explain how this can be predicted from the electronegativity of these two elements.
  d Lithium iodide forms an ionic lattice but shows some covalent character. Explain why.

1 A **2** C **3** C **4** C **5** B

6 a i Potassium: $Z = 19$; $1s^2 2s^2 2p^6 3s^2 3p^6 4s^1$

Bromine: $Z = 35$; $1s^2 2s^2 2p^6 3s^2 3p^6 3d^{10} 4s^2 4p^5$

ii

$$\overset{\bullet}{K} \ + \ \overset{\times\times}{\underset{\times\times}{\times Br \times}} \ \rightarrow \ [K]^+ \left[\overset{\times\times}{\underset{\times\times}{\bullet Br \times}}\right]^-$$

b i Carbon: $Z = 6$; $1s^2 2s^2 2p^2$
Chlorine: $Z = 17$; $1s^2 2s^2 2p^6 3s^2 3p^5$

ii

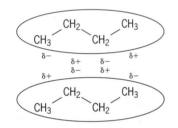

c The ions in potassium bromide are held together by strong electrostatic forces of attraction while the covalent molecules in tetrachloromethane are held together by weak van der Waals forces.

7

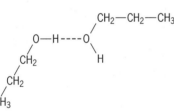

Induced dipoles on molecules of butane lead to van der Waals (induced dipole–dipole attractions) between them. These forces of attraction are relatively weak, thus the energy needed for molecules to escape each other is low.

The dipole–dipole interaction between a lone pair of electrons on the oxygen atom of one molecule of propan-1-ol and the hydrogen atom of the —OH group on another give rise to hydrogen bonding. The forces of attraction are relatively strong, thus the energy needed to overcome hydrogen bonding is much greater than that needed to overcome van der Waals forces.

8 a Electronegativity is the power of an atom to withdraw electron density from a covalent bond. A likely value for iodine is 2.4–2.6 (actual 2.5).

b The strength of the permanent dipole in the hydrogen halide molecule decreases down the group, which might suggest a decrease in boiling point as interaction between molecules becomes weaker. However, the number of electrons on the halogen atom increases, which strengthens van der Waals bonding and thus increases boiling point.

c The H—F bond is strongly polarised due to the high electronegativity of fluorine. Attraction between strong partial positive charge ($\delta+$) on hydrogen atoms and the lone pair of electrons on fluorine atoms gives rise to the formation of hydrogen bonding between molecules of hydrogen fluoride. Hydrogen bonding is significantly stronger than the permanent dipole–dipole interaction present in the other hydrogen halides.

9 a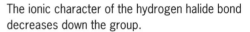

b

Bond	Difference in electronegativity	Percentage of ionic character
H—F	1.9	55
H—Cl	0.9	22
H—Br	0.7	15
H—I	0.4	10

The ionic character of the hydrogen halide bond decreases down the group.

10 a

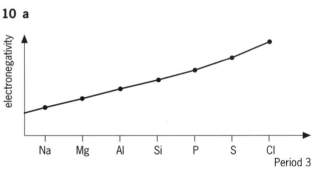

b Electronegativity decreases down Group 1 due to the increasing atomic radius. The outside of the atom is increasingly shielded from the nucleus by the increasing number of electron shells, and bonding electrons are thus less strongly attracted by the shielded nucleus.

c P_4O_{10} is a covalent compound. This could be predicted from the electronegativity of phosphorus (2.1) and oxygen (3.5). The difference in magnitude is sufficient to produce polarised covalent bonds but not to result in the electron transfer required to form ionic bonds.

d Lithium is a small positively charged ion and iodide is a large negatively charged ion. When a large negative ion is next to a small and highly charged positive ion the electron cloud around the negative ion becomes distorted so that it is no longer spherical.

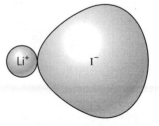

The lithium ion attracts the electron density from the iodide ion leaving the latter distorted and, in a sense, polarised. Some electron density is concentrated between the ions and the bond begins to resemble a covalent bond.

4 Properties of matter

States of matter

Substances may exist as solids, liquids and gases (Figure 4.1).

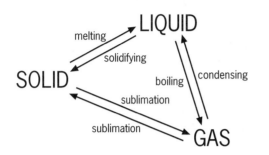

Figure 4.1 Changes of state

In **solids**, particles vibrate but are held in a fixed position (Figure 4.2). The attractive forces which hold the particles together can be ionic bonds, covalent bonds, metallic bonds, hydrogen bonds, permanent dipole–dipole forces or van der Waals forces, depending on the nature of the solid.

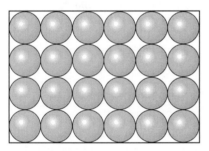

Fiigure 4.2 Particles in a solid

As the temperature of a solid increases, the particles vibrate more and more energetically. At a certain temperature, the **melting point** of the solid, the particles have sufficient energy to partially overcome, or loosen, the attractive forces that hold them together.

The energy needed to change a solid to a liquid at its melting point is called the **enthalpy of fusion**.

In **liquids**, the particles are able to move position, as well as vibrate, but are still held together by forces similar to those of a solid (Figure 4.3). At any moment in time, the arrangement of particles in a liquid resembles a disordered solid. The separation of particles in a liquid is typically 10% more than in solids.

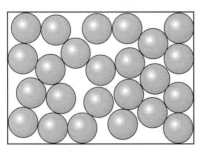

Figure 4.3 Particles in a liquid

As the temperature of a liquid increases, the particles move about more and more energetically. At a certain temperature, the **boiling point** of the liquid, the particles have sufficient energy to completely overcome the attractive forces that hold them together so that the particles are completely separate.

The energy needed to change a liquid into a vapour at its boiling point is called the **enthalpy of vaporisation**. For a given mass of a substance, it needs more energy to change a liquid into a gas than a solid into a liquid.

In **gases** the particles are completely separated and move with rapid, random motion (Figure 4.4). The degree to which gases can be compressed is a result of the large distances between particles.

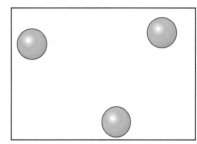

Figure 4.4 Particles in a gas

Kinetic theory of gases

The **kinetic theory** of gases is a mathematical model of an **ideal gas**. Such a gas does not exist but the model can be used to explain the behaviour of real gases.

The kinetic theory of gases makes five assumptions about a gas.

1   A gas consists of particles of negligible volume.
2   The particles are in continuous random motion.
3   The particles exert no attractive forces on each other.
4   All collisions between particles are perfectly elastic; no kinetic energy is lost.
5   The average kinetic energy of the particles is proportional to the absolute temperature.

The following equation forms the basis of the kinetic theory of gases and follows from the above assumptions:

$$pV = \tfrac{1}{3}Nmc^2$$

where p is the pressure of the gas, V is the volume of the gas, N is the number of gas particles, m is the mass of a gas particle and c^2 is the average of the speed squared of the particles. This equation can be used to account for the various laws associated with gases, such as Boyle's law and Charles's law, and the ideal gas equation (see topic 8).

Real gases behave in a similar way to an ideal gas when at room temperature and atmospheric pressure. Deviation of a real gas from ideal gas behaviour becomes pronounced at high pressures and at low temperatures since, under such conditions, the volume of the gas particles and the attractive forces between them become significant.

Types of solids

In a solid, the particles are often bonded together in an ordered pattern, forming a regular crystalline shape. The type of bonding between the particles gives rise to solids with different properties.

Metals

Metals consist of a lattice of positively charged ions surrounded by a 'sea' of mobile electrons (Figure 4.5).

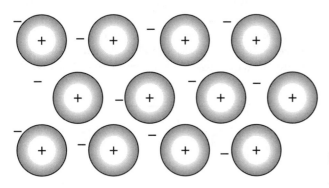

Figure 4.5 Metallic bonding

The mobility of the free electrons is responsible for the good electrical and heat conductivity of metals.

The electrostatic attractive forces between the positive ions and the electrons are described as **metallic bonds**. A large amount of energy is needed to overcome these forces, thus metals usually have high melting points and boiling points. Figure 4.6 shows how the melting point of the metals changes across a period and down a group.

Li 454 K		
Na 371 K	Mg 922 K	Al 933 K
K 336 K		
Rb 312 K		
Cs 302 K		

Figure 4.6 Changes in melting point

This can be explained in terms of the density of the sea of electrons.

- Passing across the period there is an increase in melting point. The outermost electrons are in the same main energy level and are thus a similar distance from the nucleus of the atom. As the number of electrons increases so does the electron density.
- Passing down a group there is a decrease in electron density. The outer electron is held in an orbital that is progressively further from the nucleus of the atom so the electron density decreases.

Ionic compounds

The ions in an ionic crystal are held together by strong electrostatic forces making the crystal hard and brittle (Figure 4.7).

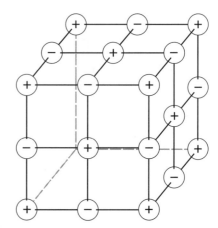

Figure 4.7 Ionic bonding

Ionic compounds have high melting points as a large amount of energy is needed to overcome these forces of attraction.

As the ions in a crystal are unable to move freely, ionic compounds do not conduct electricity when solid. However, in the liquid state, and in aqueous solution, the ions are free to move and can carry charge.

Giant covalent substances or macromolecules

Macromolecules consist of a continuous structure in which atoms are joined together by covalent bonds.

In a diamond crystal, each carbon atom is joined to four other carbon atoms (Figure 4.8). The result is a hard, brittle, crystalline structure.

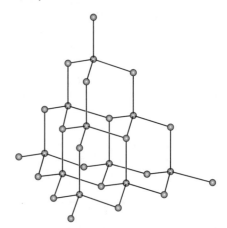

Figure 4.8 Structure of diamond

Macromolecules have very high melting points as the covalent bonds that join the atoms are very strong and a large amount of energy is needed to break them. With the exception of graphite, macromolecules are non-conductors of electricity as they have no free electrons or ions to carry charge.

In graphite the carbon atoms are arranged in layers (Figure 4.9). Within a layer, each carbon atom is joined to three other carbon atoms by strong covalent bonds, giving a continuous series of hexagons. The unbonded electron on each carbon atom is delocalised and able to move within the overlapping orbitals above and below the carbon atoms, forming weak bonds between the layers.

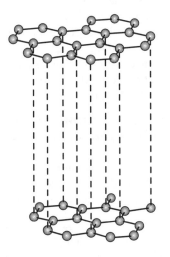

Figure 4.9 Structure of graphite

Graphite is soft, since the layers can slip over each other. It also conducts electricity, as the delocalised electrons between the layers are able to carry charge.

Simple molecular substances

Many covalent compounds exist as single **molecules** (Figure 4.10). The atoms in a molecule are held together by strong covalent bonds (intramolecular forces) but the forces of attraction between molecules (intermolecular forces) are much weaker (hydrogen bonding, permanent dipole–dipole forces or van der Waals forces).

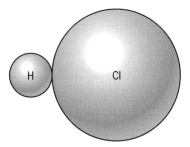

Figure 4.10 Simple molecular bonding

In the solid state, the molecules of a simple molecular substance, such as iodine, are arranged in a lattice, forming a **molecular crystal** (Figure 4.11).

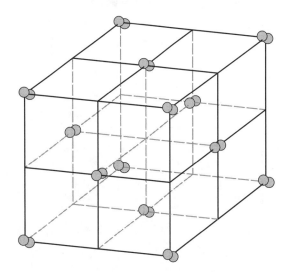

Figure 4.11 Structure of iodine

The molecules in a molecular crystal are held together by weak forces of attraction such as van der Waals forces or hydrogen bonding. Molecular crystals are soft and are non-conductors. The compounds have low melting points.

Summary

Table 4.1 summarises the physical properties of different types of solids.

Table 4.1 The physical properties of solids

Solid	Relative melting point and boiling point	Electrical conductivity when solid (particle)	Electrical conductivity when molten (particle)
metal	high	good (electron)	good (electron)
ionic	high	non-conductor	good (ion)
giant covalent	very high	non-conductor [except graphite (electron)]	non-conductor
simple molecular	low	non-conductor	non-conductor

In general terms, ionic substances tend to be soluble in water and simple molecular covalent compounds tend to be insoluble; however, this should not be given as a typical property of ionic or simple covalent compounds. There are many ionic compounds that are all but insoluble in water and many simple covalent compounds which are very soluble.

1 Which of the following best represents the particles in a solid crystal lattice?

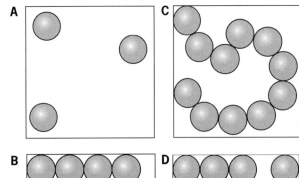

A

C

B

D

2 A substance has a high melting point and does not conduct electricity either as a solid or when molten. The substance is likely to be
  A ionic
  B macromolecular
  C metallic
  D simple covalent

3 Which of the following gases is likely to give the best approximation to ideal gas behaviour at room temperature and atmospheric pressure?
  A argon
  B carbon dioxide
  C methane
  D sulphur dioxide

4 Which of the following decreases as a solid is heated to become a liquid and finally a gas?
  A attractive forces between particles
  B motion of the particles
  C size of the particles
  D space between the particles

5 Which of the following graphs shows how the average kinetic energy of the particles varies with absolute temperature for an ideal gas?

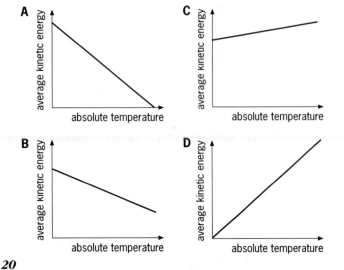

A

C

B

D

6 a Explain the terms **intermolecular forces** and **intramolecular forces**.
  b Comment on the strength of these forces in a named simple covalent compound.

7 Explain the following observations as fully as you can.
  a Graphite and diamond are two forms of the element carbon. Graphite is able to conduct electricity but diamond is not.
  b Molten sodium chloride is able to conduct electricity but solid sodium chloride is not.
  c Silicon dioxide and sulphur dioxide are covalent compounds of similar formula mass. The melting point of silicon dioxide (1726 °C) is much higher than that of sulphur dioxide (−73 °C).

8 a Define the terms **enthalpy of fusion** and **enthalpy of vaporisation** and state which is greater for a given mass of a substance.
  b Explain why 1 g of steam occupies a much greater volume than 1 g of ice.
  c Describe the motion of a single particle in a gas.

9 a In the kinetic theory of gases state three of the assumptions made about gases.
  b Under what conditions does a real gas behave like an ideal gas?
  c Explain why deviation from ideal gas behaviour becomes pronounced at high pressures and low temperatures.

10 When a solid is heated it first becomes a liquid and then a gas. Describe the changes that take place to the:
  a order of the particles
  b motion of the particles
  c space between particles
  d attractive forces between particles.

1 B **2** B **3** A **4** A **5** D

6 a Intermolecular forces are the forces between molecules; intramolecular forces are the forces between atoms within a molecule.

b In a simple covalent compound, such as methane, the intramolecular forces between atoms are covalent bonds, which are relatively strong. The intermolecular forces are van der Waals forces, which are relatively weak.

7 a In diamond each carbon atom is bonded to four other carbon atoms in a rigid three-dimensional structure. All of the outer shell of electrons are used in forming bonds so none is available to carry an electric charge.

In graphite each carbon atom is bonded to three other carbon atoms, forming a hexagonal planar structure. The remaining electron from the outer shell exists in a p orbital above and below the plane. The hexagonal planes are held together by weak forces of attraction; however, the delocalised electrons in the p orbitals are sufficiently mobile to carry an electric charge.

b Sodium chloride exists as ions both in the solid and molten phases. However, when solid, the sodium ions and chloride ions are held in position in a lattice and are unable to carry an electric charge. When sodium chloride is molten the ions are mobile and can carry charge.

c Sulphur dioxide is a simple covalent compound and exists as separate SO_2 molecules held together by weak intermolecular forces.

Silicon dioxide is a giant covalent structure in which each atom is covalently bonded to other atoms.

8 a The enthalpy of fusion is the energy change when a solid changes into a liquid at its melting point. The enthalpy of vaporisation is the energy change when a liquid changes into a gas at its boiling point. For a given mass of a substance, the enthalpy of vaporisation is greater than the enthalpy of fusion.

b The water molecules in steam are separated from each other while the water molecules in ice are held together in a rigid lattice.

c The gas particle moves rapidly in straight lines for short distances in random directions.

9 a Any three from:

- A gas consists of particles of negligible volume.
- The particles are in continuous random motion.
- The particles exert no attractive forces on each other.
- All collisions between particles are perfectly elastic; no kinetic energy is lost.
- The average kinetic energy of the particles is proportional to the absolute temperature.

b At room temperature and atmospheric pressure, or even more so at high temperatures and low pressures.

c At high pressures and low temperatures the volume of the particles is no longer negligible and the attractive forces between particles cannot be ignored.

10 a The particles become less ordered.

b The motion of the particles increases.

c The space between particles increases.

d The attractive forces between particles decrease.

5 Shape

Metallic crystals

For the purposes of considering their structure, the metal ions in metallic crystals can be regarded as spheres which, in solid metals, are packed together as closely as possible.

When arranged in a single layer, the most efficient way (wasting the least space) of packing the spheres is in the form of a hexagon in which each sphere is surrounded by six others. A square arrangement is less efficient because more space is wasted between spheres (Figure 5.1).

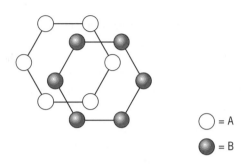

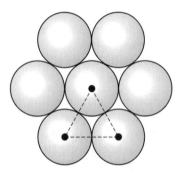

hexagonal packing

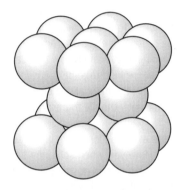

Figure 5.2 Hexagonal close packing (ABAB. . .)

In **face-centred cubic packing (cubic close packing)** the third layer does not sit directly above either the first or second layers. The pattern of layers is repeated after three layers (Figure 5.3), giving rise to an ABCABCABC. . . arrangement.

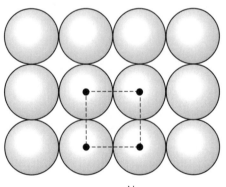

square packing

Figure 5.1 Packing in metallic crystals

The situation becomes more complex when layers are placed on top of each other. Hexagonal layers can be stacked on top of each other in two different ways.

In **hexagonal close packing** a second layer is positioned in such a way that each sphere in the second layer is in contact with three spheres of the first layer (and each sphere in the first layer is in contact with three spheres of the second layer). The third layer is placed directly above the first, and the fourth directly above the second (Figure 5.2). The result is sometimes represented as an ABABAB. . . arrangement.

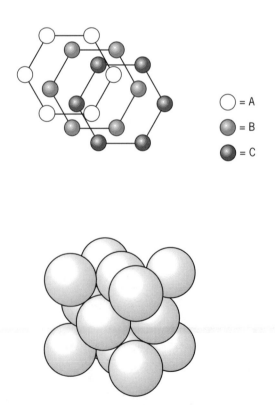

Figure 5.3 Face-centred cubic packing (ABCABC. . .)

Both hexagonal close packing and face-centred cubic packing may be considered as efficient packing since spheres occupy 74% of the available space. In both arrangements, each sphere is in contact with twelve others and is said to have a coordination number of 12.

In **body-centred cubic packing** the layers are formed from spheres arranged in squares. The second layer is positioned in such a way that each sphere in the second layer is in contact with four spheres in the first layer. The third layer sits directly above the first layer (Figure 5.4), giving rise to an ABABAB... arrangement.

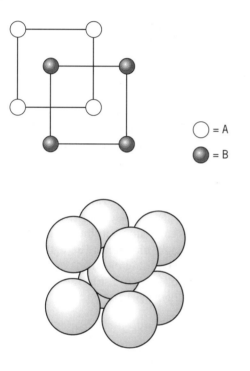

$\bigcirc$ = A
$\bullet$ = B

Figure 5.4 Body-centred cubic packing (ABAB...)

Body-centred cubic packing is less efficient than the previously described structures. Spheres occupy only 68% of the available space. Each sphere is in contact with eight others (four in the layer above and four in the layer below) and therefore has a coordination number of 8.

Table 5.1 contains examples of metals that show these structures.

Table 5.1 The types of metallic crystals

Metals showing hexagonal close packing	Metals showing face-centred cubic packing	Metals showing body-centred cubic packing
cobalt	aluminium	Group 1 metals
magnesium	calcium	barium
titanium	copper	chromium
zinc	lead	iron
	nickel	vanadium

Ionic crystals

The ions in an ionic crystal are arranged similar to those in a metallic crystal; however, each type of ion in an ionic lattice has its own coordination number. The lattice structure is, in the main, determined by two factors:

- the ratio of the number of cations to anions
- the ratio of the radii of the ions (r_A/r_B) – as this increases the coordination number of the ionic lattice increases.

The radius ratio in the sodium chloride lattice is 0.57. The ions are arranged in a face-centred cubic structure (Figure 5.5). The coordination number is 6.

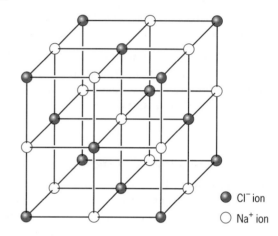

$\bullet$ Cl$^-$ ion
$\bigcirc$ Na$^+$ ion

Figure 5.5 Sodium chloride lattice – face-centred cubic

In caesium chloride the radius ratio is 0.94 (due to the larger caesium ion). The ions are arranged in a body-centred cubic structure with a coordination number of 8 (Figure 5.6).

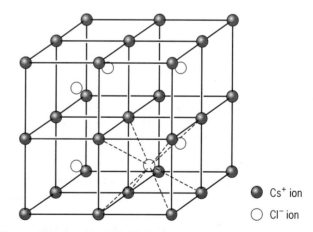

$\bullet$ Cs$^+$ ion
$\bigcirc$ Cl$^-$ ion

Figure 5.6 Caesium chloride lattice – body-centred cubic

In general, the higher the value of the radius ratio, the higher the coordination number of the lattice (Table 5.2).

Table 5.2 Radius ratio and coordination number

Radius ratio	Coordination number	Example
0.22–0.41	4	zinc sulphide
0.42–0.73	6	sodium chloride
>0.73	8	caesium chloride

Covalent structures

Electrons repel each other since they carry the same charge. According to the **valence-shell electron-pair repulsion (VSEPR) theory** the spatial arrangement of atoms in a molecule is determined by repulsion between electron pairs. Each electron pair tries to get as far away as possible from all other electron pairs.

The shape of molecules and ions that contain only single covalent bonds can be predicted from the total number of electron pairs in the outside shell of the central atom. This total number includes:

- the number of electrons originally in the outside shell of the central atom
- additional shared electrons in single covalent bonds
- any electrons lost or gained if an ion is formed.

The final shape will also depend on the presence of non-bonding pairs (lone pairs) of electrons. Lone pairs of electrons are more compact than bonding pairs of electrons and repel more strongly.

Predicting the shapes of molecules

Boron trifluoride (BF₃)

There are three electrons in the outer electron shell of boron, plus three shared electrons from the fluorine atoms (one from each atom) in the covalent bonds, giving a total of six electrons. The six electrons form three pairs of bonding electrons and no lone pairs. The bonding pairs are kept as far from each other as possible by adopting a trigonal planar shape (Figure 5.7).

Figure 5.7 Shape of BF₃

The boron–fluorine bonds are directed towards the corners of an equilateral triangle. The angle between each pair of bonds is 120°.

Methane (CH₄)

There are four electrons in the outer electron shell of carbon, plus four shared electrons from the hydrogen atoms (one from each atom) in the covalent bonds, giving a total of eight electrons. The eight electrons form four pairs of bonding electrons and no lone pairs. The bonding pairs are kept as far from each other as possible by adopting a tetrahedral shape (Figure 5.8).

Figure 5.8 Shape of CH₄

The carbon–hydrogen bonds are directed towards the corners of a tetrahedron. The angle between each pair of bonds is 109.5°.

Ammonia (NH₃)

There are five electrons in the outer electron shell of nitrogen, plus three shared electrons from the hydrogen atoms (one from each atom) in the covalent bonds, giving a total of eight electrons. The eight electrons form three pairs of bonding electrons and one lone pair. The lone pair of electrons distorts the molecule, giving a trigonal pyramidal shape (Figure 5.9).

Figure 5.9 Shape of NH₃

In ammonia repulsion between the lone pair of electrons and the bonding pairs of electrons forces the nitrogen–hydrogen bonds slightly closer to each other, resulting in a bond angle of 107°.

Water (H₂O)

There are six electrons in the outer electron shell of oxygen, plus two shared electrons from the hydrogen atoms (one from each atom) in the covalent bonds, giving a total of eight electrons. The eight electrons form two pairs of bonding electrons and two lone pairs. The lone pairs again distort the molecule, giving it a bent shape (Figure 5.10).

Figure 5.10 Shape of H₂O

In water there are two lone pairs of electrons which repel each other more strongly than bonding pairs of electrons. Repulsion between these and the bonding electrons reduces the angle between the oxygen–hydrogen bonds to 104.5°.

Phosphorus pentachloride (PCl₅)

There are five electrons in the outer electron shell of phosphorus, plus five shared electrons from the chlorine atoms (one from each atom) in the covalent bonds, giving a total of ten electrons. The ten electrons form five pairs of bonding electrons and no lone pairs. The bonding pairs are kept as far from each other as possible by adopting a trigonal bipyramidal shape (Figure 5.11).

Figure 5.11 Shape of PCl₅

The phosphorus–chlorine bonds in the central plane of the molecule are at 120° to each other while those bonds above and below are at an angle of 90° to the central plane.

Sulphur hexafluoride (SF_6)

There are six electrons in the outer electron shell of sulphur, plus six shared electrons from the fluorine atoms (one from each atom) in the covalent bonds, giving a total of 12 electrons. The 12 electrons form six pairs of bonding electrons and no lone pairs. The bonding pairs are kept as far from each other as possible by adopting an octahedral shape (Figure 5.12).

Figure 5.12 Shape of SF_6

The sulphur–fluorine bonds in the central plane of the molecule are at 90° to each other and the bonds above and below are also at an angle of 90° to the central plane.

Xenon tetrafluoride (XeF_4)

There are eight electrons in the outer electron shell of xenon, plus four shared electrons from the fluorine atoms (one from each atom) in the covalent bonds, giving a total of 12 electrons. The 12 electrons form four pairs of bonding electrons and two lone pairs. The bonding pairs of electrons and lone pairs of electrons are kept as far from each other as possible by adopting a square planar shape (Figure 5.13).

Figure 5.13 Shape of XeF_4

The xenon–fluorine bonds in the central plane of the molecule are at 90° to each other and the lone pairs above and below are also at an angle of 90° to the central plane.

Predicting the shapes of molecules containing double bonds

The same principles used in the previous section apply.

Carbon dioxide (CO_2)

There are four electrons in the outer electron shell of carbon, plus four shared electrons from the oxygen atoms (two from each atom) in the covalent bonds, giving a total of eight electrons. The eight electrons

form four pairs of bonding electrons (as two double bonds) and no lone pairs. The bonding pairs of electrons are kept as far from each other as possible by adopting a linear shape (Figure 5.14).

Figure 5.14 Shape of CO_2

The double bonds between carbon and oxygen atoms are at an angle of 180°.

Sulphur dioxide (SO_2)

There are six electrons in the outer electron shell of sulphur, plus four shared electrons from the oxygen atoms (two from each atom) in the covalent bonds, giving a total of ten electrons. The ten electrons form four pairs of bonding electrons (as two double bonds) and one lone pair. The bonding pairs of electrons and lone pair of electrons are kept as far from each other as possible by adopting a bent shape (Figure 5.15).

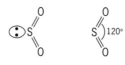

Figure 5.15 Shape of SO_2

The double bonds between sulphur and oxygen atoms are at an angle of 120°.

Predicting the shapes of ions

The shapes of ions can be predicted in a similar way to those of molecules.

Ammonium ion (NH_4^+)

There are five electrons in the outer electron shell of nitrogen, plus four shared electrons from the hydrogen atoms (one from each atom) in the covalent bonds, with one electron removed to form a positive ion, giving a total of eight electrons. The eight electrons form four pairs of bonding electrons and no lone pairs. The bonding pairs of electrons are kept as far from each other as possible by adopting a tetrahedral shape (Figure 5.16).

Figure 5.16 Shape of NH_4^+ ion

The nitrogen–hydrogen bonds are directed towards the corners of a tetrahedron. The angle between each pair of bonds is 109.5°.

Nitrate ion (NO₃⁻)

There are five electrons in the outer electron shell of nitrogen, plus six shared electrons from the oxygen atoms (two from each atom) in the covalent bonds, with one electron added to form a negative ion, giving a total of 12 electrons. The 12 electrons form six pairs of bonding electrons (as three double bonds) and no lone pairs. The bonding pairs are kept as far from each other as possible by adopting a trigonal planar shape (Figure 5.17). The nitrogen–oxygen bonds are directed towards the corners of a triangle. The angle between each pair of bonds is 120°.

Figure 5.17 Shape of NO₃⁻ ion

Sulphate ion (SO₄²⁻)

There are six electrons in the outer electron shell of sulphur, plus eight shared electrons from the oxygen atoms (two from each atom) in the covalent bonds, with two electrons added to form a negative ion, giving a total of 16 electrons. The 16 electrons form eight pairs of bonding electrons (as four double bonds) and no lone pairs. The bonding pairs are kept as far from each other as possible by adopting a tetrahedral shape (Figure 5.18).

Figure 5.18 Shape of SO₄²⁻ ion

The angle between the sulphur–oxygen bonds is 109.5°.

Phosphorus hexachloride ion (PCl₆⁻)

There are five electrons in the outer electron shell of phosphorus, plus six shared electrons from the chlorine atoms (one from each atom) in the covalent bonds, with one electron added to form a negative ion, giving a total of 12 electrons. The 12 electrons form six pairs of bonding electrons and no lone pairs. The bonding pairs of electrons are kept as far from each other as possible by adopting an octahedral shape (Figure 5.19).

Figure 5.19 Shape of PCl₆⁻ ion

All of the chlorine–phosphorus bonds in the molecule are at 90° to each other.

1 Which of the following describes hexagonal close packing?

	Arrangement	Coordination number
A	ABABAB...	8
B	ABABAB...	12
C	ABCABC...	8
D	ABCABC...	12

2 In crystalline form, potassium iodide shows 6 coordination. The radius ratio of the potassium ion to the iodide ion is:

A <0.22   B 0.22–0.41
C 0.42–0.73   D >0.73

3 The angle between the bonds in the covalent compound silicon(IV) chloride ($SiCl_4$) is
A 104.5°   B 107°
C 109.5°   D 120°

4 Which of the following is most similar in shape to NH_3?
A BF_3 B $FeCl_3$
C GaI_3 D PBr_3

5 The shape of the ammonium ion (NH_4^+) is
A trigonal planar
B tetrahedral
C trigonal pyramidal
D trigonal bipyramidal

6 The following diagram shows the crystal structures of three metals.

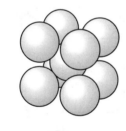

a potassium

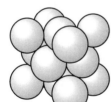

b aluminium

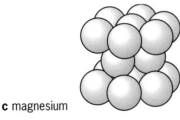

c magnesium

For each metal:
i name the structure
ii state the coordination number of the atoms
iii explain how the layers are arranged on top of each other to form a three-dimensional structure.

7 Tetrachloromethane is a covalent compound of carbon and chlorine, and has the formula CCl_4.
  a Give the electron configurations of carbon and chlorine.
  b Draw a diagram to show the shape of the CCl_4 molecule and give the value(s) of the bond angles.
  c Explain the shape of the molecule in terms of electron pair repulsion.
  d State, with reasons, whether you would expect the covalent compound NCl_3 to have the same shape, or a different shape, to that of CCl_4.

8 a State two factors that determine the lattice structure of an ionic compound.
  b The following table gives the ionic radii of some ions.

Ion	Br^-	Cs^+	I^-	Na^+
Ionic radius/nm	0.195	0.170	0.215	0.102

  Calculate the ratio of the radii of the ions in the compounds sodium bromide (NaBr) and caesium iodide (CsI) and explain why these compounds have different crystal structures.
  c Draw diagrams to show the arrangement of ions in sodium bromide and in caesium iodide and in each case state the coordination number.

9 a With the help of dot–cross diagrams explain why the O=C=O angle in carbon dioxide is 180° but the O=S=O angle in sulphur dioxide is only 119.5°.
  b Predict the size of the F—O—F angle in F_2O and explain the reasoning behind your answer.

10 Predict the shape of the following and explain your reasoning in each case.
  a PH_3
  b SF_6
  c $AlCl_4^-$
  d PH_4^+
  e CO_3^{2-}
  f SO_3

1 B **2** C **3** C **4** D **5** B

6 a i Body-centred cubic packing.

ii Coordination number 8

iii The layers are formed from spheres arranged in squares. Each sphere in the second layer is positioned in such a way as to be in contact with four atoms below it in the first layer and four atoms above it in the third layer. The third layer is immediately above the first, giving an ABABAB... arrangement.

b i Face-centred cubic packing.

ii Coordination number 12

iii The layers are formed from atoms arranged in hexagons. Each sphere in the second layer is in contact with three atoms in the first layer and three atoms in the third layer. The third layer does not sit directly above the first or second. The pattern of layers is repeated after three layers, giving an ABCABC... arrangement.

c i Hexagonal close packing.

ii Coordination number 12

iii Like aluminium, the layers are formed from atoms arranged in hexagons. Each sphere in the second layer is in contact with three atoms in the first layer and three atoms in the third layer. Unlike aluminium, the third layer sits directly above the first, giving an ABABAB... arrangement.

7 a Carbon $1s^2 2s^2 2p^2$
Chlorine $1s^2 2s^2 2p^6 3s^2 3p^5$

b

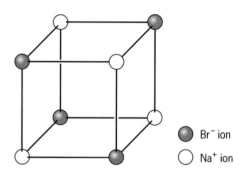

c There are four pairs of bonding electrons around the carbon atom which repel equally. A tetrahedral shape allows the pairs of electrons to get as far away from each other as possible.

d In NCl_3 there are three bonding pairs of electrons and one lone pair of electrons. The lone pair of electrons will repel more strongly. The result is that the structure will be distorted to give a trigonal pyramidal shape.

8 a The ratio of the number of cations to anions and the ratio of the radii of the ions.

b For NaBr: $r_A/r_B = \dfrac{0.102}{0.195}$

$= 0.52$

For CsI: $r_A/r_B = \dfrac{0.170}{0.215}$

$= 0.79$

The lattice structure is determined by the ratio of the number of cations to anions, which in both compounds is $1:1$, and the ratio of the radii of the ions. In caesium iodide this ratio is significantly greater. In NaBr, the Br^- ion is too big to allow 8 to pack around the small Na^+ ion.

c

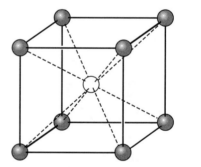

The coordination number is 6.

The coordination number is 8.

9 a

$$\overset{..}{\underset{..}{O}} \overset{\times}{\underset{\times}{:}} C \overset{\times}{\underset{\times}{:}} \overset{..}{\underset{..}{O}}$$

In carbon dioxide the four electrons (in two double bonds) in the outer shell of the carbon atom are all used in forming bonds. The linear shape allows the bonding electrons to move as far as possible away from each other.

$$\overset{..}{\underset{..}{O}} \overset{\times\times}{\underset{\times\ \times}{\cdot\ S\ \cdot}} \overset{..}{\underset{..}{O}}$$

In sulphur dioxide only four of the six electrons in the outer shell of the sulphur atom are used in forming two double bonds. The remaining two electrons form a lone pair. A trigonal planar shape allows the bonding electrons and the lone pair to move as far as possible away from each other, hence the smaller bond angle of 120°.

b In F_2O the oxygen atom has six electrons in the outer shell. Two of these are involved in making covalent bonds with the fluorine atoms while the remaining four electrons form two lone pairs.

In order to keep as far as possible from each other we would expect the two pairs of bonding electrons and two lone pairs of electrons to be distributed in the shape of a tetrahedron about the oxygen atom. However, the lone pairs repel more strongly than the bonding pairs of electrons so the structure will be distorted into a trigonal pyramidal shape in which the F—O—F bond will be a little less than 109.5°, rather like H_2O. The actual value is 103°.

10 a In PH_3 there are five electrons in the outside shell plus three shared electrons in the covalent bonds – a total of eight electrons.

The eight electrons form three pairs of bonding electrons and one lone pair of electrons.

The bonding pairs of electrons are kept as far from each other as possible by adopting a trigonal pyramidal shape.

b In SF_6 there are six electrons in the outside shell plus six shared electrons in the covalent bonds – a total of 12 electrons.

The 12 electrons form six pairs of bonding electrons and no lone pairs of electrons.

The bonding pairs are kept as far from each other as possible by adopting an octahedral shape.

c In $AlCl_4{}^-$ there are three electrons in the outside shell plus four shared electrons in the covalent bonds plus one electron added to form a negative ion – a total of eight electrons.

The eight electrons form four pairs of bonding electrons and no lone pairs of electrons.

The bonding pairs are kept as far from each other as possible by adopting a tetrahedral shape.

d In $PH_4{}^+$ there are five electrons in the outside shell plus four shared electrons in the covalent bonds less one electron removed to form a positive ion – a total of eight electrons.

The eight electrons form four pairs of bonding electrons and no lone pairs of electrons.

The bonding pairs are kept as far from each other as possible by adopting a tetrahedral shape.

e In $CO_3{}^{2-}$ there are four electrons in the outside shell of carbon plus six shared electrons from the oxygens (two from each atom) in the covalent bonds plus two electrons added to form a negative ion – a total of 12 electrons.

The 12 electrons form six pairs of bonding electrons (as three double bonds) and no lone pairs of electrons.

The bonding pairs are kept as far from each other as possible by adopting a trigonal planar shape.

f In $SO_3{}^{2-}$ there are six electrons in the outside shell of sulphur plus six shared electrons from the oxygens (two from each atom) in the covalent bonds plus two electrons added to form a negative ion – a total of 14 electrons.

The 14 electrons form six pairs of bonding electrons (as three double bonds) and one lone pair of electrons.

The electron pairs are kept as far from each other as possible by adopting a trigonal pyramidal shape.

6 The Periodic Table

The modern Periodic Table is based on a suggested ordering of elements by the Russian chemist Dmitri Mendeleev and subsequent work carried out by Henry Moseley. In it, elements are arranged in order of ascending atomic number.

Within the Periodic Table, **metallic** elements occupy positions to the left and centre, while **non-metallic** elements are to be found on the right. Hydrogen is an exception to this pattern. The atomic structure of hydrogen would indicate that it belongs at the top left of the table; however, it is usually shown separate from the remainder of Group 1 as it is a non-metal and has very different properties to the other elements in this group.

Traditionally a stepped diagonal line shows the boundary between metals and non-metals; however, the change from metal to non-metal is not so well defined. Elements in the boundary region show a mixture of properties, some typical of metals and some typical of non-metals. These boundary elements are often referred to as **semi-metals** or **metalloids**.

Group	1	2												3	4	5	6	7	0
Period 1	1.0 H 1																		4.0 He 2
2	6.9 Li 3	9.0 Be 4												10.8 B 5	12.0 C 6	14.0 N 7	16.0 O 8	19.0 F 9	20.2 Ne 10
3	23.0 Na 11	24.3 Mg 12												27.0 Al 13	28.1 Si 14	31.0 P 15	32.1 S 16	35.5 Cl 17	39.9 Ar 18
4	39.1 K 19	40.1 Ca 20	45.0 Sc 21	47.9 Ti 22	50.9 V 23	52.0 Cr 24	54.9 Mn 25	55.8 Fe 26	58.9 Co 27	58.7 Ni 28	63.5 Cu 29	65.4 Zn 30		69.7 Ga 31	72.6 Ge 32	74.9 As 33	79.0 Se 34	79.9 Br 35	83.8 Kr 36
5	85.5 Rb 37	87.6 Sr 38	88.9 Y 39	91.2 Zr 40	92.9 Nb 41	95.9 Mo 42	(98) Tc 43	101.1 Ru 44	102.9 Rh 45	106.4 Pd 46	107.9 Ag 47	112.4 Cd 48		114.8 In 49	118.7 Sn 50	121.8 Sb 51	127.6 Te 52	126.9 I 53	131.3 Xe 54
6	132.9 Cs 55	137.3 Ba 56		178.5 Hf 72	180.9 Ta 73	183.9 W 74	186.2 Re 75	190.2 Os 76	192.2 Ir 77	195.1 Pt 78	197.0 Au 79	200.6 Hg 80		204.4 Tl 81	207.2 Pb 82	209.0 Bi 83	(209) Po 84	(210) At 85	(222) Rn 86
7	(223) Fr 87	226.0 Ra 88		(261) Rf 104	(262) Db 105	(263) Sg 106	(264) Bh 107	(265) Hs 108	(266) Mt 109	(269) Uun 110	(272) Uuu 111	(277) Uub 112			(285) Uuq 114		(289) Uuh 116		(293) Uno 118

138.9 La 57	140.1 Ce 58	140.9 Pr 59	144.2 Nd 60	(145) Pm 61	150.4 Sm 62	152.0 Eu 63	157.3 Gd 64	158.9 Tb 65	162.5 Dy 66	164.9 Ho 67	167.3 Er 68	168.9 Tm 69	173.0 Yb 70	175.0 Lu 71
227.0 Ac 89	232.0 Th 90	231.0 Pa 91	238.0 U 92	237.0 Np 93	(244) Pu 94	(243) Am 95	(247) Cm 96	(247) Bk 97	(251) Cf 98	(252) Es 99	(257) Fm 100	(258) Md 101	(259) No 102	(260) Lr 103

key

 metals

semi-metals

non-metals

Figure 6.1 Periodic Table

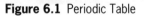

Metals

Metals exist as giant structures in which a 'sea' of free electrons is able to move through a lattice of positively charged ions (cations).

- Metals are good conductors of electricity and heat.
- With the exception of mercury, metals are solids at room temperature.
- Metals form positively charged ions (cations) by the loss of one or more electrons.

$$M \rightarrow M^{n+} + ne^-$$

Non-metals

Unlike metals, non-metals show considerable variation in structure.

- At room temperature:
  - the elements of Group 0, which includes neon (Ne) and argon (Ar), are all gases existing as separate atoms
  - the elements of Groups 5, 6 and 7 that are gases, such as nitrogen (N_2), oxygen (O_2) and chlorine (Cl_2), exist as separate molecules, each containing two atoms (diatomic)
  - bromine (Br_2), from Group 7, is a liquid, existing as separate molecules
  - sulphur (S_8), from Group 6, and iodine (I_2), from Group 7, are solids, existing as lattices of separate molecules
  - carbon (two forms – graphite and diamond) and silicon from Group 4 are solids, existing as giant molecular structures.
- Non-metals are generally poor conductors of electricity and heat.
- Non-metals form negatively charged ions (anions) by gaining one or more electrons.

$$X + ne^- \rightarrow X^{n-}$$

Semi-metals

The semi-metals are a cluster of elements that occur in the Periodic Table at the boundary between metals and non-metals. The classification is not definitive but usually includes boron, silicon, germanium, arsenic and tellurium, and sometimes selenium, antimony, bismuth and polonium.

As the name suggests, these elements exhibit properties that are intermediate between typical metals and typical non-metals. For example, in conducting electricity:

- metals are good **conductors** of electricity
- semi-metals are poor conductors of electricity and are often called electrical **semi-conductors**
- non-metals do not conduct an electric current and are called **insulators**.

Carbon displays both semi-metal and non-metal properties in its different forms. Graphite is a poor conductor and therefore is a semi-metal, while diamond is an insulator and is therefore a non-metal.

Group

The vertical columns of elements in the Periodic Table are called **groups**. Each group has a group number.

All of the elements within a particular group have similarities in their electronic configuration; they have the same number of electrons in the outermost sub-shell. As the chemistry of an element is determined by its electronic configuration, the elements within each group have similar chemical properties.

Passing down a group, there is a definite trend in reactivity as the metallic character of the elements increases with the increasing size of the atom.

In groups that contain metallic elements, the increase in metallic character is responsible for a general increase in reactivity. Metal–metal bonds decrease in strength as atoms increase in size and bonding electrons are less strongly attracted as non-bonding electrons increasingly shield nuclei.

Conversely, in groups that contain non-metallic elements, the increase in metallic character is responsible for a general decrease in reactivity.

Period

The horizontal rows of elements in the Periodic Table are called **periods**. Each period has a period number. Reading across a period from left to right the metallic nature of the elements decreases. Elements to the left are metals, those to the right are non-metals while those in the centre exhibit a mixture of properties and are thus described as semi-metals.

There is a significant variation in the properties of the elements across a period. Table 6.1 (page 32) contains information about the elements in Period 3.

The variation in melting point is a result of both bond strength and structure of the elements.

- Across sodium, magnesium and aluminium, which are metals, there is increasing nuclear charge and increasing density of mobile electrons which increases the strength of the metal–metal bonds.
- Silicon is a macromolecule in which each atom is linked to other atoms in three dimensions by strong covalent bonds.
- Phosphorus (P_4), sulphur (S_8) and chlorine (Cl_2) are simple molecular substances. Their melting points are determined by the size of the van der Waals forces between molecules which, in turn, are related to the sizes of the molecules.

Periodic trends of oxides and chlorides

Periodic trends are also evident in compounds of the elements in a period.

Period 3 oxides

The Period 3 oxides that are ionic lattices or giant molecular solids have strong forces between particles; thus they have high melting points (Table 6.2). Those that are simple molecular compounds have weak forces between molecules; hence they have low melting points.

The reaction of these oxides with water changes across the period:

- sodium oxide and magnesium oxide react to give alkaline solutions; thus they are described as **alkaline oxides**:

$$Na_2O(s) + H_2O(l) \rightarrow 2Na^+(aq) + 2OH^-(aq)$$
$$MgO(s) + H_2O(l) \rightarrow Mg^{2+}(aq) + 2OH^-(aq)$$

- aluminium oxide and silicon dioxide are insoluble
- phosphorus pentoxide and sulphur dioxide react to give acidic solutions; thus they are described as **acidic oxides**:

$$P_4O_{10}(s) + 6H_2O(l) \rightarrow 4H_3PO_4(aq)$$
phosphoric acid

$$SO_2(g) + H_2O(l) \rightarrow H_2SO_3(aq)$$
sulphurous acid

Across the period, ionic oxides react with water to form alkaline solutions while soluble covalent oxides react to form acidic solutions.

Period 3 chlorides

The Period 3 chlorides show a similar trend to the oxides (Table 6.3). At room temperature both aluminium chloride and phosphorus pentachloride have unusual properties that are worthy of comment.

- Aluminium chloride exists as a **dimer**. This is a structure in which two molecules of $AlCl_3$ link together to form a single molecule, Al_2Cl_6.
- Phosphorus pentachloride exists as an ionic solid containing the ions PCl_4^+ and PCl_6^-. However, at 162°C it **sublimes** to form a simple molecular gas, PCl_5.

The reaction of these chlorides with water changes across the period:

- sodium chloride and magnesium chloride dissolve to form neutral solutions
- silicon tetrachloride, phosphorus pentachloride and sulphur dichloride react to give acidic solutions.

Across the period, ionic chlorides dissolve in water to form neutral solutions while covalent chlorides are hydrolysed by water to form acidic solutions.

Table 6.1 Period 3 elements

Element	Na	Mg	Al	Si	P (white)	S (rhombic)	Cl	Ar
Melting point/°C	98	650	660	1410	44	115	−101	−189
Boiling point/°C	900	1100	2520	3200	280	445	−34	−186
State at 20°C	solid	solid	solid	solid	solid	solid	gas	gas
Electrical conductivity		good		poor		very poor		
Type of element	metal	metal	metal	semi-metal	non-metal	non-metal	non-metal	non-metal
Type of structure		giant metallic		macro-molecular		simple molecular		

Table 6.2 Period 3 oxides

Oxide	Na_2O	MgO	Al_2O_3	SiO_2	P_4O_{10}	SO_2	Cl_2O
Melting point/°C	1130	2862	2054	1726	358(sub)	−73	−111
Type of bonding	ionic	ionic	ionic/covalent	covalent	covalent	covalent	covalent
Type of structure		lattice		giant molecular		simple molecular	

Table 6.3 Period 3 chlorides

Chloride	NaCl	$MgCl_2$	$AlCl_3$	$SiCl_4$	PCl_5	SCl_2	Cl_2
Melting point/°C	801	708	181(sub)	−69	162(sub)	−78	−101
Type of bonding	ionic	ionic	covalent	covalent	covalent	covalent	covalent
Type of structure		lattice		simple molecular			

1 Which of the following is a characteristic of most non-metals?
  A forms an ion by gaining one or more electrons
  B good conductor of electricity
  C good conductor of heat
  D solid at room temperature

2 Which of the following Group 2 elements is the most reactive?
  A barium
  B calcium
  C magnesium
  D strontium

3 Which of the following Group 6 elements is the least reactive?
  A oxygen
  B selenium
  C sulphur
  D tellurium

4 Which of the following Period 2 elements exists as a giant covalent structure?
  A carbon
  B fluorine
  C lithium
  D nitrogen

5 Which of the following elements may be classified as a semi-metal?
  A argon
  B boron
  C magnesium
  D sulphur

6 The Periodic Table devised by Mendeleev was one of several attempts made by early chemists to arrange the elements into groups. In 1864 the English chemist John Newlands noticed that if the elements were arranged in order of increasing atomic weight the eighth element, starting from a particular element, was 'a sort of repetition' of that element – rather like an octave of music. He called this pattern the law of octaves. This is part of the arrangement.

H	Li	Be	B	C	N	O
F	Na	Mg	Al	Si	P	S
Cl	K	Ca	Cr	Ti	Mn	Fe

  a What was the basis of the 'sort of repetition' observed by Newlands?
  b From the above table, give one example of a group of elements which are:
    i found in the same group in the modern Periodic Table
    ii not found in the same group in the modern Periodic Table.
  c Suggest why Newlands' octaves were not well received by fellow chemists of his day.

7 The following graph shows the melting points of the Period 3 elements.

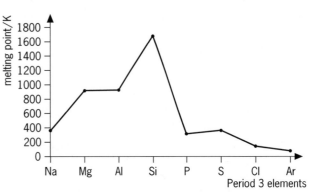

Explain, in terms of structure and bonding:
  a the increase in melting point between sodium and aluminium
  b why silicon has the highest melting point
  c the variation in melting point between phosphorus and chlorine.

8 Describe and explain the trends across the period by the reaction of water with:
  a Period 3 oxides
  b Period 3 chlorides.

9 The following graph shows the atomic radii of the elements in Group 1.

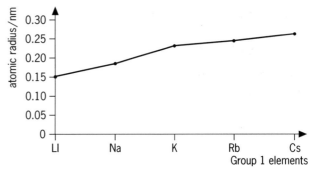

  a Why does the atomic radius increase down the group?
  b Why does reactivity increase down the group?
  c Why does melting point decrease down the group?

10 The following table shows some of the Group 7 elements.

F
Cl
Br
I

Describe how the:
  a elements form ions
  b state of the elements, at room temperature and atmospheric pressure, changes down the group
  c reactivity changes down the group
  d atomic radius changes down the group.

1 A

2 A

3 D

4 A

5 B

6 a Similarities in chemical reactions.
  b i For example, Li, Na, K
    ii For example, H, F, Cl
  c Some of the elements were placed in groups with others exhibiting similar chemistry; however, there were too many instances where elements with obviously different chemistry were grouped together in order to make them appear to fit the pattern.

7 a The atoms have an increasing nuclear charge and an increasing number of mobile electrons which results in an increase in the strength of the metal–metal bonds.
  b Silicon has a giant three-dimensional molecular structure, similar to diamond, in which each atom is bonded to four other atoms by strong covalent bonds. A large amount of energy is needed to break these bonds.
  c The elements phosphorus, sulphur and chlorine exist as simple covalent structures. The melting point of each is determined by the strength of the van der Waals forces between molecules. Each of the elements has a low melting point; however, the sulphur molecule (S_8) is the biggest and most liable to polarisation. The van der Waals forces are greatest in sulphur, thus it has the highest melting point of the three elements.

8 a The oxides of sodium and magnesium are ionic and react with water to give alkaline solutions. Sodium oxide is more soluble than magnesium oxide and forms a more alkaline solution.
  Aluminium oxide is ionic with a high degree of covalent character while silicon dioxide is covalent. Both of these oxides are effectively insoluble in water.
  The oxides of phosphorus, sulphur and chlorine are covalent and react with water to give acidic solutions.
  b The chlorides of sodium and magnesium are ionic. They dissolve in water to form neutral solutions.
  The chlorides of aluminium, silicon, phosphorus and sulphur are covalent. They react with water to form acidic solutions containing HCl.

9 a The number of electron shells increases down the group.
  b As the atomic radius increases the outside of the atom is increasingly shielded from the nucleus by an increasing number of inner shells of electrons. Less energy is needed to remove the outermost electron, thus elements become progressively more reactive.
  c As the atoms increase in size ionic charge increases and the outer electron is held in an orbit that is progressively further away from the nucleus so electron density decreases. The melting points decrease down the group because the metal–metal bonds decrease in strength.

10 a Ions are formed by the gain of a single electron to give an X^- ion.
  b Fluorine and chlorine are gases, bromine is a liquid and iodine is a solid.
  c Reactivity decreases down the group.
  d The atomic radius increases down the group.

7 Formulae and equations

Empirical formula and molecular formula

The formula of a substance is determined by the number and type of atoms of which it is made:

- the **empirical formula** represents the simplest ratio of atoms in a compound
- the **molecular formula** represents the actual number of atoms in a molecule.

For ethane, the empirical formula is CH_3 and the molecular formula is C_2H_6.

$$
\begin{array}{c}
\quad H \quad H \\
\quad | \quad\ | \\
H-C-C-H \\
\quad | \quad\ | \\
\quad H \quad H
\end{array}
$$
ethane

For ethanol, the empirical formula is C_2H_6O; this is the simplest ratio of atoms. The molecular formula is also C_2H_6O.

$$
\begin{array}{c}
\quad H \quad H \\
\quad | \quad\ | \\
H-C-C-O-H \\
\quad | \quad\ | \\
\quad H \quad H
\end{array}
$$
ethanol

Since ionic compounds exist as giant ionic structures, it makes no sense to talk about 'ionic molecules'. The formula of an ionic compound is really an empirical formula, reflecting the ions present in their simplest ratio.

Calculating empirical formulae

The empirical formula of a substance can be calculated from data that gives the percentage composition, by mass, of each element in the compound.

Example 1
An organic compound is found to consist of 40.0% carbon, 6.7% hydrogen and 53.3% oxygen.
Atoms of carbon, hydrogen and oxygen differ in mass.

- To find the ratio of carbon atoms to hydrogen atoms to oxygen atoms we must divide each percentage by the relative atomic mass of the element.

$$
C:H:O = \frac{40.0}{12} : \frac{6.7}{1} : \frac{53.3}{16}
$$
$$
= 3.3 : 6.7 : 3.3
$$

- The ratio is obtained in its simplest form by dividing through by the smallest number.

$$
C:H:O = \frac{3.3}{3.3} : \frac{6.7}{3.3} : \frac{3.3}{3.3}
$$
$$
= 1:2:1
$$

- The empirical formula of the compound is CH_2O.

Sometimes the ratio may involve a number that is not a whole number. For example, aluminium oxide is found to consist of 52.9% aluminium and 47.1% oxygen. If we calculate the empirical formula as we did above:

$$
Al:O = \frac{52.9}{27} : \frac{47.1}{16}
$$
$$
= 1.96 : 2.94
$$

Dividing through by the smallest number:

$$
Al:O = \frac{1.96}{1.96} : \frac{2.94}{1.96}
$$
$$
= 1:1.5
$$

In order to obtain the empirical formula we must multiply through by 2 so $1:1.5$ becomes $2:3$. The empirical formula of aluminium oxide is Al_2O_3.

Calculating molecular formulae

In order to calculate the molecular formula from the empirical formula, the relative molecular mass, M_r, of the compound is needed. This value is often obtained from a mass spectrum (see topic 23) of the substance.

Example 2
In the case of CH_2O in the previous example, M_r is 60.

- The empirical formula mass of CH_2O is $12 + 2 + 16 = 30$.

$$
M_r : \text{empirical formula mass} = 60:30
$$
$$
= 2:1
$$

- The molecular formula must be 2 times the empirical formula and is therefore $C_2H_4O_2$.

Note that the molecular formula is always a whole number times the empirical formula. Sometimes the empirical formula and molecular formula are the same.

Structural and display formulae

The molecular formula of a compound gives the number and type of atoms it contains but it gives no details of the compound's structure – the spatial arrangement of the atoms.

The **structural formula** gives some indication of how the atoms are arranged in a molecule and may be written in a variety of ways. For example, the structural formula of ethanoic acid may be written:

$CH_3-CO-OH$ CH_3-COOH $CH_3.CO.OH$ $CH_3.COOH$

Structural formulae can also show the arrangement of atoms, or groups of atoms, in space.

ethanoic acid

A structural formula in which all the covalent bonds are shown is sometimes called a **display formula** or a **graphical formula**.

A structural formula may be simplified by using brackets where groups of atoms are repeated. For example, the structural formula for hexane, $CH_3-CH_2-CH_2-CH_2-CH_2-CH_3$, may be written as $CH_3-(CH_2)_4-CH_3$.

Balancing equations

In a mathematical equation the expression given to the left of the '=' sign is equal to that on the right. The same thing is true of a balanced chemical equation except that the '=' sign is replaced by '→' or, in the case of a reversible reaction, by '⇌'.

In a balanced chemical equation the same number of atoms of each element as appear on the left-hand side of the arrow must appear on the right-hand side.

Example 3
The following equation is unbalanced.

$$CuCO_3 + HCl \rightarrow CuCl_2 + H_2O + CO_2 \qquad \text{(unbalanced)}$$

- Counting the number of atoms on each side of the equation:

Left-hand side	Right-hand side
$1 \times Cu$	$1 \times Cu$
$1 \times C$	$1 \times C$
$3 \times O$	$3 \times O$
$1 \times H$	$2 \times H$
$1 \times Cl$	$2 \times Cl$

- For the equation to be balanced $1 \times H$ and $1 \times Cl$ must be added to the left-hand side and the equation becomes:

$$CuCO_3 + 2HCl \rightarrow CuCl_2 + H_2O + CO_2 \qquad \text{(balanced)}$$

In order to balance some equations several steps may be necessary.

$$C_3H_8 + O_2 \rightarrow CO_2 + H_2O \qquad \text{(unbalanced)}$$

- Count the atoms.

Left-hand side	Right-hand side
$3 \times C$	$1 \times C$
$2 \times O$	$3 \times O$
$8 \times H$	$2 \times H$

- Adding $2 \times C$ to the right-hand side the equation becomes:

$$C_3H_8 + O_2 \rightarrow 3CO_2 + H_2O \qquad \text{(unbalanced)}$$

- This also changes the number of oxygen atoms on the right-hand side so:

Left-hand side	Right-hand side
$3 \times C$	$3 \times C$
$2 \times O$	$7 \times O$
$8 \times H$	$2 \times H$

- Adding $6 \times H$ to the right-hand side the equation becomes:

$$C_3H_8 + O_2 \rightarrow 3CO_2 + 4H_2O \qquad \text{(unbalanced)}$$

- Again, this changes the number of oxygen atoms on the right-hand side so:

Left-hand side	Right-hand side
$3 \times C$	$3 \times C$
$2 \times O$	$10 \times O$
$8 \times H$	$8 \times H$

- Adding $8 \times O$ to the left-hand side the equation is finally balanced.

$$C_3H_8 + 5O_2 \rightarrow 3CO_2 + 4H_2O \qquad \text{(balanced)}$$

Ionic equations

Iron displaces copper(II) ions from an aqueous solution of copper(II) sulphate, according to the following equation.

$$CuSO_4 + Fe \rightarrow Cu + FeSO_4$$

If the equation is written in terms of ions:

$$Cu^{2+} + SO_4^{2-} + Fe \rightarrow Cu + Fe^{2+} + SO_4^{2-}$$

it is clear that the sulphate ion, SO_4^{2-}, plays no part in the reaction. Ions that do not take part in reactions are called **spectator ions** and should not be included in ionic equations.

The ionic equation for this reaction is:

$$Cu^{2+} + Fe \rightarrow Cu + Fe^{2+}$$

The reactions that occur during electrolysis are often represented by pairs of **half-equations** involving ions and electrons.

The following pair of equations represents the reactions that occur during the electrolysis of an aqueous solution of sodium chloride.

- at the anode: $\qquad 2Cl^- - 2e^- \rightarrow Cl_2$
- at the cathode: $\qquad 2H^+ + 2e^- \rightarrow H_2$

State symbols

The following symbols should normally be written after each substance in an equation to indicate its **state**.

- (s) = solid
- (l) = liquid
- (g) = gas
- (aq) = aqueous solution

$$2K(s) + H_2O(l) \rightarrow 2KOH(aq) + H_2(g)$$

The equation indicates that solid potassium reacts with liquid water to produce an aqueous solution of potassium hydroxide and hydrogen gas.

1 The relative molecular mass of a compound is 60. Its molecular formula could be
   A CH_3CONH_2
   B $HOCH_2CH_2OH$
   C H_2NCONH_2
   D CH_3CH_2SH

2 The empirical formula of a compound is C_2H_4O. The compound could be
   A CH_3COOH
   B $HOCH_2CH_2CH_2CH_2OH$
   C $CH_3CH_2CH_2COOH$
   D $HOCH_2CH_2CH_2CH_2CH_2CH_2OH$

3 A compound consists of 38.71% carbon, 9.68% hydrogen and 51.61% oxygen. The compound could be
   A CH_3OH
   B $HOCH_2CH_2OH$
   C CH_3COCH_3
   D CH_3CH_2COOH

4 Which one of the following equations is **not** balanced?
   A $C_3H_8 + 5O_2 \rightarrow 3CO_2 + 4H_2O$
   B $C_2H_5OH + 3O_2 \rightarrow 2CO_2 + 3H_2O$
   C $CH_3COOH + O_2 \rightarrow 2CO_2 + 2H_2O$
   D $CH_3COCH_3 + 4O_2 \rightarrow 3CO_2 + 3H_2O$

5 Which of the following display formulae correctly represents the compound with the molecular formula CH_3CH_2CHO?

   A
   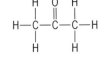

   B

   C
   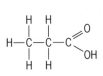

   D
   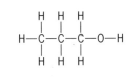

6 Rewrite the following equations so that they balance and include state symbols.
   a $Na + H_2O \rightarrow NaOH + H_2$
   b $KNO_3 \rightarrow KNO_2 + O_2$
   c $C_3H_6 + O_2 \rightarrow CO_2 + H_2O$
   d $Al_2O_3 + H_2SO_4 \rightarrow Al_2(SO_4)_3 + H_2O$
   e $CuCO_3 + HCl \rightarrow CuCl_2 + CO_2 + H_2O$

7 Analysis of an organic compound, **X**, shows it to have the following composition by mass: carbon, 53.33%; hydrogen, 11.11%; oxygen, 35.56%. The mass spectrum of compound **X** shows it to have a relative molecular mass of 90.
   a Calculate the empirical formula of compound **X**.
   b Calculate the molecular formula of compound **X**.
   c Compound **X** contains two —OH groups bonded to different carbon atoms on a straight carbon chain. Draw two possible display formulae for this compound.

8 An inorganic compound is known to have a formula of the form $Na_2X.nH_2O$ where X is an inorganic anion and n is a whole number.
   Analysis of the compound shows it to have the following composition by mass: sodium, 18.25%; sulphur, 12.70%; oxygen, 63.49%; hydrogen, 5.56%.
   a Find the formula of the inorganic compound.
   b Calculate the expected analysis of the anhydrous form of this compound, in which nH_2O have been removed.

9 The empirical formula of an organic compound, **Y**, is the same as its molecular formula. Analysis of the compound shows it to have the following composition by mass: carbon, 38.10%; hydrogen, 7.41%; oxygen, 16.93%; chlorine 37.57%.
   a Calculate the molecular formula of compound **Y**.
   b The compound contains an —OH group and a —Cl group. Draw two possible display formulae for this compound in which these groups are attached to different carbon atoms.

10 Write balanced equations, including state symbols, for the following. Ionic equations should be given if appropriate.
   a Solid zinc carbonate decomposes on heating to form zinc oxide.
   b Iron(III) oxide powder reacts with dilute sulphuric acid to form a solution of iron(III) sulphate.
   c When added together, solutions of barium nitrate and magnesium sulphate produce a white precipitate of barium sulphate.
   d Electrolysis of dilute sulphuric acid produces hydrogen at the cathode and oxygen at the anode.
   e When chlorine gas is bubbled through a solution of potassium bromide the solution goes red-brown due to the release of bromine.

1 C **2** C **3** B **4** C **5** B

6 a $2Na_{(s)} + 2H_2O_{(l)} \rightarrow 2NaOH_{(aq)} + H_{2(g)}$
 b $2KNO_{3(s)} \rightarrow 2KNO_{2(s)} + O_{2(g)}$
 c $C_3H_{6(g)} + \frac{9}{2}O_{2(g)} \rightarrow 3CO_{2(g)} + 3H_2O_{(g)}$
 d $Al_2O_{3(s)} + 3H_2SO_{4(aq)} \rightarrow Al_2(SO_4)_{3(aq)} + 3H_2O_{(l)}$
 e $CuCO_{3(s)} + 2HCl_{(aq)} \rightarrow CuCl_{2(aq)} + CO_{2(g)} + H_2O_{(l)}$

7 a $C:H:O = \dfrac{53.33}{12} : \dfrac{11.11}{1} : \dfrac{35.56}{16}$

$= 4.44 : 11.11 : 2.22$

$= \dfrac{4.44}{2.22} : \dfrac{11.11}{2.22} : \dfrac{2.22}{2.22}$

$= 2:5:1$

The empirical formula of compound **X** is C_2H_5O.
 b A compound of formula C_2H_5O would have a relative molecular mass of $(2 \times 12) + (5 \times 1) + (1 \times 16) = 45$.
   The relative atomic mass of compound **X** is 90 so its

formula must be $\dfrac{90}{45} = 2 \times C_2H_5O$

$= C_4H_{10}O_2$

 c Compound **X** could be any of the following.

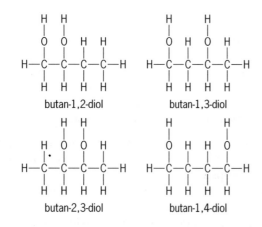

butan-1,2-diol   butan-1,3-diol

butan-2,3-diol   butan-1,4-diol

8 a Two atoms of sodium have a relative atomic mass of $2 \times 23 = 46$. As this accounts for 18.25% of the mass of the compound the relative formula mass must be:

$46 \times \dfrac{100}{18.25} = 252$

Sulphur: $\dfrac{12.70}{100} \times 252 = 32$

The compound contains $\dfrac{32}{32} = 1$ sulphur atom

Oxygen: $\dfrac{63.49}{100} \times 252 = 160$

The compound contains $\dfrac{160}{16} = 10$ oxygen atoms

Hydrogen: $\dfrac{5.56}{100} \times 252 = 14$

The compound contains $\dfrac{14}{1} = 14$ hydrogen atoms

14 hydrogen atoms will combine with 7 oxygen atoms to give $7H_2O$.
   Remaining are $(2 \times Na) + (1 \times S) + (3 \times O)$, hence the compound is $Na_2SO_3.7H_2O$.

 b The relative formula mass of Na_2SO_3 is $(2 \times 23) + 32 + (3 \times 16) = 126$

% sodium $= \dfrac{46 \times 100}{126} = 36.51\%$

% sulphur $= \dfrac{32 \times 100}{126} = 25.40\%$

% oxygen $= \dfrac{48 \times 100}{126} = 38.10\%$

9 a $C:H:O:Cl = \dfrac{38.10}{12} : \dfrac{7.41}{1} : \dfrac{16.93}{16} : \dfrac{37.57}{35.5}$

$= 3.18 : 7.41 : 1.06 : 1.06$

$= \dfrac{3.18}{1.06} : \dfrac{7.41}{1.06} : \dfrac{1.06}{1.06} : \dfrac{1.06}{1.06}$

$= 3:7:1:1$

The empirical formula of compound **Y** is C_3H_7OCl.

 b Compound **Y** could be any of the following.

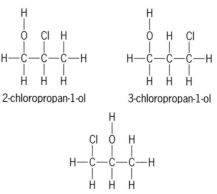

2-chloropropan-1-ol   3-chloropropan-1-ol

1-chloropropan-2-ol

10 a $ZnCO_{3(s)} \rightarrow ZnO_{(s)} + CO_{2(g)}$
 b $Fe_2O_{3(s)} + 3H_2SO_{4(aq)} \rightarrow Fe_2(SO_4)_{3(aq)} + 3H_2O_{(l)}$
 c $Ba(NO_3)_{2(aq)} + MgSO_{4(aq)} \rightarrow BaSO_{4(s)} + Mg(NO_3)_{2(aq)}$
 d Cathode reaction: $2H^+_{(aq)} + 2e^- \rightarrow H_{2(g)}$
   Anode reaction: $4OH^-_{(aq)} - 4e^- \rightarrow O_{2(g)} + 2H_2O_{(l)}$
 e $Cl_{2(g)} + 2Br^-_{(aq)} \rightarrow 2Cl^-_{(aq)} + Br_{2(aq)}$

8 Amount of substance

The mole

Atoms, ions and molecules have very small masses. It would be impossible to count them or weigh them individually.

The amounts of substances involved in chemical reactions are scaled up into moles using the **Avogadro constant** (L), which has the value 6.023×10^{23} particles mol^{-1}.

A mole is a quantity of particles; 1 mole contains 6.023×10^{23} particles.

The significance of this number is that it scales the mass of a particle in atomic mass units exactly into grams.

- The mass of 1 atom of carbon-12 is 12 amu.
- The mass of 1 mole of carbon-12 atoms is 12 g.
- The mass of 1 mole of a molecule is the sum of 1 mole of each of the atoms it contains. For example, the mass of 1 mole of methane (CH_4) is $12 g + (4 \times 1 g) = 16 g$.

Chemical equations usually imply that the quantities are in moles.

$$CH_4 \;+\; 2O_2 \;\rightarrow\; CO_2 \;+\; 2H_2O$$

1 mol	2 mol	1 mol	2 mol
16 g	64 g	44 g	36 g

Ideal gas equation

Boyle's law states that gas volume (V) varies inversely with pressure (p) at constant temperature – if you increase the pressure on a gas its volume gets less.

As $V \propto 1/p$ the product of volume and pressure (pV) is constant.

Charles's law states that gas volume (V) varies directly with temperature (T) at constant pressure – if you heat up a gas its volume gets larger.

As $V \propto T$ the ratio of volume to temperature (V/T) is constant.

These two laws can be combined to give the general gas law:

$$\frac{pV}{T} = \text{constant}$$

The exact value of the constant depends on the amount of gas. When the amount of gas is 1 mole the constant is known as the **molar gas constant**, R, which has the value $8.314\,J\,K^{-1}\,mol^{-1}$.

The ideal gas equation for n moles of gas is:

$$pV = nRT$$

The equation can also be written as:

$$pV = \frac{m}{M} RT$$

where m = mass and M = molar mass.

When using the ideal gas equation T represents **absolute temperature**. This is the temperature of the gas compared to absolute zero, the lowest temperature that is theoretically attainable. Absolute temperature is always expressed in kelvins, K.

Absolute zero is at 0 K, which is $-273\,°C$, so to convert from degrees Celsius to kelvins we simply add 273. For example, if room temperature is taken to be $20\,°C$ this also equals $20 + 273 = 293\,K$. Notice that the symbol ° is not used with K.

We can use the ideal gas equation to calculate the relative molar mass of a gas.

Example 1

$350\,cm^3$ of gas at atmospheric pressure and $30\,°C$ has a mass of $0.42\,g$. Calculate the molar mass of the gas.

- $p = 101\,325\,Pa$ (the pressure must be expressed in Pa)
- $V = 350\,cm^3 = 350 \times 10^{-6}\,m^3$ (the volume must be expressed in m^3)
- $T = 30\,°C = 303\,K$ (the temperature must be expressed in K)
- $R = 8.314\,J\,K^{-1}\,mol^{-1}$
- $m = 0.42\,g$

$$M = \frac{m \times R \times T}{p \times V}$$

$$= \frac{0.42\,g \times 8.314\,J\,K^{-1}\,mol^{-1} \times 303\,K}{101\,325\,Pa \times 350 \times 10^{-6}\,m^3}$$

$$= 29.8 \approx 30$$

The molar mass of the gas is 30.

Molar gas volume

1 mole of any gas occupies $22.4\,dm^3$ at standard temperature and pressure (s.t.p.), which is $0\,°C$ and atmospheric pressure.

As a rough approximation, the molar gas volume may be taken as $24\,dm^3\,mol^{-1}$ at room temperature and atmospheric pressure.

Partial pressure

When dealing with gases it is often more convenient and more meaningful to consider partial pressures rather than concentrations.

In a mixture of gases each gas contributes to the total pressure the same pressure that it would exert if the gas was present alone in a vessel of the same volume as that occupied by the mixture. The pressure exerted by the gas is called its **partial pressure**.

This principle is sometimes referred to as **Dalton's law of partial pressures**. Strictly speaking, it is only true for ideal gases; however, it provides a very good approximation for the behaviour of real gases.

Example 2

A mixture of gases contains 32 g of oxygen, 56 g of nitrogen and 10 g of hydrogen at a pressure of 100 kPa. What is the partial pressure of each gas?

The gases are in the molar ratio of

$$32/32 : 56/28 : 10/2 = 1 : 2 : 5$$

Since 1 mole of any gas occupies the same volume at a constant temperature the partial pressures of the gases will be in the same ratio as the number of moles of them.

Total pressure = 100 kPa therefore:

$$P_{\text{partial oxygen}} = \frac{(1 \times 100\,\text{kPa})}{8}$$
$$= 12.5\,\text{kPa}$$

$$P_{\text{partial nitrogen}} = \frac{(2 \times 100\,\text{kPa})}{8}$$
$$= 25.0\,\text{kPa}$$

$$P_{\text{partial hydrogen}} = \frac{(5 \times 100\,\text{kPa})}{8}$$
$$= 62.5\,\text{kPa}$$

Molarity

Molarity is concerned with the **concentration** of a solution. It indicates the number of particles in $1\,\text{dm}^3$ of solution.

A 1 molar solution contains 1 mole of a substance dissolved in water, or some other solvent, to make $1\,\text{dm}^3$ of solution.

Example 3

How much magnesium sulphate ($MgSO_4.7H_2O$) is needed to make $1\,\text{dm}^3$ of a $1.00\,\text{mol}\,\text{dm}^{-3}$ solution?

The mass of 1 mol of $MgSO_4.7H_2O$ is

$$24 + 32 + (4 \times 16) + 7 \times (2 + 16) = 246\,\text{g}$$

246 g of magnesium sulphate is needed to make $1\,\text{dm}^3$ of a $1.00\,\text{mol}\,\text{dm}^{-3}$ solution.

Example 4

What mass of magnesium sulphate is there in $50\,\text{cm}^3$ of a $1.00\,\text{mol}\,\text{dm}^{-3}$ solution?

$$\frac{50}{1000} = 0.05\text{ mol of magnesium sulphate}$$
$$= 0.05 \times 246\,\text{g}$$
$$= 12.3\text{ g of magnesium sulphate}$$

Molarity in titrations

The concentration of a solution may be found experimentally by titrating it against another solution of known concentration, sometimes using an indicator.

Example 5

Potassium hydroxide reacts with dilute sulphuric acid according to the following equation:

$$2KOH(aq) + H_2SO_4(aq) \rightarrow K_2SO_4(aq) + 2H_2O(l)$$

$25.0\,\text{cm}^3$ of aqueous potassium hydroxide of unknown concentration are exactly neutralised by $14.4\,\text{cm}^3$ of dilute sulphuric acid of concentration $0.202\,\text{mol}\,\text{dm}^{-3}$.

Find the concentration of the potassium hydroxide solution.

The number of moles of sulphuric acid is

$$0.0144\,\text{dm}^3 \times 0.202\,\text{mol}\,\text{dm}^{-3} = 0.0029\,\text{mol}$$

From the equation, 2 mol of potassium hydroxide are neutralised exactly by 1 mol of sulphuric acid. So the number of moles of potassium hydroxide is

$$2 \times 0.0029\,\text{mol} = 0.0058\,\text{mol}$$

$25.0\,\text{cm}^3$ of the solution contains $0.0058\,\text{mol}$ of potassium hydroxide. So $1\,\text{dm}^3$ ($1000\,\text{cm}^3$) of the solution contains

$$\frac{1000}{25} \times 0.0058\,\text{mol} = 0.232\,\text{mol}$$

The concentration of the potassium hydroxide solution is $0.232\,\text{mol}\,\text{dm}^{-3}$.

Example 6

Calcium carbonate reacts with dilute hydrochloric acid according to the following equation.

$$CaCO_3(s) + 2HCl(aq) \rightarrow CaCl_2(aq) + CO_2(g) + H_2O(l)$$

What volume of dilute hydrochloric acid, of concentration $1.03\,\text{mol}\,\text{dm}^{-3}$, is needed to completely react with $17.5\,\text{g}$ of calcium carbonate and what volume of gas, measured at s.t.p., is produced?

The relative formula mass of calcium carbonate is

$$40 + 12 + (3 \times 16) = 100$$

1 mol of calcium carbonate has a mass of $100\,\text{g}$, therefore $17.5\,\text{g}$ is

$$\frac{17.5}{100} = 0.175\,\text{mol}$$

According to the equation, 1 mol of calcium carbonate reacts with 2 mol of hydrochloric acid. So 0.175 mol of calcium carbonate reacts with

$$2 \times 0.175 = 0.350\text{ mol of hydrochloric acid}$$

$1\,\text{dm}^3$ of hydrochloric acid contains 1.03 mol, therefore the volume which contains 0.350 mol is

$$\frac{0.350}{1.03} \times 1\,\text{dm}^3 = 0.340\,\text{dm}^3$$

From the equation, 1 mol of calcium carbonate produces 1 mol of carbon dioxide gas, so 0.175 mol of calcium carbonate produces 0.175 mol of carbon dioxide.

At s.t.p. 1 mole of gas occupies $22.4\,\text{dm}^3$ so the volume of 0.175 mol of carbon dioxide will be

$$0.175 \times 22.4 = 3.92\,\text{dm}^3$$

1 What mass of sodium carbonate ($Na_2CO_3.10H_2O$) is needed to make $250\,cm^3$ of aqueous solution of concentration $0.05\,mol\,dm^{-3}$?

A $1.325\,g$
B $1.550\,g$
C $3.575\,g$
D $14.300\,g$

2 $1\,dm^3$ of nitrogen contains x atoms. How many atoms are there in $1\,dm^3$ of argon?

A $\dfrac{x}{4}$

B $\dfrac{x}{2}$

C x

D $2x$

3 How many moles of H^+ ions are present in $18.2\,cm^3$ of dilute hydrochloric acid of concentration $0.035\,mol\,dm^{-3}$?

A 6.37×10^{-4}
B 1.27×10^{-3}
C 1.92×10^{-3}
D 3.85×10^{-3}

4 Calcium carbonate reacts with dilute nitric acid according to the following equation.

$$CaCO_3(s) + 2HNO_3(aq) \rightarrow Ca(NO_3)_2(s) + CO_2(g) + H_2O(l)$$

What volume of $0.25\,mol\,dm^{-3}$ dilute nitric acid is required to completely react with $0.50\,g$ of calcium carbonate?

A $10\,cm^3$
B $20\,cm^3$
C $40\,cm^3$
D $80\,cm^3$

5 A mixture of gases contains $44\,g$ of carbon dioxide ($M_r = 44$), $48\,g$ of methane ($M_r = 16$) and $112\,g$ of nitrogen ($M_r = 28$). The pressure of the mixture is $200\,kPa$.

The partial pressure of the methane, in kPa, is

A 25
B 50
C 75
D 100

6 Complete combustion of $1\,mol$ of an organic compound **X** gives $2\,mol$ of carbon dioxide and $3\,mol$ of water. $0.15\,g$ of compound **X** vapour occupies $112\,cm^3$ at s.t.p. ($1\,mol$ of a gas occupies $22.4\,dm^3$ at s.t.p.).

a What is meant by s.t.p.?
b What is the relative molecular mass of compound **X**?
c What is the molecular formula of compound **X**?
d Write an equation, including state symbols, for the complete combustion of compound **X**.

7 a State the ideal gas equation.
b When vaporised, $0.17\,g$ of a volatile liquid occupies a volume of $58.1\,cm^3$ at a pressure of $101\,kPa$ and a temperature of $80\,°C$. Calculate the molar mass of the liquid.

8 A Group 2 metal carbonate has the formula MCO_3. It reacts with dilute hydrochloric acid according to the following equation.

$$MCO_3(s) + 2HCl(aq) \rightarrow MCl_2(aq) + CO_2(g) + H_2O(l)$$

$0.517\,g$ of MCO_3 was neutralised by $29.20\,cm^3$ of hydrochloric acid of concentration $0.240\,mol\,dm^{-3}$.

a Calculate the number of moles of hydrochloric acid used.
b Calculate the number of moles of MCO_3 that reacted.
c Calculate the relative formula mass of MCO_3.
d Calculate the relative atomic mass of M and identify it.

9 In a titration, $23.40\,cm^3$ of dilute sulphuric acid, of concentration $0.103\,mol\,dm^{-3}$, was required to neutralise $25.0\,cm^3$ of sodium hydroxide solution.

a Write an equation for the reaction between dilute sulphuric acid and aqueous sodium hydroxide.
b Calculate the number of moles of sulphuric acid used in the titration.
c Calculate the concentration of the sodium hydroxide solution in $mol\,dm^{-3}$.

10 $0.84\,g$ of a divalent metal oxide, MO, was reacted with $200\,cm^3$ of hydrochloric acid of concentration $7.30\,g$ per $1000\,cm^3$. The excess acid was neutralised by $80\,cm^3$ of sodium hydroxide solution containing $5.0\,g\,dm^{-3}$.

a Write equations for the two reactions.
b Calculate the number of moles of sodium hydroxide used to neutralise the excess hydrochloric acid.
c Calculate the number of moles of hydrochloric acid used in total.
d Calculate the number of moles of hydrochloric acid used to react with the metal oxide.
e Find the relative formula mass of the metal oxide and deduce the identity of the metal.

1 C

2 B

3 A

4 C

5 C

6 a Standard temperature and pressure, $0\,°C$ and atmospheric pressure.

   b Relative molecular mass $= 0.15 \times \dfrac{22\,400}{112} = 30$

   c C_2H_6

   d $C_2H_{6(g)} + \frac{7}{2}O_{2(g)} \rightarrow 2CO_{2(g)} + 3H_2O_{(g)}$

7 a The ideal gas equation for n moles of gas is $pV = nRT$
   where p = pressure of gas in Pa
   V = volume of gas in m^3
   T = temperature of gas in K
   R = molar gas constant in $JK^{-1}mol^{-1}$

   b Using the equation $M = \dfrac{m \times R \times T}{p \times V}$

   $m = 0.17\,g$
   $p = 101\,kPa = 101\,000\,Pa$
   $V = 58.1\,cm^3 = 58.1 \times 10^{-6}\,m^3$
   $T = 80\,°C = 353\,K$
   $R = 8.314\,JK^{-1}mol^{-1}$

   $M = \dfrac{0.17\,g \times 8.314\,JK^{-1}mol^{-1} \times 353\,K}{101\,000\,Pa \times 58.1 \times 10^{-6}\,m^3}$

   $= 85.02 \approx 85$

   You may have noticed that the units given in the equation do not balance in this form. In order to achieve this it is necessary to use Jm^{-3} in place of Pa ($Pa = Nm^{-2}$; $J = Nm$ therefore $N = Jm^{-1}$ therefore $Pa = Jm^{-3}$). However this complication is unnecessary.

8 a $0.240 \times \dfrac{29.20}{1000} = 0.0070\,mol$

   b 2 moles of HCl react with 1 mole of MCO_3

   $\dfrac{0.007}{2}\,mol = 0.0035\,mol$

   c $\dfrac{0.517}{0.0035} = 147.7$

   d Relative formula mass of $CO_3^{2-} = 60$
   Relative atomic mass of $M = 147.7 - 60 = 87.7$
   Therefore M = strontium, Sr

9 a $2NaOH_{(aq)} + H_2SO_{4(aq)} \rightarrow Na_2SO_{4(aq)} + 2H_2O_{(l)}$

   b Number of moles of H_2SO_4

   $= 0.103\,mol\,dm^{-3} \times \dfrac{23.40}{1000}\,dm^3$

   $= 0.00241\,mol$

c 2 moles of NaOH react with 1 mole of H_2SO_4

   $0.00241\,mol \times 2 = 0.00482\,mol$

   Concentration of NaOH

   $= 0.00482\,mol \times \dfrac{1000}{25}\,dm^{-3}$

   $= 0.193\,mol\,dm^{-3}$

10 a $MO_{(s)} + 2HCl_{(aq)} \rightarrow MCl_{2(aq)} + H_2O_{(l)}$
   $HCl_{(aq)} + NaOH_{(aq)} \rightarrow NaCl_{(aq)} + H_2O_{(l)}$

   b Concentration of of NaOH

   $= \dfrac{5}{40}$

   $= 0.125\,mol\,dm^{-3}$

   Number of moles of NaOH used

   $= 0.125\,mol\,dm^{-3} \times \dfrac{80}{1000}\,dm^{-3}$

   $= 0.010\,mol$

   c Concentration of HCl

   $= \dfrac{7.30}{36.5}$

   $= 0.20\,mol\,dm^{-3}$

   Number of moles of HCl used in total

   $= 0.20\,mol\,dm^{-3} \times \dfrac{200}{1000}\,dm^3$

   $= 0.040\,mol$

   d Moles of HCl used to dissolve the metal oxide
   $= (0.040 - 0.010)\,mol$
   $= 0.030\,mol$

   e 1 mol of MO reacts with 2 mol of HCl

   Number of moles of MO $= \dfrac{0.030}{2}\,mol$

   $= 0.015\,mol$

   Relative formula mass of MO $= \dfrac{0.84}{0.015}$

   $= 56$

   Relative atomic mass of $M = 56 - 16$
   $= 40$
   Therefore M = calcium, Ca

9 Organic chemistry

Carbon compounds

Organic chemistry is the study of the covalent compounds of carbon. However, it does not include carbonates, carbon monoxide and carbon dioxide. Organic chemicals are compounds of carbon and other elements, typically hydrogen, oxygen, nitrogen and halogens.

Carbon exhibits the unusual property of **catenation**. In its compounds, carbon atoms are able to join together forming chains of varying length. These chains may be straight, branched or cyclic, giving rise to millions of possible compounds.

Organic compounds are classified on the basis of the functional group which is present. The **functional group** is the reactive part of the molecule. In the main, this determines the chemical and physical properties of the molecule.

A group of compounds that contain the same functional group is called a **homologous series** and can often be represented by a general formula. Members of a homologous series have similar general properties since they have the same functional group. Physical properties, such as melting point and boiling point, show a gradual change as the carbon chain gets longer.

Table 9.2 Suffixes for homologous series

Homologous series	Functional group	Suffix
alkanes	$-\overset{\overset{\displaystyle H}{\mid}}{\underset{\underset{\displaystyle H}{\mid}}{C}}-H$	-ane
alkenes	$\overset{}{>}C=C\overset{}{<}$	-ene
alcohols	$-\overset{\mid}{\underset{\mid}{C}}-O-H$	-anol
aldehydes	$-C\overset{\displaystyle \nearrow O}{\searrow_{\displaystyle H}}$	-anal
ketones	$-\overset{\mid}{\underset{\mid}{C}}-\overset{\overset{\displaystyle O}{\parallel}}{C}-\overset{\mid}{\underset{\mid}{C}}-$	-anone
carboxylic acids	$-C\overset{\displaystyle \nearrow O}{\searrow_{\displaystyle O-H}}$	-anoic acid
esters	$-C\overset{\displaystyle \nearrow O}{\searrow_{\displaystyle O-\overset{\mid}{\underset{\mid}{C}}-}}$	-anoate

Naming organic compounds

In simplest terms the name of an organic compound consists of:

- a prefix, which indicates how many carbon atoms are present
- a suffix, which indicates the functional group present in the molecule.

The same prefixes are used in naming members of all homologous series (Table 9.1).

Table 9.1 Prefixes for homologous series

Number of carbon atoms in chain	Prefix
1	meth-
2	eth-
3	prop-
4	but-
5	pent-
6	hex-
7	hept-
8	oct-
9	non-
10	dec-

Table 9.2 lists the suffixes for the main functional groups.

Names are obtained by combining prefix and suffix.

$$CH_3-CH_2-CH_2-CH_2-CH_2-CH_3$$

This compound contains a chain of six carbon atoms and it is an alkane, so it is hexane.

$$CH_3-CH_2-CH_2-COOH$$

This compound contains a chain of four carbon atoms and is a carboxylic acid, so it is butanoic acid.

Alkanes and alkenes

In many compounds the carbon skeleton is branched. The names given to the side-chains also depend on the number of carbon atoms they contain. Methyl, ethyl and propyl indicate one, two and three carbon atoms in a side-chain, respectively.

The position of side-chains is indicated by numbering the carbon atoms in the longest carbon chain of the molecule.

$$\overset{\displaystyle CH_3}{\underset{5 \quad\quad 4 \quad\quad 3 \quad\quad 2 \quad\quad 1}{CH_3-CH_2-CH_2-\overset{\mid}{CH}-CH_3}}$$

The longest carbon chain in this molecule is five atoms long, so its name is based on pentane. There is a side-chain containing one carbon atom so it is a methylpentane. The methyl side-chain is on the second carbon atom of the longest chain, so it is 2-methylpentane.

Notice that the longest chain is numbered so that the side-chain is on the lowest numbered carbon atom of the longest chain. Numbering the longest chain in the above example from the opposite end would have given 4-methylpentane, which is incorrect.

When side-chains of equal length are present the prefixes 'di', 'tri', and so on, are used.

$$CH_3-CH_2-CH_2-\overset{\overset{\displaystyle CH_3}{|}}{CH}-\overset{\overset{\displaystyle CH_3}{|}}{CH}-CH_3$$
$$\;\;\;6\quad\;\;5\quad\;\;4\quad\;3\quad\;2\quad\;1$$

This compound is 2,3-dimethylhexane.

When side-chains of unequal length are present they are given in alphabetical order.

$$CH_3-\overset{\overset{\displaystyle CH_3}{|}}{CH}-\overset{\overset{\displaystyle CH_2CH_3}{|}}{CH}-CH_2-CH_2-CH_2-CH_3$$
$$\;\;1\quad\;\;2\quad\;\;3\quad\;\;4\quad\;\;5\quad\;\;6\quad\;\;7$$

This compound is 3-ethyl-2-methylheptane.

In alkenes, the carbon–carbon double bond is assigned to the lowest numbered carbon atom.

$$CH_3-CH_2-CH{=}CH-CH_3$$
$$\;\;5\quad\;\;4\quad\;\;3\quad\;\;2\quad\;\;1$$

This compound is pent-2-ene and not pent-3-ene.

The prefix 'cyclo' is used to indicate a cyclic compound. The name is determined by the number of carbon atoms in the ring.

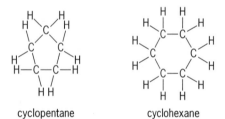

cyclopentane   cyclohexane

Cyclopentane is a ring of five carbon atoms while cyclohexane is a ring of six.

The carbon atoms in a ring are always numbered in such a way that side-chains are on the lowest numbered carbon atom.

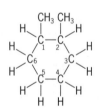

This compound is 1,2-dimethylcyclohexane.

Alcohols, aldehydes, ketones, carboxylic acids and esters

In aldehydes and carboxylic acids the functional group can only be at the end of a carbon chain. The end carbon atom is therefore numbered 1, although the number is not included in the name.

$$\overset{\overset{\displaystyle Cl}{|}}{CH_3-CH-CH_2-CHO}$$
$$\;\;4\quad\;\;3\quad\;\;2\quad\;\;1$$

This compound is 3-chlorobutanal.

In alcohols the functional group can be anywhere on the carbon chain and so the carbon atom that carries the functional group needs to be identified. In ketones the functional group cannot be at the end of the chain.

In the ketones derived from pentane the functional group may be on carbon-2 or carbon-3 (but not on carbon-1 since this would be an aldehyde), giving two possible compounds.

$$CH_3-CH_2-CH_2-\overset{\overset{\displaystyle O}{\|}}{C}-CH_3 \qquad CH_3-CH_2-\overset{\overset{\displaystyle O}{\|}}{C}-CH_2-CH_3$$
pentan-2-one   pentan-3-one

In the alcohols derived from pentane the functional group may be on carbon-1, carbon-2 or carbon-3, giving three possible compounds.

$$HO-CH_2-CH_2-CH_2-CH_2-CH_3 \qquad CH_3-\overset{\overset{\displaystyle OH}{|}}{CH}-CH_2-CH_2-CH_3$$
pentan-1-ol   pentan-2-ol

$$CH_3-CH_2-\overset{\overset{\displaystyle OH}{|}}{CH}-CH_2-CH_3$$
pentan-3-ol

The following compound is both an alcohol and a carboxylic acid.

$$HOCH_2-CH_2-COOH$$

Where two groups that are usually shown by suffixes are present, the order of precedence is:

carboxylic acid > aldehyde   or   ketone > alcohol

The above compound is therefore 3-hydroxypropanoic acid.

$$CH_3-CO-CH_2-COOH$$

This compound is 3-oxobutanoic acid.

Common names are often used where systematic names become complicated and cumbersome. Glucose is rather easier on the tongue than 2,3,4,5,6-penta-hydroxyhexanal!

$$HOC-CHOH-CHOH-CHOH-CHOH-CH_2OH$$

Esters are formed from carboxylic acids and alcohols. The name of the ester is formed from the prefix of the alcohol followed by the carboxylic acid with the ending changed to '-oate'.

This ester is formed from methanol and ethanoic acid. Its name is therefore methyl ethanoate.

1 Which of the following compounds is an aldehyde?
  A $CH_3-CH_2-CH_3$
  B $CH_3-CH_2-CH_2-CH_2-OH$
  C CH_3-CHO
  D $CH_3-CO-CH_3$

2 The compound $CH_3-CH_2-CH_2-COOH$ is
  A an alcohol
  B an aldehyde
  C a carboxylic acid
  D a ketone

3 Which of the following is 3-bromopentanoic acid?
  A $BrCH_2-CH_2-CH_2-CH_2-COOH$
  B $CH_3-CHBr-CH_2-CH_2-COOH$
  C $CH_3-CH_2-CHBr-CH_2-COOH$
  D $CH_3-CH_2-CH_2-CHBr-COOH$

4 The compound $CH_3-CHBr-CH_2Cl$ is
  A 1-chloro-2-bromopropane
  B 1-bromo-2-chloropropane
  C 2-bromo-1-chloropropane
  D 2-bromo-3-chloropropane

5 The compound $CH_3-CO-CH_2-CH_2-OH$ is
  A 1-hydroxybutan-3-one
  B 2-oxobutan-4-ol
  C 3-oxobutan-1-ol
  D 4-hydroxybutan-2-one

6 Explain the meaning of the following:
  a catenation
  b functional group
  c homologous series
  d the prefix 'hept-'
  e the suffix '-anal'.

7 A compound has the following structure.

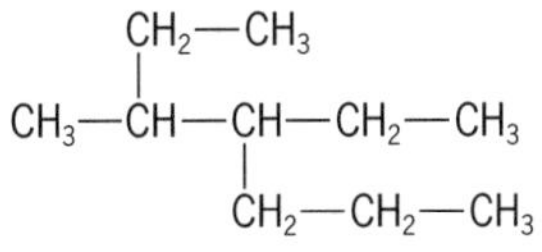

$$CH_3-\underset{\underset{\displaystyle CH_2-CH_2-CH_3}{|}}{\overset{\overset{\displaystyle CH_2-CH_3}{|}}{CH}}-CH-CH_2-CH_3$$

  a To which homologous series does this compound belong?
  b How many carbon atoms are there in the longest carbon chain of the molecule?
  c What side-chains are present in the molecule?
  d What is the systematic name for this compound?

8 Draw a display formula for each of the following compounds:
  a 3-ethylhexane
  b 2,3,4-trimethylpentane
  c 3-methylpent-1-ene
  d 3-chloropentanal
  e 3-oxopentanoic acid
  f 1,2,3-trimethylcyclopentane.

9 Each of the following compounds is incorrectly named. In each case, explain why the name is wrong and give the correct name.
  a 2-ethylpropane

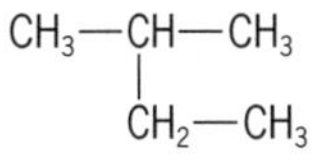

$$CH_3-\underset{\underset{\displaystyle CH_2-CH_3}{|}}{CH}-CH_3$$

  b 3,4-dibromopentane

$$CH_3-CH_2-\overset{\overset{\displaystyle Br}{|}}{CH}-\overset{\overset{\displaystyle Br}{|}}{CH}-CH_3$$

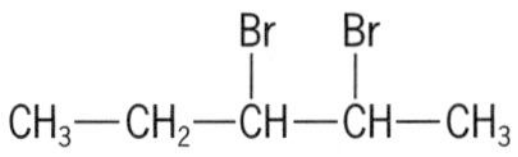

  c 3-methyl-4-ethyloctane

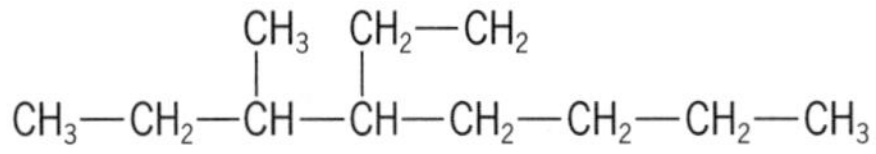

$$CH_3-CH_2-\overset{\overset{\displaystyle CH_3}{|}}{CH}-\overset{\overset{\displaystyle CH_2-CH_2}{|}}{CH}-CH_2-CH_2-CH_2-CH_3$$

  d 2-oxopentan-4-ol

$$CH_3-\overset{\overset{\displaystyle O}{||}}{C}-CH_2-\overset{\overset{\displaystyle OH}{|}}{CH}-CH_3$$

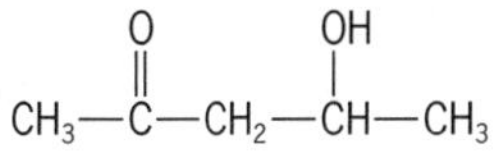

10 For each of the following molecular formulae, draw display formulae for two compounds which contain different functional groups.
  Name the functional group present in each compound and name the compound.
  a C_4H_8O
  b $C_4H_8O_2$

ANSWERS ANSWERS ANSWERS ANSWERS ANSWERS ANSWERS ANSWERS ANSWERS ANSWERS

1 C **2** C **3** C **4** C **5** A

6 a The formation of chains of atoms (usually of the same element) in chemical compounds.
  b The group of atoms responsible for the characteristic chemical reactions of a compound.
  c A series of related chemical compounds that have the same functional group but differ in formula only by the number of CH_2 groups in a chain. Members of a homologous group have similar chemical properties.
  d A carbon chain containing seven atoms.
  e The functional group —CHO which is present in an aldehyde.

7 a alkane
  b 7
  c an ethyl group and a methyl group
  d 4-ethyl-3-methylheptane

8 a

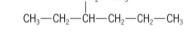
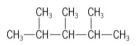

  b

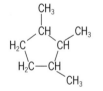

  c

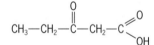

  d

  e

  f

9 a The longest chain contains four carbon atoms. The correct name is 2-methylbutane.
  b The carbon atoms should be numbered from the other end of the longest carbon chain so that the side-groups are on the lowest numbered carbons possible. The correct name is 2,3-dibromopentane.
  c The side-groups should be given in alphabetical order. The correct name is 4-ethyl-3-methyloctane.
  d The ketone group takes precedence over the alcohol group in determining the ending of the name. The correct name is 4-hydroxypentan-2-one.

10 a For example:

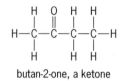

butan-2-one, a ketone

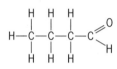

butanal, an aldehyde

  b For example:

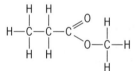

methyl propanoate, an ester

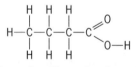

butanoic acid, a carboxylic acid

10 Alkanes and alkenes

Alkanes

Alkanes are said to be **saturated** hydrocarbons since they contain only carbon–carbon single bonds. They have the general formula C_nH_{2n+2}.

Alkanes are not very reactive; however, like all hydrocarbons, they will burn readily in air releasing a large amount of heat energy.

$$CH_4(g) + 2O_2(g) \rightarrow CO_2(g) + 2H_2O(l); \qquad \Delta H^\ominus = -890\,kJ\,mol^{-1}$$

If insufficient oxygen is present, incomplete combustion occurs, forming either carbon or the poisonous gas carbon monoxide. Incomplete combustion also results in less heat being evolved.

$$CH_4(g) + O_2(g) \rightarrow C(s) + 2H_2O(l)$$
$$CH_4(g) + \tfrac{3}{2}O_2(g) \rightarrow CO(g) + 2H_2O(l)$$

Alkanes do not react with chlorine at room temperature or in the dark, but in the presence of ultraviolet radiation a **free-radical substitution** reaction takes place. The reaction between chlorine and methane may produce an explosion.

- Initiation

$$Cl_2 \rightarrow 2Cl\cdot$$

Ultraviolet radiation provides the energy needed to separate molecules of chlorine into highly reactive atoms (known as **free radicals**). The Cl—Cl bonds are easier to break than the C—H bonds of the alkane.

- Propagation

$$Cl\cdot + CH_4 \rightarrow CH_3\cdot + HCl$$
$$CH_3\cdot + Cl_2 \rightarrow CH_3Cl + Cl\cdot$$

The chlorine free radicals collide with other molecules, creating more free radicals. Each of the propagation steps is exothermic, leading to the rapid production of energy.

- Termination

$$Cl\cdot + CH_3\cdot \rightarrow CH_3Cl$$
$$CH_3\cdot + CH_3\cdot \rightarrow CH_3CH_3$$
$$Cl\cdot + Cl\cdot \rightarrow Cl_2$$

This occurs when two free radicals combine to form a stable molecule.

This reaction produces a mixture of chloromethanes (CH_3Cl, CH_2Cl_2, $CHCl_3$, CCl_4) together with traces of hydrocarbons such as ethane. The formation of dichloromethane, for example, results from

$$CH_3Cl + Cl\cdot \rightarrow CH_2Cl\cdot + HCl$$
$$CH_2Cl\cdot + Cl_2 \rightarrow CH_2Cl_2 + Cl\cdot$$

In this reaction the number of possible chloroalkanes in the final mixture is determined by the number of hydrogen atoms on the alkane used.

The physical properties of the alkanes show a gradual change as the size of the molecules increases.

For example, as the number of carbon atoms increases the van der Waals forces between molecules becomes greater, resulting in an increase in boiling point. At normal room temperature (25 °C) alkanes containing 1–4 carbon atoms are typically gases, those with 5–16 carbon atoms are liquids and those containing more than 16 carbon atoms are solids.

Alkenes

Linear alkenes have the general formula C_nH_{2n}. They are described as **unsaturated** hydrocarbons because they contain carbon–carbon double bonds. Alkenes are far more reactive compounds than alkanes.

In a carbon–carbon double bond, one of the covalent bonds is formed by sharing a pair of electrons in the normal way. This is referred to as a σ **(sigma) bond**. Each carbon atom now has a spare p orbital containing a single electron. These p orbitals overlap giving a second pair of shared electrons that form a second bond, which is referred to as a π **(pi) bond**. The areas of overlap above and below the plane of the molecule together make up the π bond (Figure 10.1).

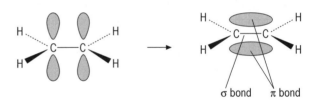

Figure 10.1 Formation of a π bond in ethene

The σ bonds around each carbon atom are at 120°. This allows the bonding pairs of electrons to keep as far apart from each other as possible.

The second bond, the π bond, results in an area of high electron density and is the reason why alkenes are so reactive. Alkenes undergo **addition reactions** in which other chemicals are added across the double bond. The reactions of ethene are typical of the group.

The **hydrogenation** of alkenes occurs at 150 °C in the presence of a finely divided nickel catalyst.

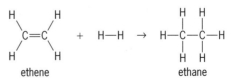

This reaction is important in the manufacture of margarines. The unsaturated vegetable oils from which margarines are made are hardened by hydrogenation. The amount of hydrogenation is controlled to produce hard or soft margarine as required.

Alkenes readily react with chlorine or bromine to form a saturated product.

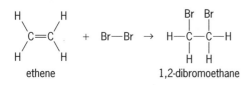

ethene   1,2-dibromoethane

The decolorisation of bromine water is used as a test for unsaturation since it distinguishes between alkanes and alkenes.

This reaction is an example of **electrophilic addition** (electrophilic = electron seeking) (Figure 10.2).

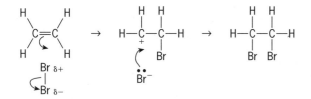

Figure 10.2 Electrophilic addition of bromine

The mechanism of this reaction involves the following.

- The high electron density of the carbon–carbon double bond induces a dipole in the bromine molecule, forming a bromine atom which is electron deficient ($\delta+$) and therefore electrophilic.
- A pair of electrons from the carbon–carbon double bond makes a new bond between one of the carbon atoms and the electrophilic bromine atom. (A curly arrow is used to represent the movement of a pair of electrons.)
- The other carbon atom remains as an electron-deficient **carbocation** (an ion containing a positively charged carbon atom) while the bromine–bromine bond breaks releasing a bromide ion.
- A lone pair of electrons from the bromide ion (and not the negative charge) forms a bond with the carbocation.

Alkenes react in a similar way with hydrogen bromide in the gas phase or as a concentrated aqueous solution to form bromoalkanes (Figure 10.3).

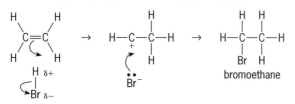

bromoethane

Figure 10.3 Electrophilic addition of hydrogen bromide

The mechanism is similar to that described above, except that the hydrogen bromide has a permanent dipole, leaving the hydrogen atom electron-deficient. In aqueous solution $H^+_{(aq)}$ ions act as electrophiles.

When a hydrogen halide is added across the carbon–carbon double bond in some alkenes two products are possible (Figure 10.4).

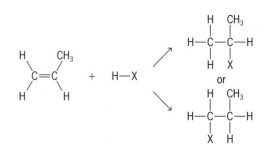

Figure 10.4 HX addition to propene – two possible products

The product from the addition of hydrogen halides to alkenes can be predicted using **Markownikoff's rule**. This rule states that 'the hydrogen atom always adds to the carbon atom bonded to the greater number of hydrogen atoms in the alkene'. It can be explained by considering the relative stability of the two possible carbocations that can be formed (Figure 10.5).

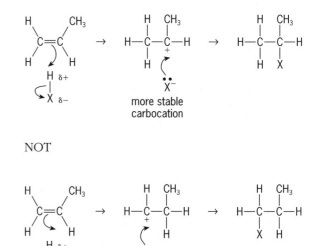

more stable carbocation

NOT

less stable carbocation

Figure 10.5 Stability of carbocations

A secondary carbocation (one in which the positive charge is on a carbon atom bonded to another carbon atom) is more stable than a primary carbocation (one in which the positive charge is on a carbon atom bonded only to hydrogen atoms).

Alkenes react with steam at 300 °C and a pressure of 60 atmospheres (6 MPa) in the presence of phosphoric acid catalyst to form alcohols.

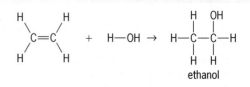

ethanol

Table 10.1 Common addition polymers

Polymer	Common name	Monomer	Repeating unit
poly(ethene)	polythene or polyethylene	$H_2C{=}CH_2$	$-(CH_2-CH_2)_n$
poly(propene)	polypropylene	$H_2C{=}CHCH_3$	$-(CH_2-CHCH_3)_n$
poly(tetrafluoroethene)	PTFE	$F_2C{=}CF_2$	$-(CF_2-CF_2)_n$
poly(chloroethene)	polyvinylchloride or PVC	$H_2C{=}CHCl$	$-(CH_2-CHCl)_n$

Addition polymerisation

Under suitable conditions, alkenes undergo **addition polymerisation**. A pair of electrons from the reactive carbon–carbon double bond of ethene forms a bond between one molecule and another. This process is repeated many times to produce a long-chain molecule.

$$nH_2C{=}CH_2 \rightarrow -(CH_2-CH_2)_n$$
ethene, monomer   poly(ethene), polymer

Many small unsaturated molecules (**monomers**) form a long saturated molecule (**polymer**). The polymer is represented as a unit repeated many times (n).

A certain amount of side-branching occurs during the polymerisation, depending on the reaction conditions. In the case of ethene, very high pressure (100–300 MPa) and moderate temperatures (420–570 K) produce a low-density form of poly(ethene). The average polymer molecule contains $4{-}40 \times 10^3$ carbon atoms and has many short branches. Such branched molecules do not pack together easily, hence the low density. Ethene can also be polymerised at low pressure (0.1–5 MPa) and low temperature (310–360 K) using a catalyst, to give a high-density form of poly(ethene) that is virtually unbranched. High-density polythene has greater strength and rigidity, and a higher melting point than the low-density form. Table 10.1 (above) gives details of common addition polymers.

Condensation polymerisation

Condensation polymerisation involves the reaction of two monomers with the loss of a small molecule which is often, but not always, water. Each monomer contains two functional groups, so they can attach to two other monomer molecules. There are several groups of condensation polymers, each characterised by a particular linkage between monomers (Table 10.2).

Table 10.2 Polymer linkages

Polymer	Linkage
polyesters	$-O-\overset{\overset{O}{\|\|}}{C}-$
polyamides	$-\overset{\overset{O}{\|\|}}{C}-NH-$
polyurethanes	$-O-\overset{\overset{O}{\|\|}}{C}-NH-$
polysaccharides	$-O-$

Terylene is a polyester made from benzene-1,4-dicarboxylic acid and ethane-1,2-diol by the loss of water. This is an example of an **esterification** reaction (considered more fully in topic 22).

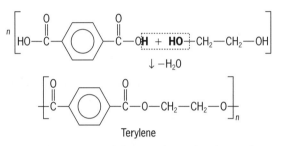

Terylene

Nylon is the name used for a group of polyamides made from diamines and dicarboxylic acid chlorides by the loss of HCl. Nylon-6,6 is made from hexane-1,6-diamine and hexane-1,6-dioyl dichloride.

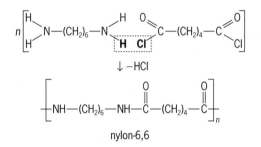

nylon-6,6

There are many examples of naturally occurring condensation polymers. Polysaccharides, such as cellulose, consist of many glucose molecules joined by glycosidic links (—O—) with the loss of water. Where monomer molecules have a complex structure the polymers they form are often presented as block diagrams. For example, Figure 10.6 represents the formation of a polysaccharide.

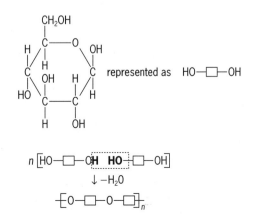

Figure 10.6 Formation of a polysaccharide

1 Which of the following free-radical reactions can be described as a propagation reaction?
A $CH_3 \cdot + CH_3-CH_2-CH_2 \cdot \rightarrow CH_3-CH_2-CH_2-CH_3$
B $CH_3-CH_2-CH_3 \rightarrow CH_3 \cdot + CH_3-CH_2 \cdot$
C $CH_3-CH_2-CH_2 \cdot + CH_3-CH_2-CH_2 \cdot \rightarrow$
$CH_3-CH_2-CH_2-CH_2-CH_2-CH_3$
D $CH_3 \cdot + CH_3-CH_2-CH_3 \rightarrow CH_4 + CH_3-CH_2-CH_2 \cdot$

2 How many different chloroalkanes are there likely to be in the mixture of products formed when ethane reacts with chlorine in the presence of sunlight?
A 6
B 7
C 8
D 9

3 Which of the following is the correct mechanism for the first step of the electrophilic addition of chlorine to an alkene?

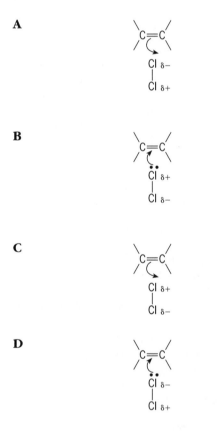

4 Poly(phenylethene) has the following structure.

From which of the following is poly(phenylethene) made?

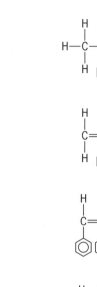

5 Starch is a condensation polymer in which glucose molecules are joined together by —O— linkages. To which group of condensation polymers does starch belong?
A polyamides
B polyesters
C polysaccharides
D polyurethanes

6 Draw and describe the mechanism for the reaction of bromine with propene. Your description should include the terms: electrophilic, dipole, carbocation and lone pair.

7 a Describe addition polymerisation, using the polymerisation of chloroethene ($H_2C=CHCl$) as an example.
b The first synthetic rubber was made from 2-chlorobuta-1,3-diene.

$$n\left[CH_2=CH-\underset{\underset{Cl}{|}}{C}=CH_2\right] \rightarrow \left[CH_2-CH=\underset{\underset{Cl}{|}}{C}-CH_2\right]_n$$

SBR is an important modern synthetic rubber. It is a co-polymer, made from styrene (phenylethene) and butadiene.

$$\underset{\underset{H}{|}}{\overset{\overset{\bigcirc}{|}}{C}}=\underset{\underset{H}{|}}{\overset{\overset{H}{|}}{C}} + CH_2=CH-CH=CH_2$$

Suggest a structure for the repeating unit in SBR rubber.

8 The following diagram shows part of a protein molecule.

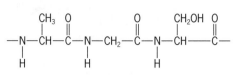

a To which group of condensation polymers do proteins belong?

b Draw the link that joins the amino acids in a protein molecule.

c Draw the three amino acids which polymerised to form the part of the protein molecule shown above.

d Name one other group of condensation polymers which are naturally occurring and give one example.

9 Alkynes are a series of hydrocarbons that contain a carbon–carbon triple bond. The first one in the series is ethyne, H—C≡C—H.

a Give a balanced equation for the complete combustion of ethyne.

b In ethyne, the carbon atoms are joined by a σ bond, as in ethene, but each carbon atom has two spare p orbitals, set at right angles to each other, with which to form two π bonds. Draw a diagram to show the structure of an ethyne molecule and label the bonds between the carbon atoms.

c Suggest a value for the H—C≡C bond angle.

d Explain whether you expect the chemistry of ethyne to be more like ethane or ethene. Give one example of a reaction of ethyne you would expect, apart from combustion.

10 a Use Markownikoff's rule to predict the product of the following reactions. Draw the structure of each product.

i CH_3—CH=CH_2 + H—I

ii $(CH_3)_2$C=CH_2 + H—Cl

iii $(CH_3)_2$C=CHCH$_3$ + H—Br

b The reaction of hydrogen halides and alkenes involves electrophilic addition. Alkyl groups stabilise carbocations. Use this information to account for Markownikoff's rule in terms of the stability of the intermediate species in reactions **ai** and **aii**.

1 D

2 D

3 C

4 B

5 C

6

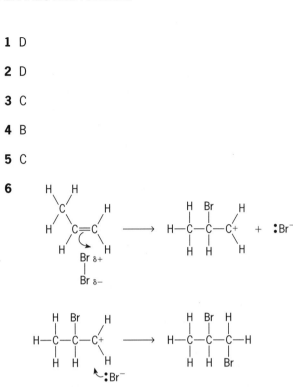

An electrophile is something that seeks, or is attracted to, electrons. There is a high electron density around the carbon–carbon double bond in propene.

The electron density causes the bonding electrons in the bromine molecule to become unevenly distributed. A dipole is induced in which the bonding electrons move away from the carbon–carbon double bond, leaving one end of the bromine molecule slightly positive and the other slightly negative.

A pair of electrons from the carbon–carbon double bond forms a bond between one of the carbon atoms and one of the bromine atoms, with the remaining bromine atom being released as a bromide ion. This leaves the other carbon atom carrying a positive charge, thus the molecule becomes a carbocation.

A lone pair of electrons from the bromide ion forms a bond between it and the positively charged carbon atom.

7 a Addition polymerisation involves the addition of a large number of alkene monomers. It is possible because alkenes contain a carbon–carbon double bond. The pair of electrons from one of these bonds provides the bond between one alkene molecule and the next.

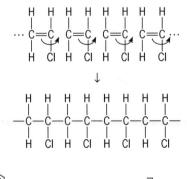

b

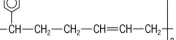

8 a Polyamides
b —CO—NH—
c H_2N—$CH(CH_3)$—COOH, H_2N—CH_2—COOH, H_2N—$CH(CH_2OH)$—COOH
d e.g. polysaccharides, cellulose

9 a $C_2H_2(g) + \frac{5}{2} O_2(g) \rightarrow 2CO_2(g) + H_2O(g)$
b

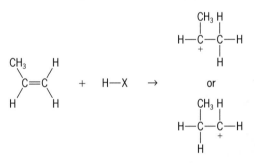

c 180°, which allows the bonding electrons to stay as far apart as possible.
d Expect the chemistry of ethyne to be similar to ethene since they both have reactive π bonds and will readily undergo addition reactions.

For example, H—C≡C—H + Br—Br →
BrH—C=C—HBr + Br—Br → Br_2H—C—C—HBr_2

10 a i CH_3—CHI—CH_3
   ii $(CH_3)_2$ClC=CH_3
   iii $(CH_3)_2$BrC—CH_2CH_3
b Addition of a hydrogen halide to propene, for example, could result in two possible intermediate carbocations.

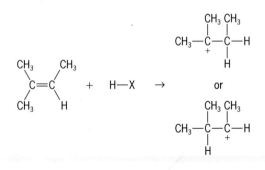

The intermediate carbocation in which the positive charge is positioned on carbon-2 will be the more stable since it is stabilised by two alkyl groups rather than one, so we would predict the halide to add at the 2 position.

Addition of a hydrogen halide to 2-methylbut-2-ene could also result in two possible intermediate carbocations.

The intermediate carbocation in which the positive charge is positioned on carbon-2 will be the more stable since it is stabilised by three alkyl groups rather than two, so we would predict the halide to add at the 2 position.

11 Crude oil

Primary distillation

Crude oil is a complex mixture of hydrocarbons (mainly alkanes), together with small amounts of other chemicals containing sulphur, nitrogen and oxygen. Alkenes and alkynes are not present but are made from other hydrocarbons by various **reforming** processes.

The mixture is first separated into a series of fractions by **fractional distillation** (Figure 11.1).

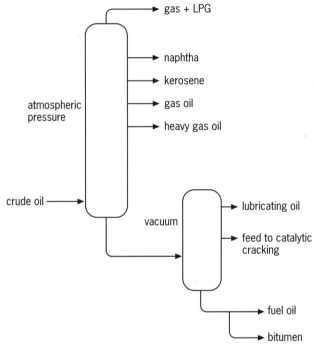

Figure 11.1 Primary distillation of crude oil

Crude oil is heated in a furnace before passing into a fractional distillation column. The temperature in the column gradually changes from 400 °C at the bottom to 30–40 °C at the top. Most of the chemicals that make up crude oil enter the column as gases. As they pass up the column they cool and change to liquids. These liquids collect at different levels and pass out of the column.

The material that remains at the bottom of the column is called the **residue**. It consists of chemicals that have very high boiling points. This residue passes into another column where a further distillation is carried out under reduced pressure.

Primary distillation separates crude oil into a series of **fractions**. A fraction consists of a group of different chemicals that have similar boiling points. Each fraction is a useful source of chemicals either used directly, or after further processing.

The composition of crude oil varies, depending on its source; however, it is always the case that the proportions of the various fractions produced do not match the demand for them. In general, there is insufficient of the smaller molecules, such as those required for petrol, and an excess of the larger molecules typical of heavy gas oils. Large hydrocarbon molecules are converted into smaller ones by the process of **cracking**.

Cracking

The mechanism for cracking depends on the method used. **Steam (thermal) cracking** involves breaking carbon–carbon and carbon–hydrogen bonds to form free radicals. **Catalytic cracking** involves the formation of carbocation intermediates. Cracked molecules may break up in different ways.

Large alkanes can be cracked to produce smaller ones plus alkenes.

$$CH_3-(CH_2)_8-CH_3 \rightarrow CH_3-(CH_2)_6-CH_3 + CH_2{=}CH_2$$

Cracking removes the side-chains from aromatic compounds.

Cracking opens the rings of cyclic alkanes.

Breaking bonds requires a large amount of energy. Cracking is carried out at high temperatures, either in the presence of steam, to produce a high yield of alkenes, or in the presence of a catalyst to give molecules suitable for use in petrol.

In steam cracking a preheated hydrocarbon vapour is mixed with steam in a cracking furnace at a temperature in the region of 800 °C.

The use of a catalyst allows the reaction to be carried out at a lower temperature, typically 500 °C. In fluidised catalytic cracking, a mixture of the silica/alumina catalyst powder and hydrocarbon vapour is fed into a cracking reactor. The catalyst is regenerated by burning off the carbon which is deposited on it during the cracking process.

In both cracking processes the mixture of products formed is subsequently separated by fractional distillation.

Desulphurisation

Crude oil contains small amounts of sulphur in the form of organic sulphur compounds.

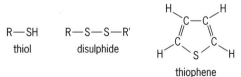

If these compounds were allowed to remain in the crude oil during refining they would damage the catalysts used in some processes. Any sulphur compounds remaining in fuels would produce acidic sulphur dioxide on combustion, which may damage engine parts and pollute the atmosphere.

One method of removing sulphur compounds involves three stages:

1 hydrofining – sulphur compounds are converted into hydrocarbons and hydrogen sulphide using hydrogen under high pressure
2 gas purification – hydrogen sulphide is removed from the hydrocarbon residue
3 sulphur recovery – hydrogen sulphide is oxidised to elemental sulphur.

The sulphur is a valuable by-product that is used in the vulcanisation of rubber and in the manufacture of sulphuric acid.

Fuels and the environment

When sulphur dioxide (SO_2) and nitrogen oxides (NO_x) are released into the atmosphere they increase the acidity of the rain.

This so called 'acid rain' causes damage to structures such as the erosion of limestone buildings. It also reduces the pH of soil, causing the release of heavy metal cations. These ions may subsequently be carried away into water courses where they can have a devastating effect on aquatic life.

Sulphur compounds are removed from fuels such as petrol and diesel, so sulphur dioxide is not a problem in this respect.

Nitrogen oxides are not derived from nitrogen compounds in the fuel but by the oxidation of atmospheric nitrogen during combustion. The amount of nitrogen oxides produced by an engine depends upon its design and the temperature at which the fuel burns: reducing the temperature reduces the amount of nitrogen oxides. Many manufacturers are now fitting 'low NO_x' engines to their cars. These reduce, but do not eliminate, the production of nitrogen oxides.

Many modern cars are also fitted with **catalytic converters** that reduce the emission of harmful gases. These devices contain a metal catalyst, such as rhodium, on a ceramic matrix (Figure 11.2). When the exhaust mixture passes through the converter nitrogen oxides, carbon monoxide and unburnt hydrocarbons are converted to nitrogen, carbon dioxide and water.

Figure 11.2 Catalytic converter

The efficient running of a petrol engine depends upon the smooth combustion of the fuel. If the fuel is not combusted smoothly the engine runs irregularly in a way that is described as 'knocking'. In the past, engine knocking was avoided by adding an anti-knock agent to the petrol in the form of lead tetraethyl,

$Pb(C_2H_5)_4$. In order to prevent an accumulation of lead in the engine cylinders, bromoethane was also added to the petrol and the lead was expelled in the exhaust gas as lead(II) bromide. However, long-term studies have shown that the lead bromide in the atmosphere presents a serious health hazard. If inhaled, some of the lead is retained by the body, where it acts as a cumulative poison.

Lead tetraethyl has been eliminated in modern fuels and leaded fuels are no longer sold in the UK. This has been possible by substituting alternative anti-knock agents, such as methyl tert-butyl ether (MTBE), which present no health risk, and by increasing the anti-knock value of petrol by reforming.

The process of **reforming** is concerned with changing the shape of hydrocarbon molecules, but not their size.

• Cyclic alkanes are converted to aromatic compounds.

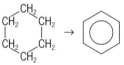

• Alkanes are converted into cyclic alkanes.

$$CH_3-CH_2-CH_2-CH_2-CH_2-CH_3 \rightarrow$$

• Straight-chain alkanes are converted into branched-chain alkanes.

$$CH_3-CH_2-CH_2-CH_2-CH_2-CH_3 \rightarrow CH_3-CH_2-CH_2-CH-CH_3$$
$$| $$
$$CH_3$$

Increasing the proportion of aromatic compounds and branched-chain alkanes in petrol increases its anti-knock value and thus reduces the dependence on anti-knock agents.

1 Primary distillation sorts out the compounds in crude oil into fractions on the basis of
 A boiling point
 B melting point
 C number of carbon atoms in the molecule
 D relative molecular mass

2 The process of converting sulphur compounds to hydrocarbons and hydrogen sulphide is called
 A cracking
 B distillation
 C hydrofining
 D reforming

3 Which of the following correctly describes the mechanism involved in catalytic cracking and in steam cracking?

	Catalytic cracking	Steam cracking
A	free radical	free radical
B	free radical	carbocation intermediates
C	carbocation intermediates	carbocation intermediates
D	carbocation intermediates	free radical

4 Which of the following would not be produced by reforming hexane ($CH_3CH_2CH_2CH_2CH_2CH_3$)?
 A $CH_3CH_2CH_2CH(CH_3)_2$
 B $CH_3C(CH_3)_2CH(CH_3)CH_3$
 C $CH_3CH(CH_3)CH(CH_3)CH_3$
 D $CH_3C(CH_3)_2CH_2CH_3$

5 Which of the following compounds is an example of a thiol?

 A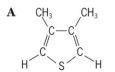

 B $CH_3—S—CH_2—CH_3$

 C $CH_3—CH_2—CH_2—SH$

 D $CH_3—S—S—CH_3$

6 **a** Describe the action of a catalytic converter on the exhaust from a motor car.
 b Explain why removing nitrogen oxides from exhaust gas is a benefit to the environment.

7 **a** Describe what happens during reforming.
 b Why is reforming a benefit to the environment?

8 Read the following passage carefully and answer the questions that follow.

Cracking is an added-value process in which large hydrocarbon molecules from oil are broken, or cracked, into smaller ones. The yield of products depends on several factors, including the quality of the feedstock.

In general, density is a good indicator of quality. A density of less than $0.9\,g\,cm^{-3}$ indicates that the feedstock is readily crackable and will produce a high percentage of high-value products. Refractive index is also a good indicator of the proportion of aromatic compounds in the feedstock. A value of around 1.48 indicates a good quality feedstock.

 a Why is cracking described as an added-value process?
 b Suggest likely products when the following undergo cracking.

 i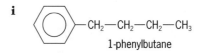
 1-phenylbutane

 ii $CH_3—CH_2—CH_2—CH_2—CH_2—CH_3$
 hexane

 iii $CH_3—CH_2—CH=CH_2$
 but-1-ene

 c What are the characteristics of a good feedstock?

9 **a** Why is it desirable to remove organic sulphur compounds from petroleum fuels?
 b The following table contains information about water and hydrogen sulphide.

	Water	Hydrogen sulphide
Relative molecular mass	18	34
Melting point/°C	0	−86
Boiling point/°C	100	−63
Bond angle	104.5°	93.3°

 i The RMM of hydrogen sulphide is almost twice that of water yet at room temperature hydrogen sulphide is a gas while water is a liquid. Explain why.
 ii Explain why the bond angle in hydrogen sulphide is less than in water.
 c Hydrogen sulphide is converted into sulphur in the Claus process. This involves two steps.

$$6H_2S(g) + 5O_2(g) \rightarrow 2SO_2(g) + 2S_2(g) + 6H_2O(g)$$
(equation 1)
$$4H_2S(g) + 2SO_2(g) \rightarrow 4H_2O(g) + 3S_2(g)$$
(equation 2)

Describe what is happening in equations 1 and 2 in terms of oxidation and reduction.

10 Steam cracking is a free-radical process. A molecule of propene forms an allyl radical by the loss of a hydrogen atom. The following diagram shows two forms of the allyl radical. Notice that the movement of single electrons is shown using half-arrows.

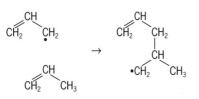

a Suggest a mechanism for the reaction of an allyl radical with a molecule of propene.

b Suggest how the product from part **a** might further react to form cyclopentene and a methyl radical (this involves cyclic intermediates).

1 A **2** C **3** D **4** B **5** C

6 a In a catalytic converter harmful gases are converted to harmless gases. Nitrogen oxides, carbon monoxide and unburnt hydrocarbons pass over a catalyst and are converted into nitrogen, carbon dioxide and water vapour.

b Nitrogen oxides contribute to acid rain which causes damage to limestone buildings and iron structures. Acid rain also reduces the pH of soil which releases heavy metal ions into water courses. The drop in pH and the heavy metal ions in water courses kills aquatic life.

7 a Straight-chain alkanes are converted to branched-chain alkanes, alkanes are converted to cyclic alkanes, and cyclic alkanes to aromatic compounds.

b The products of reforming have a higher anti-knock value than the starting materials. This allows petrol to burn more smoothly in a car engine and reduces the need for harmful anti-knock agents, such as lead tetraethyl, which pollute the atmosphere and are absorbed by the body.

8 a The products have more commercial value than the starting material.

b i For example, benzene and butene.
ii For example, butane and ethene.
iii For example, ethene and ethene.

c Density of less than $0.9\,g\,cm^{-3}$ and refractive index around 1.48.

9 a If sulphur compounds are not removed they will be oxidised to sulphur dioxide during combustion. This gas is acidic and will corrode engine parts. Once released to the atmosphere in exhaust gas, it will contribute to acid rain.

b i The intermolecular forces between molecules of water are much stronger (hydrogen bonding) than between molecules of hydrogen sulphide.
ii A sulphur atom is larger than an oxygen atom. There is a greater repulsion between the hydrogen atoms and the lone pairs of electrons on the sulphur atom in hydrogen sulphide than is the case with the oxygen atom in water.

c In equation 1, hydrogen sulphide is oxidised to sulphur, and to sulphur dioxide. In equation 2, hydrogen sulphide is oxidised to sulphur while sulphur dioxide is reduced to sulphur.

10 a

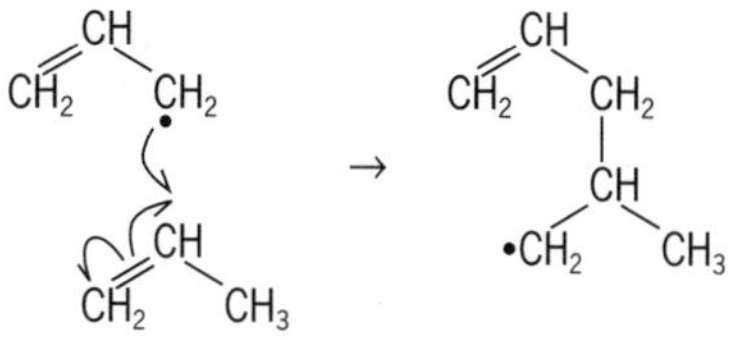

b

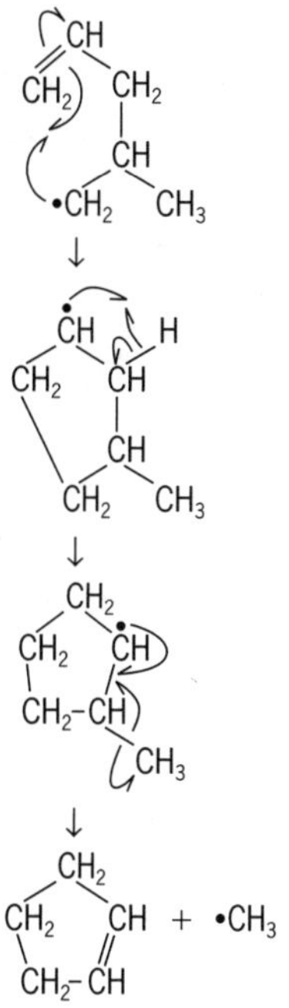

12 Isomerism and structure

Isomerism exists where molecules have the same molecular formula but their atoms are arranged in different ways. There are two main types of isomerism: structural isomerism and stereoisomerism.

Structural isomerism

Isomerism is largely associated with organic compounds. Compounds that are structural isomers have the same molecular formula but they have different structures. There are three ways in which this can occur.

Chain isomerism

Chain isomerism occurs where there is more than one way of arranging the carbon skeleton of a molecule. For example, the carbon chain of an alkane corresponding to the formula C_4H_{10} can be drawn as a straight chain, giving butane, or a branched chain, giving 2-methylpropane.

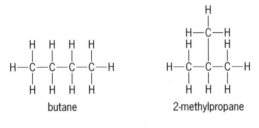
butane 2-methylpropane

Increasing the length of the carbon chain increases the number of possible isomers. There are three isomers with the formula C_5H_{12}:

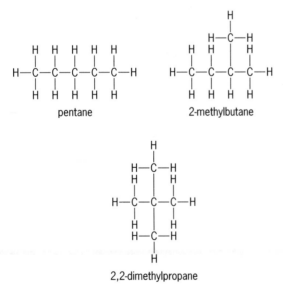
pentane 2-methylbutane

2,2-dimethylpropane

The number of possible chain isomers increases sharply with increasing numbers of atoms in the carbon skeleton (Table 12.1).

Table 12.1 Number of isomers of alkanes

Number of atoms in the carbon skeleton	Number of isomers
1	1
2	1
3	1
4	2
5	3
10	75
20	366 319
30	4 111 846 763
40	62 491 178 805 831

Chain isomers have similar chemical properties but the differences in structure give rise to some small variation in physical properties. Branched chains have less surface area so the attractive van der Waals forces between molecules are weaker. As a result, branched-chain compounds tend to have lower melting points and boiling points than their straight-chain isomers (Table 12.2).

Table 12.2 Isomers of C_4H_{10}

Isomer of C_4H_{10}	Melting point/°C	Boiling point/°C
butane	−138.4	−0.5
2-methylpropane	−159.6	−11.7

Positional isomerism

Positional isomers have the same carbon skeleton and the same functional group. What makes them different is that the functional group is attached to different positions on the carbon skeleton.

There are two isomers of chloropropane:

$CH_3CH_2CH_2Cl$ $CH_3CHClCH_3$
1-chloropropane   2-chloropropane

There are three isomers of pentanol:

$CH_3CH_2CH_2CH_2CH_2OH$ $CH_3CH_2CH_2CHOHCH_3$
pentan-1-ol   pentan-2-ol

$CH_3CH_2CHOHCH_2CH_3$
pentan-3-ol

The functional group may be part of the carbon skeleton. There are two isomers of butene:

$CH_2{=}CHCH_2CH_3$ $CH_3CH{=}CHCH_3$
but-1-ene   but-2-ene

Positional isomers have similar chemistry, since they have the same functional group; however, the positioning of the functional group sometimes results in differences in properties.

Functional isomerism

Functional isomers have the same molecular formula but different functional groups.

The compound with the formula C_2H_6O may be:

  CH₃CH₂OH   or   CH₃OCH₃
  ethanol (an alcohol)   methoxymethane (an ether)

The compound with the formula C_3H_6O may be:

  CH₃CH₂CHO   or   CH₃COCH₃
  propanal (an aldehyde)   propanone (a ketone)

The compound with the formula $C_4H_8O_2$ may be:

  CH₃CH₂CH₂COOH   or   CH₃COOCH₂CH₃
  butanoic acid   ethyl ethanoate
  (a carboxylic acid)   (an ester)

Functional isomers have different functional groups and thus have different chemical and physical properties.

Stereoisomerism

Stereoisomerism is concerned with compounds that have the same structural formulae but whose bonds are arranged differently in space. There are two ways in which this can occur: geometric and optical.

Geometric isomerism

This type of isomerism is found in alkenes. The carbon–carbon double bond in alkenes consists of a central σ bond in the plane of the molecule and the parts of a π bond above and below. These π electron clouds prevent the rotation of the carbon–carbon double bond. If each of the carbon atoms in the double bond has two different atoms or groups attached to it, two configurations are possible. When groups are on the same side of the carbon–carbon double bond the name of the compound carries the prefix *cis*, and when they are on opposite sides, the prefix *trans*.

For example, there are two forms of but-2-ene, *cis*-but-2-ene and *trans*-but-2-ene.

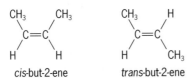

cis-but-2-ene   trans-but-2-ene

Geometric isomers have different physical and chemical properties. *Trans* isomers tend to pack together rather better so they have higher melting points. However, *cis* isomers tend to be more polar and thus have higher boiling points.

Butenedioic acid exhibits *cis–trans* isomerism and thus exists in two forms. *Cis*-butenedioic acid is readily converted to its anhydride because of the proximity of the carboxylic acid groups. *Trans*-butenedioic acid can be converted to the same anhydride but requires heating to a much higher temperature (Figure 12.1).

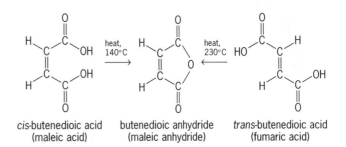

cis-butenedioic acid   butenedioic anhydride   trans-butenedioic acid
(maleic acid)   (maleic anhydride)   (fumaric acid)

Figure 12.1 *cis* and *trans* compounds react differently

Optical isomerism

When four different atoms or groups are attached to a carbon atom, the resulting molecule has no symmetry. This means that it is different from its mirror image and cannot be superimposed on it. The molecule is said to be **chiral** and the carbon atom at the centre is described as **asymmetric**.

1-bromo-1-chloroethane is such a molecule and exists in two forms that differ only in the way the bonds are arranged in space (Figure 12.2). These two forms are called **enantiomers**.

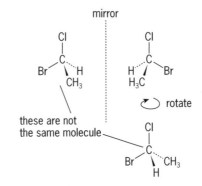

Figure 12.2 Optical isomerism

Chiral molecules are said to be **optically active** since they rotate the plane of polarised light. If polarised light is passed through a solution containing only one of the enantiomers, the plane of the light will be rotated either to the right, dextro- or (+) rotatory, or to the left, laevo- or (−) rotatory. A similar solution containing only the other enantiomer will rotate the plane of the light in the opposite direction.

A solution containing equal amounts of the enantiomers is optically inactive since the two effects cancel out. Such a solution is called a **racemic mixture**, or a **racemate**.

Many naturally occurring molecules are optically active. Enantiomers have identical physical properties and identical chemical properties except in reactions with other optically active substances. Enzymes are **stereospecific** – they can distinguish between enantiomers. Enzymes will catalyse reactions involving one enantiomer but not the other.

1 How many structural isomers are there with the formula C_6H_{14}?
   A 4
   B 5
   C 6
   D 7

2 Which pair of compounds are positional isomers?
   A $CH_3CH_2CH_2Br$ and $CH_3CH_2CH_2Cl$
   B $CH_3CH_2CH_2OH$ and $CH_3CHOHCH_3$
   C $CH_3CH_2CH_2NH_2$ and $CH_3CH_2CH_2CH_2NH_2$
   D $CH_3CH_2CH_2OH$ and $CH_3OCH_2CH_3$

3 Which of the following is a functional isomer of CH_3CH_2COOH?
   A $CH_3CH_2OOCH_3$
   B $CH_3CH_2CH_2COOH$
   C $HOCH_2CH_2CH_2OH$
   D $HOCH_2CH_2CHO$

4 Which of the following compounds exhibits *cis–trans* isomerism?

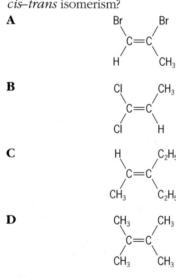

5 Which of the following compounds is optically active?

6 a The following diagram represents two of the possible nine chain isomers of C_7H_{16}.

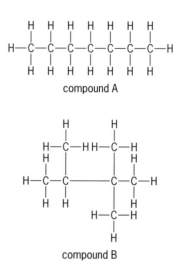

compound A

compound B

   i Name compounds A and B.
   ii Explain how you would expect the physical properties of these compounds to differ.
   b Draw the carbon skeletons of the other seven possible chain isomers of C_7H_{16}.

7 Draw the structural formulae and name the two positional isomers of
   a chlorobutane
   b pentene
   c hexanone.

8 Give the functional isomers indicated for each of the following formulae.
   a $C_4H_8O_2$ (a carboxylic acid and an ester)
   b C_4H_8O (an aldehyde and a ketone)
   c C_5H_{10} (a cyclic alkane and an alkene)
   d C_3H_8O (an alcohol and an ether)

9 a Give the structural formulae and name the *cis* and *trans* forms of each of the following compounds.
   i pent-2-ene $CH_3CH_2CH=CHCH_3$
   ii 1,2-dichloroethene $HClC=CClH$
   iii but-2-ene-1,4-diol $CH_2OHCH=CHCH_2OH$
   b Draw all the possible *cis–trans* isomers of 1,4-dibromo-1,3-butadiene,

   $CHBr=CH-CH=CHBr$

10 a Explain the terms:
   i chiral
   ii enantiomer
   iii optically active
   iv dextro-rotatory
   v racemic mixture.
   b Draw the enantiomers of 2-aminopropanoic acid, CH_3CHNH_2COOH, and label the asymmetric carbon on one of the structures.

1 B

2 B

3 D

4 A

5 C

6 a i Compound A = heptane
Compound B = 2,2,3-trimethylbutane
ii Expect 2,2,3-trimethylbutane to have a lower boiling point as it has a smaller surface area so the van der Waals forces between molecules will be weaker.
b The carbon frameworks are:

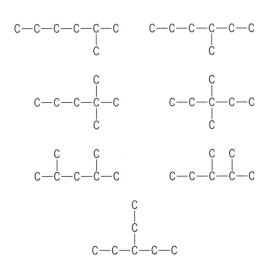

7 a CH₃CH₂CH₂CH₂Cl — 1-chlorobutane
CH₃CH₂CHClCH₃ — 2-chlorobutane
b CH₃CH₂CH₂CH=CH₂ — pent-1-ene
CH₃CH₂CH=CHCH₃ — pent-2-ene
c CH₃CH₂CH₂CH₂COCH₃ — hexan-2-one
CH₃CH₂CH₂COCH₂CH₃ — hexan-3-one

8 a CH₃CH₂CH₂COOH and CH₃CH₂COOCH₃ or CH₃COOCH₂CH₃
b CH₃CH₂CH₂CHO and CH₃CH₂COCH₃
c

CH₃—CH₂—CH₂—CH=CH₂ and or CH₃—CH₂—CH=CH—CH₃
d CH₃CH₂CH₂OH or CH₃CHOHCH₃ and CH₃OCH₂CH₃

9 a

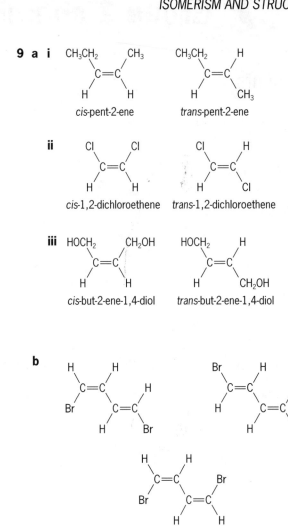

i cis-pent-2-ene   trans-pent-2-ene
ii cis-1,2-dichloroethene   trans-1,2-dichloroethene
iii cis-but-2-ene-1,4-diol   trans-but-2-ene-1,4-diol

b

10 a i A molecule is chiral if it cannot be superimposed on its mirror image.
ii Mirror images of a molecule that cannot be superimposed on each other are called enantiomers.
iii Enantiomers are optically active because they rotate the plane of polarised light.
iv A dextro-rotatory compound rotates the plane of polarised light to the right.
v A racemic mixture contains equal amounts of enantiomers and is optically inactive.

b

CH₃
H—C···COOH
NH₂

CH₃
H—C···NH₂
COOH

asymmetric carbon atom

13 Group 2 chemistry

Elements

Group 2 elements are sometimes called the **alkaline earth metals**. Group 2 consists of the metallic elements shown in Table 13.1.

Table 13.1 Group 2

Element	Symbol
beryllium	Be
magnesium	Mg
calcium	Ca
strontium	Sr
barium	Ba
radium	Ra

Group 2 elements are light silver metals which form ions with a charge of +2. They form ionic compounds which are white or colourless.

The elements show common group behaviour and there are clearly discernible trends in properties passing up and down the group.

Radium is a radioactive element and it is generally not included in the study of this group of elements.

Beryllium does not react like a typical Group 2 metal. Its compounds show a high degree of covalent character due, in part, to its small atomic radius.

Physical properties of the elements

The elements in Group 2 have low melting points and boiling points (Table 13.2) when compared with the transition metals (see topic 17). However, like all metals, they are good conductors of both heat and electricity.

The atoms of each metal have two outer electrons. They form ions by losing these two electrons and thus achieving the electronic configuration of a noble gas.

$$Mg \rightarrow Mg^{2+} + 2e^-$$
$$1s^2 2s^2 2p^6 3s^2 \quad 1s^2 2s^2 2p^6$$

As atomic number increases down the group the nuclear charge increases. This might reasonably be expected to make it more difficult for electrons to be lost. However, an increasing number of complete electron shells will shield the outer electrons from the nuclear charge. Also, the attraction between the positively charged nucleus and the negatively charged electrons will diminish as the distance between them increases. These latter two factors make it easier for electrons to be lost.

Overall, there is a decrease in electronegativity passing down the group, making it easier for atoms to lose electrons. As the elements form ions by the loss of electrons, the result is an increase in reactivity as you go down the group.

least reactive	Mg
↓	Ca
	Sr
most reactive	Ba

Chemical reaction with air and oxygen

When left exposed to air, Group 2 metals react quickly with oxygen, becoming covered with a white film of metal oxide.

The metals react vigorously when heated with oxygen, forming white solid oxides. (The symbol M in the following equations represents any of the metals.)

$$2M + O_2 \rightarrow 2MO$$

These oxides do not decompose on heating.

Chemical reaction with water

The metals react with water or steam to produce the metal oxide and hydrogen.

$$M + H_2O \rightarrow MO + H_2$$

Magnesium is the least reactive of the metals and reacts very slowly with water but more vigorously with steam. Calcium, strontium and barium react vigorously with water.

Table 13.2 Physical properties of the Group 2 elements

Element	Atomic number	Outer electrons	Melting point /°C	Boiling point /°C	Metallic radius /nm	Ionic radius /nm	Electro-negativity
Mg	12	$3s^2$	650	1100	0.160	0.072	1.2
Ca	20	$4s^2$	840	1490	0.197	0.100	1.0
Sr	38	$5s^2$	770	1390	0.215	0.113	1.0
Ba	56	$6s^2$	730	1640	0.224	0.136	0.9

Chemical reaction with acids

The metals react with acids to give the metal salt and hydrogen. For example, with hydrochloric acid

$$M + 2HCl \rightarrow MCl_2 + H_2$$

The reaction with sulphuric acid is not a good method of preparing the sulphates of calcium, strontium or barium. The sulphates of these three metals are not very soluble. When pieces of one of these metals are reacted with dilute sulphuric acid they become coated with a layer of the sulphate and this prevents further reaction.

Chemical reaction of metal oxides with water

The Group 2 metal oxides are bases which react with water to produce alkaline solutions of hydroxides.

$$MO + H_2O \rightarrow M(OH)_2$$

Magnesium oxide is not very soluble in water and gives only a weakly alkaline solution.

Calcium oxide gives off a great deal of heat when it reacts with water. The resulting solution of calcium hydroxide is better known as lime water. It is used to test for carbon dioxide.

Solubility of hydroxides and sulphates

Solubility in water is often given as a typical characteristic of ionic compounds; however, the situation is more complex than this. There are many ionic compounds which are effectively insoluble in water. When an ionic solid dissolves in water two processes take place.

1 The ionic lattice breaks. This process requires energy and is thus endothermic. The amount of energy needed depends on the size of the ions and the charge they carry. The energy needed to break the lattice is smallest for large ions carrying a small charge.
2 After the ionic lattice breaks down the ions become surrounded by water molecules. This process is called **hydration** and it is exothermic, giving out energy. The larger the ions the less energy is given out by this process.

Simplistically, we might expect an ionic compound to be soluble if the energy given out from the hydration of its ions is greater than or equal to the energy required to break the ionic lattice.

In Group 2 **hydroxides**, the energy released by hydration increases in comparison to the energy needed to break down the ionic lattice and so the compounds become more soluble passing down the group (Table 13.3).

Table 13.3 Solubility of Group 2 hydroxides and sulphates

Element	Solubility of hydroxide /mol per 100 g water at 298 K	Solubility of sulphate /mol per 100 g water at 298 K
magnesium	2.00×10^{-5}	1.8×10^{-1}
calcium	1.53×10^{-3}	1.10×10^{-3}
strontium	3.37×10^{-3}	7.11×10^{-5}
barium	1.50×10^{-2}	9.43×10^{-7}

Conversely, in Group 2 **sulphates** the energy released by hydration decreases in comparison to the energy needed to break down the ionic lattice and so the compounds become less soluble passing down the group.

However, in considering solubility changes, **entropy** must also be taken into account. Entropy is a measure of the amount of order in a system and is considered more fully in topic 25. For now, we can say that the highly ordered arrangement of ions in the solid becomes a much less ordered arrangement of ions within the water. When an ionic substance dissolves its entropy increases.

Thermal stability of salts

Group 2 **carbonates** decompose when heated to give the oxide and carbon dioxide.

$$MCO_3 \rightarrow MO + CO_2$$

The thermal stability of the carbonates increases down the group. Magnesium carbonate decomposes at 350 °C while barium carbonate does not decompose until a temperature of 1350 °C.

Group 2 **nitrates** decompose when heated to give the metal oxide, nitrogen dioxide and oxygen.

$$2M(NO_3)_2 \rightarrow 2MO + 4NO_2 + O_2$$

It is interesting to note that lithium carbonate decomposes in the same way as the Group 2 carbonates, while the other Group 1 carbonates do not decompose on heating. Similarly, lithium nitrate decomposes in the same way as a Group 2 nitrate, whereas the remaining Group 1 nitrates decompose on strong heating, giving nitrites and oxygen.

Group 2 **hydroxides** decompose when heated to give the metal oxide and water.

$$M(OH)_2 \rightarrow MO + H_2O$$

Uses of Group 2 elements and their compounds

Magnesium burns with a bright white light. It is used for flares and incendiary bombs. It is also a powerful reducing agent and is widely used as a sacrificial anode to protect steel structures, such as ships and bridges, from rusting. The low density of magnesium makes it useful in low-density alloys, such as duralumin, which are used in the construction of aircraft.

Magnesium oxide has a very high melting point (2800 °C) and is used as a lining material for furnaces. Magnesium hydroxide is a weak alkali and is used in indigestion remedies and toothpaste to neutralise acid.

Calcium oxide is widely used in agriculture in order to reduce the acidity of soil. It is used in the manufacture of cement, plaster and mortar, and plays an important role in the extraction of iron, in the blast furnace, where it combines with acidic impurities which would otherwise damage the furnace lining.

Calcium hydroxide solution is more commonly called lime water. It is used in the laboratory to demonstrate the presence of carbon dioxide.

An insoluble form of calcium sulphate, commonly called plaster of Paris, is used in modelling and to hold broken bones in position.

Barium sulphate is insoluble and opaque to X-rays. When swallowed by patients, subsequent X-ray examination may show up imperfections in the soft tissue of the digestive system. Barium compounds are toxic but the very low solubility of barium sulphate makes it safe to use in the body.

1 Which of the following statements about the Group 2 elements is true?
   A All form an ion which carries a charge of +2
   B Cationic radius is always larger than metallic radius
   C Electronegativity increases down the group
   D Covalent character in compounds of the elements increases down the group

2 Which of the following have the same electron configuration?
   A Be^{2+} and Ne
   B Mg and F^-
   C Sr^{2+} and Rb^+
   D Ba and Sr

3 The formula of crystalline magnesium sulphate may be written $MgSO_4.xH_2O$. Heating 4.3 g of crystalline magnesium sulphate to constant mass produced 2.1 g of the anhydrous salt. The value of x is
   A 2
   B 5
   C 7
   D 10

4 Which of the following compounds of magnesium is chemically unchanged by heating?
   A magnesium carbonate
   B magnesium hydroxide
   C magnesium nitrate
   D magnesium oxide

5 Which of the following is **not** likely to be a property of radium?
   A The electronegativity is 0.9
   B The radius of the Ra^{2+} ion is 0.140 nm
   C The melting point is 918 °C
   D The outer electron configuration is $7s^2$

6 a Explain what happens when an ionic solid dissolves in water.
   b The following table summarises the solubility of two compounds of some Group 2 elements.

Element	Hydroxide	Sulphate
magnesium	sparingly soluble	very soluble
calcium	reasonably soluble	reasonably soluble
barium	very soluble	sparingly soluble

Explain why the solubilities of the two groups of compounds show opposite trends.

7 a Give general equations for the thermal decomposition of Group 2 carbonates and nitrates.
   b What group trend is shown by the thermal decomposition of Group 2 carbonates?

c In the Periodic Table, elements which are diagonally next to each other sometimes show similarities in their compounds. Comment on whether such a relationship exists between lithium and magnesium by considering the thermal stability of their carbonates and nitrates.

8 The following table gives the first and second ionisation energies for the Group 2 elements.

Element	First ionisation energy/kJ mol^{-1}	Second ionisation energy/kJ mol^{-1}
magnesium	738	1451
calcium	590	1145
strontium	550	1064
barium	503	965

a Give equations for the first and second ionisations of a metal, M.
b Explain why the second ionisation energy is greater than the first for all of these elements.
c Explain why the first (and second) ionisation energy decreases down the group.
d The following graph shows the pattern of successive ionisation energies for magnesium. Account for the shape of this graph.

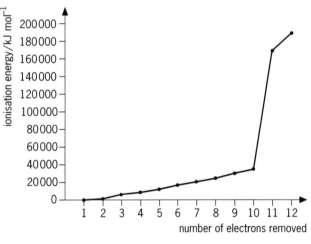

9 Explain each of the following statements as fully as you can, giving equations where appropriate.
   a When reacted with water, sodium oxide dissolves to give a strong alkali but magnesium oxide gives a weak alkali.
   b When carbon dioxide is bubbled through aqueous calcium hydroxide the solution initially goes cloudy but subsequently clears.

c Reacting calcium with dilute sulphuric acid is not a satisfactory way of making calcium sulphate.

d Barium sulphate is used to show up soft tissue for X-ray examination.

e Barium is stored under oil but magnesium is not.

10 Magnesium is widely distributed in minerals. Two important sources are dolomite ($CaCO_3.MgCO_3$) and carnallite ($MgCl_2.KCl.6H_2O$).

a Calculate the percentage by mass of magnesium in each of these ores.

b Dolomite is converted to a mixture of calcium hydroxide and magnesium hydroxide. Suggest how this can be done.

c Calcium is removed by addition of magnesium ions. Comment on the equilibrium position in this reaction.

$$Ca(OH)_2.Mg(OH)_2(s) + Mg^{2+}(aq) \rightleftharpoons 2Mg(OH)_2(s) + Ca^{2+}(aq)$$

d Magnesium hydroxide is converted to magnesium oxide, and then to magnesium chloride by heating with coke and chlorine gas.

$$2MgO(s) + C(s) + 2Cl_2(g) \rightarrow 2MgCl_2(s) + CO_2(g)$$

Which reactants are oxidised and which are reduced in this reaction?

e Magnesium is made by the electrolysis of molten magnesium chloride. Suggest why molten magnesium chloride and not molten magnesium oxide is used. Why does the magnesium chloride have to be molten and not solid? Write half-equations for the reaction at each electrode.

1 A

2 C

3 C

Formula mass of $MgSO_4 = 24.3 + 96 = 120.3$
Formula mass of $MgSO_4.xH_2O = 120.3 + 18x$

$$\frac{4.3}{120.3 + 18x} = \frac{2.1}{120.3}$$

$x = 7$

4 D

5 C

6 a The ionic lattice breaks down, separating the ions. The ions then become surrounded by water molecules in a process called hydration.

b Breaking down the ionic lattice requires energy and is an endothermic process. Conversely, hydration gives out energy and is an exothermic process. The solubility of an ionic compound depends on the energy given out during hydration being greater than or equal to the energy needed to break down the lattice.

In the case of the hydroxides, the energy released by hydration becomes progressively greater compared to the energy needed to break the ionic lattice so solubility increases down the group.

The opposite situation exists with sulphates. The energy released by hydration becomes progressively less compared to the energy needed to break the ionic lattice so solubility decreases down the group.

7 a Group 2 carbonates decompose to give the metal oxide and carbon dioxide.

$$MCO_3 \rightarrow MO + CO_2$$

Group 2 nitrates decompose to give the metal oxide, nitrogen dioxide and oxygen.

$$2M(NO_3)_2 \rightarrow 2MO + 4NO_2 + O_2$$

b Group 2 carbonates increase in stability passing down the group. Magnesium carbonate is the least stable, decomposing at $350\,°C$, while barium carbonate is the most stable and does not decompose until a temperature of $1350\,°C$.

c For the most part Group 1 carbonates and nitrates differ in their properties to Group 2 carbonates and nitrates. Most Group 1 carbonates do not decompose on heating and nitrates decompose to nitrites with the loss of oxygen only.

Lithium is the exception. Its carbonate and nitrate decompose in the same way as a Group 2 element, thus showing similarities with the compounds of magnesium.

8 a First ionisation: $M \rightarrow M^+ + e^-$
Second ionisation: $M^+ \rightarrow M^{2+} + e^-$

b After the first electron is removed, there is an imbalance between the positive charge at the nucleus and the total negative charge of the electrons

surrounding it. The remaining electron in the outer shell experiences a greater attractive force from the nucleus so more energy is required to remove it.

c The charge on the nucleus increases down the group which might be expected to increase the ionisation energy. However, there is an increasing number of complete shells of electrons which shield the outer electrons from the charge on the nucleus and the outer electrons are further away from the nucleus, which reduces the force of attraction between them. The latter two effects are more significant than the increase in nuclear charge.

d 1 and 2 are 3s electrons, 3–8 are 2p electrons, 9 and 10 are 2s electrons, and 11 and 12 are 1s electrons. Within each orbital it becomes progressively more difficult to remove electrons as the imbalance between nuclear charge and total negative charge becomes greater. The changes of gradient represent the start of electrons being removed from a complete shell. The stability of a complete electron shell has to be overcome hence there is a step-up in the amount of energy needed.

9 a Both sodium oxide and magnesium react with water to form a hydroxide.

$$Na_2O + H_2O \rightarrow 2NaOH$$
$$MgO + H_2O \rightarrow Mg(OH)_2$$

However, magnesium hydroxide is not very soluble so it only gives a weak alkaline solution.

b Calcium hydroxide initially reacts with carbon dioxide to give insoluble calcium carbonate.

$$Ca(OH)_2 + CO_2 \rightarrow CaCO_3 + H_2O$$

Calcium carbonate is able to react with additional carbon dioxide to form soluble calcium hydrogencarbonate.

$$CaCO_3 + CO_2 + H_2O \rightarrow Ca(HCO_3)_2$$

c Calcium sulphate is not very soluble. When calcium is added to dilute sulphuric acid a layer of insoluble calcium sulphate forms around the calcium and prevents the reaction going to completion.

d Barium nitrate is opaque to X-rays. When it is swallowed by a patient details of the soft tissue show up under X-ray examination. Barium sulphate is also very insoluble so it is not absorbed by the body.

e Barium is a very reactive metal and will react with oxygen and water vapour in the air. Storing it under oil prevents it coming into contact with the air.

$$2Ba(s) + O_2(g) \rightarrow 2BaO(s)$$
$$Ba(s) + H_2O(g) \rightarrow BaO(s) + H_2(g)$$

Magnesium is less reactive than barium. Also, when exposed to the air magnesium rapidly forms a layer of oxide which covers the metal and prevents any further reaction.

$$2Mg(s) + O_2(g) \rightarrow 2MgO(s)$$

10 a Dolomite: relative formula mass

$= 40 + 60 + 24.3 + 60$

$= 184.3$

% by mass of magnesium

$= \dfrac{24.3}{184.4} \times 100$

$= 13.18\%$

Carnallite: relative formula mass

$= 24.3 + 71 + 39.1 + 35.5 + 108$

$= 277.9$

% by mass of magnesium in carnallite

$= \dfrac{24.3}{277.9} \times 100$

$= 8.74\%$

b Heating dolomite produces a mixture of calcium oxide and magnesium oxide, which form hydroxides when added to water.

$CaCO_3.MgCO_3(s) \rightarrow CaO.MgO(s) + 2CO_2(g)$
$CaO.MgO(s) + 2H_2O(l) \rightarrow Ca(OH)_2.Mg(OH)_2(s)$

c Magnesium hydroxide is much less soluble than calcium hydroxide so the equilibrium position is well to the right, favouring the removal of calcium ions.

d Magnesium oxide and chlorine are reduced. Carbon is oxidised.

e Magnesium oxide has a very high melting point so it would be most costly to melt it. Solid magnesium chloride is composed of ions but they are not mobile and so they are unable to conduct electricity.

At the cathode: $Mg^{2+}(aq) + 2e^- \rightarrow Mg(l)$
At the anode: $2Cl^-(aq) \rightarrow Cl_2(g) + 2e^-$

14 Group 7 chemistry

Elements

The elements of Group 7 are often called the **halogens**. It consists of the non-metallic elements listed in Table 14.1.

Table 14.1 Group 7

Element	Symbol	Colour	State at room temperature
fluorine	F	pale green	gas
chlorine	Cl	yellow-green	gas
bromine	Br	red-brown	liquid
iodine	I	black	solid

These elements exist as diatomic molecules (X_2) at room temperature. They form both ionic and covalent compounds and exhibit a range of **oxidation states**.

The elements show distinct similarities in behaviour and clearly discernible trends in properties passing up and down the group.

The element astatine (At) is also in Group 7 but relatively little is known of its chemistry and it is generally not included in the study of this group of elements.

Fluorine has some properties that are anomalous when compared to the behaviour of other members of Group 7.

Physical properties of the elements

The trend of increasing boiling point down the group (Table 14.2) is explained by considering the nature of the weak attractive van der Waals forces between halogen molecules. Temporary fluctuations in electron density within molecules result in these temporary induced dipole–dipole attractions between molecules. The size of the attractive force increases with the surface area of contact between molecules, thus as both atomic and molecular radii increase with increasing atomic number, so the boiling point also increases.

Electronegativity is the power of an atom to withdraw electron density from a covalent bond. The trend of decreasing electronegativity down this group may be explained by considering those factors that determine the electronegativity of an element.

- The atomic number of an atom indicates the charge on its nucleus. As atomic number increases we might expect the attraction for the pair of electrons in a covalent bond to increase.
- Conversely, as atomic number increases so does the number of shells of electrons. The outer electrons of an atom will be increasingly shielded by complete inner shells of electrons and will thus be less strongly attracted by the nucleus.
- The attraction between oppositely charged particles falls rapidly as the distance between them increases. As the radius of an atom increases the outer electrons are further from the nucleus and are attracted less strongly.

Overall, the trend in electronegativity in this group indicates that the increase in shielding and the increase in atomic radius make a greater contribution than the increase in nuclear charge.

Halogens as oxidising agents

Oxidation involves the loss of one or more electrons. In bringing about an oxidation, the oxidising agent is itself reduced.

A halogen atom has seven electrons in its outer shell and requires just one more to attain the stable electron configuration of a noble gas.

$$\tfrac{1}{2}F_2 \quad + e^- \rightarrow \quad F^-$$
$$1s^2 2s^2 2p^5 \qquad 1s^2 2s^2 2p^6$$

All of the halogens are oxidising agents; however, oxidising power decreases down the group. The reason for this involves a similar explanation to that given for decreasing electronegativity.

Table 14.2 Physical properties of the Group 7 elements

Element	Atomic number	Outer electrons	Melting point/°C	Boiling point/°C	Radius of atom/nm	Radius of ion X^-/nm	Electronegativity
F	9	$2s^2 2p^5$	−220	−188	0.071	0.133	4.0
Cl	17	$3s^2 3p^5$	−101	−34	0.099	0.180	3.0
Br	35	$4s^2 4p^5$	−7	59	0.114	0.195	2.8
I	53	$5s^2 5p^5$	114	184	0.133	0.215	2.5

As nuclear charge increases we would expect that the attraction for an extra electron would increase. Thus, on the basis of this factor alone, we would expect the oxidising power of the halogens to increase down the group. However, as atomic number increases, the number of complete electron shells between the nucleus and the outer shell increases, as does the distance between the nucleus and the outer electrons. On the basis of the increase in shielding and the reduction in attraction between the nucleus and the outer electrons we would expect the oxidising power of the halogens to decrease down the group.

It is the latter two factors which have the greater effect, leading to a decrease in the oxidising power of the halogens as atomic number increases. Fluorine is the most powerful oxidising agent and iodine the least.

$$F > Cl > Br > I$$
oxidising power

Fluorine is a very reactive chemical and is not used in normal laboratory work. The relative oxidising power of chlorine, bromine and iodine is easily demonstrated using **displacement reactions**. These involve adding a halogen to solutions containing the other two halide ions.

The halogens displaced by these reactions are in such low concentrations that it may not be possible to see if a reaction has taken place. If the aqueous solution is shaken with an organic solvent, such as cyclohexane, the halogen will be concentrated in the organic layer and the colour will become more obvious.

Chlorine displaces both bromide and iodide from solution.

$$Cl_2(aq) + 2Br^-(aq) \rightarrow 2Cl^-(aq) + Br_2(aq)$$

Br_2 forms a yellow-brown colour in cyclohexane.

$$Cl_2(aq) + 2I^-(aq) \rightarrow 2Cl^-(aq) + I_2(aq)$$

I_2 forms a pink–purple colour in cyclohexane.

Bromine displaces iodide but not chloride from solution.

$$Br_2(aq) + 2Cl^-(aq) \rightarrow \text{no reaction}$$
$$Br_2(aq) + 2I^-(aq) \rightarrow 2Br^-(aq) + I_2(aq)$$

Iodine displaces neither chloride nor bromide from solution.

$$I_2(aq) + 2Cl^-(aq) \rightarrow \text{no reaction}$$
$$I_2(aq) + 2Br^-(aq) \rightarrow \text{no reaction}$$

Halides as reducing agents

Conversely, **reduction** involves gaining one or more electrons. The reducing power of halide ions increases down the group. Iodide is the most powerful reducing agent and fluoride the least.

$$F^- < Cl^- < Br^- < I^-$$
reducing power

The trend in reducing power is demonstrated by the reaction of solid halide salts with concentrated sulphuric acid. In these reactions the oxidation state of sulphur in sulphuric acid is reduced from +6 to +4, 0 or −2 depending on the reducing power of the halide ion (oxidation state is discussed more fully in topic 17).

$H_2SO_4 + NaCl$	$\rightarrow HCl$	Displacement reaction only.
$H_2SO_4 + NaBr$	$\rightarrow HBr$	Displacement reaction.
	$\rightarrow SO_2$	Sulphuric acid is reduced. The oxidation state of sulphur in sulphuric acid is +6 while in sulphur dioxide it is only +4.
	$\rightarrow Br_2$	Bromide oxidised to bromine.
$H_2SO_4 + NaI$	$\rightarrow HI$	Displacement reaction.
	$\rightarrow SO_2$	Sulphuric acid is reduced. The oxidation state of sulphur in sulphuric acid is +6 while in sulphur dioxide it is +4, as elemental sulphur it is 0 while as hydrogen sulphide it is −2.
	$\rightarrow S$	
	$\rightarrow H_2S$	
	$\rightarrow I_2$	Iodide oxidised to iodine.

Reaction of chlorine with water and alkalis

Chlorine is commonly added to water to sterilise it. The gas kills bacteria, making the water safe to drink. Chlorine dissolves in water to form a pale-green solution. The gas reacts with water, establishing the following equilibrium:

$$Cl_2(g) + H_2O(l) \rightleftharpoons HCl(aq) + HClO(aq)$$
oxidation state $\quad 0 \qquad\qquad\qquad -1 \qquad +1$

Notice that in this reaction the chlorine is being simultaneously oxidised and reduced. This is known as **disproportionation**. If universal indicator is added to the water it first turns red, since both HCl (hydrochloric acid) and HOCl (chloric(I) acid) release H^+ in solution. Subsequently, the red colour is lost and the solution becomes colourless because chloric(I) acid is also an effective bleach.

If chlorine water is left in bright sunlight the green colour fades and oxygen is formed.

$$2Cl_2(aq) + 2H_2O(l) \rightleftharpoons 4HCl(aq) + O_2(g)$$

Under these conditions chlorine oxidises water to oxygen, itself being reduced to chloride ions.

The products of the reaction of chlorine with alkalis depend upon the temperature at which the reaction takes place. At 15 °C with cold, dilute alkali a mixture of the chloride and the chlorate(I), also known as hypochlorite, is produced.

$$Cl_2(g) + 2OH^-(aq) \rightarrow Cl^-(aq) + ClO^-(aq) + H_2O(l)$$

The chlorate(I) which is produced then decomposes to give the chloride and chlorate(v) ions.

$$3ClO^-(aq) \rightarrow 2Cl^-(aq) + ClO_3^-(aq)$$

The second reaction is slow at 15 °C but rapid at 70 °C therefore it is possible to control the products of the reaction(s) by controlling the temperature.

Chlorine forms a series of oxoacids (Table 14.3).

Table 14.3 Oxoacids of chlorine

Oxoacid of chlorine	Formula	Oxidation state of chlorine
Chloric(I) acid	HOCl	+1
Chloric(III) acid	$HClO_2$	+3
Chloric(V) acid	$HClO_3$	+5
Chloric(VII) acid	$HClO_4$	+7

As the oxidation number of the chlorine increases, both the thermal stability and the acid strength of the oxoacids increases but the oxidising strength of the acids decreases.

Iodine and thiosulphate titrations

Iodine reacts with thiosulphate ions to produce iodide and tetrathionate ions.

$$I_2(aq) + 2S_2O_3^{2-}(aq) \rightarrow 2I^-(aq) + S_4O_6^{2-}(aq)$$

It is possible to use this reaction as an accurate method of measuring the concentration of an oxidising agent, using a technique called **titration**.

An excess of iodide ions is added to an acidified solution of the oxidising agent. Some of the iodide ions will be oxidised to iodine. The amount of iodine produced will depend upon the concentration of the oxidising agent. The concentration of iodine produced is then calculated by titrating against a solution of sodium thiosulphate of known concentration.

1 Which of the following statements about the Group 7 elements is true?

A Electronegativity increases down the group

B Anionic radius is always smaller than atomic radius

C They are all gases at 25 °C

D They all form an ion which carries a charge of −1

2 Which of the following is **not** likely to be a property of astatine?

A The outer electron configuration is $6s^2 6p^5$

B The melting point is 300 °C

C The electronegativity is 2.9

D The radius of the At^- ion is 0.227 nm

3 Which of the following (in addition to sodium hydrogensulphate) is/are produced when solid sodium fluoride is heated with concentrated sulphuric acid?

A hydrogen fluoride only

B hydrogen fluoride and sulphur dioxide

C hydrogen fluoride, sulphur dioxide and fluorine

D hydrogen fluoride, sulphur dioxide, fluorine and sulphur

4 Which of the following equations represents a disproportionation reaction?

A $Cl_2(g) + 2Br^-(aq) \rightarrow 2Cl^-(aq) + Br_2(l)$

B $Cl_2(g) + 2OH^-(aq) \rightarrow Cl^-(aq) + ClO^-(aq) + H_2O(l)$

C $H_2SO_4(l) + 2HBr(g) \rightarrow SO_2(g) + 2H_2O(l) + Br_2(l)$

D $I_2(aq) + 2S_2O_3{}^{2-}(aq) \rightarrow 2I^-(aq) + S_4O_6{}^{2-}(aq)$

5 When concentrated sulphuric acid reacts with iodine to produce hydrogen sulphide, the oxidation state of the sulphur in sulphuric acid changes from

A +4 to 0

B +4 to −2

C +6 to 0

D +6 to −2

6 When chlorine is bubbled through cold dilute aqueous sodium hydroxide a solution containing the chlorate(I) ion is formed.

a Give an equation for the reaction which occurs and state the changes in the oxidation state of chlorine.

b State what happens to the chlorate(I) in terms of oxidation state when this solution is heated.

7 a i What is electronegativity?

ii What is the trend in electronegativity passing down the Group 7 elements?

iii Explain this trend.

b Aqueous bromine and a small amount of cyclohexane are added to separate solutions of potassium chloride and potassium iodide with shaking. In each case, describe what you would observe and give an equation for any reaction that takes place.

c Discuss the trend in reducing properties of the halide ions which is shown by the above experiments.

8 The trend in the reducing ability of the chloride, bromide and iodide ions can be shown by the reaction between concentrated sulphuric acid and potassium chloride, potassium bromide and potassium iodide.

a State the products obtained from each of these reactions and identify those which are formed by the reduction of concentrated sulphuric acid.

b Explain how the results of the above reactions show a trend in reducing ability of the halide ions and state this trend.

9 Copper(II) ions liberate iodine from potassium iodide solution according to the following equation:

$$2Cu^{2+}(aq) + 4I^-(aq) \rightarrow 2CuI(s) + I_2(aq)$$

The iodine may then be titrated by sodium thiosulphate.

a Give an ionic equation for the reaction between iodine and sodium thiosulphate.

b How many moles of sodium thiosulphate are needed to react with the iodine liberated by 1 mole of copper(II) ions?

c An aqueous solution of an unknown copper(II) salt was made by dissolving 1.836 g in water to make 100 cm³ of solution. An excess of potassium iodide was added to 25 cm³ of the aqueous copper(II) salt solution and the liberated iodine reacted exactly with 18.4 cm³ of aqueous sodium thiosulphate of concentration 0.10 mol dm⁻³. Calculate the relative formula mass of the copper(II) salt and deduce its formula, given that it contains five waters of crystallisation.

10 a Discuss the equilibrium which is established when chlorine gas is bubbled into water, in terms of changes to the oxidation state of chlorine. Explain what happens if a few drops of universal indicator are added to the solution.

b Describe and explain what happens if an aqueous solution of chlorine is left in bright sunlight.

c When chlorine is bubbled into hot aqueous sodium chloride solution both chloride ions and chlorate(v) ions ($ClO_3{}^-$) are produced. Write half-equations for:

chlorine → chloride ions

and

chlorine → chlorate(v) ions

Use these half-equations to deduce the overall equation for the reaction.

1 D **2** C **3** A **4** B **5** D

6 a $Cl_{2(aq)} + 2OH^-_{(aq)} \rightarrow Cl^-_{(aq)} + ClO^-_{(aq)} + H_2O_{(l)}$
The oxidation state of chlorine changes from 0 to -1 forming chloride ions, and from 0 to $+1$ forming chlorate(I) ions.

b $3ClO^-_{(aq)} \rightarrow 2Cl^-_{(aq)} + ClO_3^-_{(aq)}$
The chlorate(I) decomposes to give chloride and chlorate(v) ions. The oxidation state of the chlorate(I) ions changes from $+1$ to -1 forming chloride ions, and from $+1$ to $+5$ forming chlorate(v) ions.

7 a i The tendency of an atom to attract electrons to itself.
ii Electronegativity decreases down the group.
iii Passing down the group there is an increasing charge on the nucleus which, taken in isolation, would be expected to increase electronegativity. However, as atomic number increases so does the number of shells of electrons so the outer electrons of an atom will be increasingly shielded by complete inner shells of electrons. Also, as the radius of an atom increases, the outer electrons are further from the nucleus and the attraction is less. Overall the increase in shielding and the increase in atomic radius make a greater contribution than the increase in nuclear charge.

b No reaction with potassium chloride solution, but with potassium iodide solution bromine displaces the iodide ions. The cyclohexane turns pink-purple due to the presence of iodine.

$Br_{2(aq)} + 2I^-_{(aq)} \rightarrow 2Br^-_{(aq)} + I_{2(aq)}$

c Reduction involves gaining one or more electrons so a reducing agent is a substance which gives up electrons. In these reactions iodide is the most powerful reducing agent since it reduces bromine while chloride is the least powerful since it is unable to reduce bromine. The reducing power of halide ions increases down the group.

8 a With potassium chloride the product is:

- hydrogen chloride – only a displacement reaction occurs. The concentrated sulphuric acid is not reduced.

With potassium bromide the products are:

- hydrogen bromide – a displacement reaction occurs.
- sulphur dioxide and bromine – sulphuric acid oxidises bromide to bromine and is itself reduced to sulphur dioxide.

With potassium iodide the products are:

- hydrogen iodide – a displacement reaction occurs.
- sulphur dioxide, sulphur, hydrogen sulphide and iodine – sulphuric acid oxidises iodide to iodine and is itself reduced to sulphur dioxide, sulphur and hydrogen sulphide.

b These reactions show an increase in reducing power of the halide ions passing down the group. This is evident by the degree to which sulphuric acid is reduced in the reactions.
The chloride ion is not a sufficiently strong reducing agent to reduce sulphuric acid. The bromide ion is able to reduce sulphuric acid such that the oxidation state of the sulphur changes from $+6$ to $+4$ in the product formed. The iodide ion reduces the sulphuric acid such that the oxidation state of the sulphur changes from $+6$ to $+4$, 0 and -2 in the products formed.

9 a $I_2 + 2S_2O_3^{2-} \rightarrow 2I^- + S_4O_6^{2-}$
b 1
c $18.4\,cm^3$ of $0.10\,mol\,dm^{-3}$ thiosulphate ions $=$ 0.00184 mol of thiosulphate ions
$25\,cm^3$ contains 0.459 g of the copper(II) salt
Since 1 mol of copper(II) ions react with 1 mol of thiosulphate ions
$0.459\,g \equiv 0.00184$ mol of the copper(II) salt
1 mol of the copper(II) salt has a relative formula mass of

$$\frac{0.459}{0.00184} = 249.5$$

By comparison with likely soluble copper(II) salts the unknown salt is copper(II) sulphate pentahdyrate, $CuSO_4.5H_2O$.

10 a Chlorine reacts with water establishing the following equilibrium:

$$Cl_2 + H_2O \rightleftharpoons HCl + HClO$$

Chlorine has an oxidation state of 0. In forming HClO, chlorine is oxidised and its oxidation state changes from 0 to $+1$. Conversely, in forming HCl chlorine is reduced and its oxidation state changes from 0 to -1.
When universal indicator is added to the water it first turns red because both HCl and HOCl are acidic in aqueous solution. The red colour is then lost and the solution becomes colourless because HOCl is also a bleach.

b When chlorine water is left in bright sunlight the green colour fades and oxygen gas is formed.

$$2Cl_{2(aq)} + 2H_2O_{(l)} \rightleftharpoons 4HCl_{(aq)} + O_{2(g)}$$

c Chlorine to chloride:
$\frac{1}{2}Cl_2 + e^- \quad\quad \rightarrow Cl^-$
Chlorine to chlorate(v):
$\frac{1}{2}Cl_2 + 6OH^- \quad \rightarrow ClO_3^- + 3H_2O + 5e^-$
Overall:
$\frac{5}{2}Cl_2 + 5e^- \quad \rightarrow 5Cl^-$
$\frac{1}{2}Cl_2 + 6OH^- \quad \rightarrow ClO_3^- + 3H_2O + 5e^-$
$3Cl_{2(g)} + 6OH^-_{(aq)} \rightarrow 5Cl^-_{(aq)} + ClO_3^-_{(aq)} + 3H_2O_{(l)}$

15 Energetics

Enthalpy changes

Enthalpy is the heat content of a system. The enthalpy change of a system, indicated by the symbol ΔH, is the amount of heat taken in or given out when a system reacts at constant pressure.

A reaction in which heat energy is given out to the surroundings is said to be **exothermic**. Heat energy is lost by the system so the change in enthalpy, ΔH, is negative (Figure 15.1).

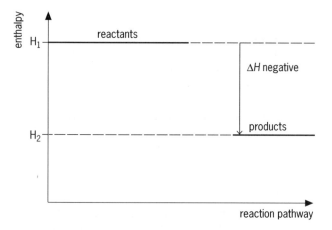

Figure 15.1 Reaction pathway for an exothermic reaction

Conversely, a reaction in which heat energy is taken in from the surroundings is said to be **endothermic**. Heat energy is gained by the system so the change in enthalpy, ΔH, is positive (Figure 15.2).

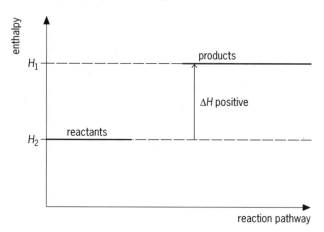

Figure 15.2 Reaction pathway for an endothermic reaction

Enthalpy change is measured in $kJ\,mol^{-1}$. The enthalpy change for a reaction is affected by the temperature, the pressure, the quantities of the reactants and the physical state of the reactants and products. In order to avoid confusion, standard molar enthalpy change values are often used.

The **standard molar enthalpy change**, $\Delta H^{\ominus}$, is the enthalpy change of 1 mole of reactants measured at 298 K (25 °C) and 100 kPa (1 bar) pressure. The symbol $^{\ominus}$ indicates that the enthalpy change is measured under standard conditions. The chemicals should also be in their standard state at 298 K, for example water should be liquid, carbon should be graphite.

Standard enthalpy of combustion

The standard molar enthalpy change of combustion $\Delta H_c^{\ominus}$ is the enthalpy change when 1 mole of a compound is completely burned in oxygen under standard conditions (298 K and 100 kPa), all reactants and products being in their standard states.

The enthalpy change is usually given alongside an equation with state symbols. Here is such an equation for the complete combustion of methane under standard conditions.

$$CH_4(g) + 2O_2(g) \rightarrow CO_2(g) + 2H_2O(l);$$
$$\Delta H_c^{\ominus} = -890\,kJ\,mol^{-1}$$

Accurate enthalpies of combustion may be determined experimentally using a bomb calorimeter.

Standard enthalpy of formation

The standard molar enthalpy change of formation $\Delta H_f^{\ominus}$ is the enthalpy change when 1 mole of a compound is formed from its elements under standard conditions, all reactants and products being in their standard states.

$$Ca(s) + C(graphite) + \tfrac{3}{2}O_2(g) \rightarrow CaCO_3(s);$$
$$\Delta H_f^{\ominus} = -1206\,kJ\,mol^{-1}$$

The standard molar enthalpy change of formation of an element in its standard state is always zero, by definition, since an element cannot be formed from anything else.

Most compounds have negative enthalpies of formation, and are said to be **exothermic compounds**. However, a few, such as benzene, have positive enthalpies of formation and are called **endothermic compounds**.

Enthalpies of formation are usually calculated indirectly using Hess's law.

Hess's law (law of constant heat summation)

Hess's law states that the enthalpy change of a reaction depends only on the initial and final states of the reaction and is independent of the route taken or the number of steps involved.

It follows from this definition that the enthalpy change of a reaction is the sum of the individual enthalpy changes for each step. This is particularly useful for working out molar enthalpy changes of formation where it is not possible to react the elements directly. For example,

$$C(s) + 2H_2(g) \rightarrow CH_4(g)$$

It is not possible to produce methane by reacting carbon and hydrogen directly. However, it is possible to calculate the standard enthalpy of formation of methane using standard enthalpies of combustion since the reactants and products in the above equation burn to give carbon dioxide and water.

An energy level diagram can be drawn (Figure 15.3).

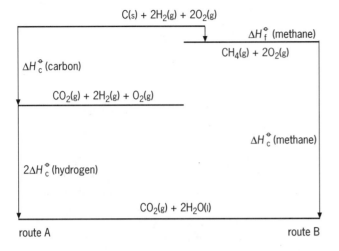

route A   route B

Figure 15.3 Energy level diagram for methane

From data tables:

$\Delta H_c^{\ominus}$(carbon) $= -393.5\,\text{kJ}\,\text{mol}^{-1}$
$\Delta H_c^{\ominus}$(hydrogen) $= -285.8\,\text{kJ}\,\text{mol}^{-1}$
$\Delta H_c^{\ominus}$(methane) $= -890.3\,\text{kJ}\,\text{mol}^{-1}$

Using Hess's law:

$\Sigma\Delta H$ (steps in route A) $= \Sigma\Delta H$ (steps in route B)

$-393.5\,\text{kJ}\,\text{mol}^{-1} + 2(-285.8\,\text{kJ}\,\text{mol}^{-1})$
$\qquad\qquad = \Delta H_f^{\ominus}\text{(methane)} + (-890.3\,\text{kJ}\,\text{mol}^{-1})$
$\Delta H_f^{\ominus}$(methane)
$= -393.5\,\text{kJ}\,\text{mol}^{-1} + 2(-285.8\,\text{kJ}\,\text{mol}^{-1}) - (-890.3\,\text{kJ}\,\text{mol}^{-1})$
$= -74.8\,\text{kJ}\,\text{mol}^{-1}$

Standard enthalpy of atomisation

The standard molar enthalpy change of atomisation $\Delta H_a^{\ominus}$ of an element is the enthalpy change when 1 mole of its atoms, in the gaseous state, is formed from the element under standard conditions.

$\tfrac{1}{2}H_2(g) \rightarrow H(g); \qquad \Delta H_a^{\ominus} = +218\,\text{kJ}\,\text{mol}^{-1}$

This should not be confused with the standard bond dissociation enthalpy, $\Delta H_d^{\ominus}$, for the same change.

$H_2(g) \rightarrow 2H(g); \qquad \Delta H_d^{\ominus} = +436\,\text{kJ}\,\text{mol}^{-1}$

Bond enthalpy and mean bond enthalpy

The standard molar enthalpy change of bond dissociation $\Delta H_d^{\ominus}$ is the energy change when 1 mole of bonds is broken, the molecules and resulting fragments being in the gaseous state at 298 K and a pressure of 100 kPa.

A bond dissociation energy refers to a specific bond in a molecule; however, the exact value depends on the local environment of the bond. If the C—H bonds in methane are broken, one after another, each will have a different bond dissociation enthalpy.

$CH_4(g) \rightarrow CH_3(g) + H(g); \qquad \Delta H_{d1}^{\ominus} = +425\,\text{kJ}\,\text{mol}^{-1}$
$CH_3(g) \rightarrow CH_2(g) + H(g); \qquad \Delta H_{d2}^{\ominus} = +470\,\text{kJ}\,\text{mol}^{-1}$
$CH_2(g) \rightarrow CH(g) + H(g); \qquad \Delta H_{d3}^{\ominus} = +416\,\text{kJ}\,\text{mol}^{-1}$
$CH(g) \rightarrow C(g) + H(g); \qquad \Delta H_{d4}^{\ominus} = +335\,\text{kJ}\,\text{mol}^{-1}$

For this reason, in a polyatomic molecule, it is more useful to know the average amount of energy needed to break a particular bond.

$CH_4(g) \rightarrow C(g) + 4H(g)$

The total amount of energy needed to bring about the above change is $+425 + 470 + 416 + 335 = +1646\,\text{kJ}\,\text{mol}^{-1}$, so the average amount of energy needed to break a C—H bond in methane is $+1646 \div 4 = +412\,\text{kJ}\,\text{mol}^{-1}$.

The average amount of energy needed to break a C—H bond in all types of molecules, including the methane molecule, is $+413\,\text{kJ}\,\text{mol}^{-1}$. This is called the **mean bond enthalpy**, or simply the **bond energy**. Values for different bonds are found in data tables.

Bond enthalpies can be used to estimate the enthalpy changes of reactions ($\Delta H_r^{\ominus}$). Breaking bonds requires energy, while making bonds releases energy.

$\Delta H_r^{\ominus} = \Sigma \Delta H_d^{\ominus}\text{(bonds broken)} - \Sigma \Delta H_d^{\ominus}\text{(bonds formed)}$

The complete combustion of propane is shown as:

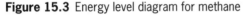

$$
\begin{array}{c}
\text{H} \quad \text{H} \quad \text{H} \\
| \quad\;\; | \quad\;\; | \\
\text{H—C—C—C—H} \quad + \quad 5(\text{O}={\text{O}}) \\
| \quad\;\; | \quad\;\; | \\
\text{H} \quad \text{H} \quad \text{H}
\end{array}
$$
$$\rightarrow \quad 3(\text{O}={\text{C}}={\text{O}}) \quad + \quad 4(\text{H—O—H})$$

The bonds broken are:

- $8 \times$ C—H
- $2 \times$ C—C
- $5 \times$ O=O

The bonds formed are:

- $6 \times$ C=O
- $8 \times$ H—O

$\Delta H_r^{\ominus} = 8\Delta H_d^{\ominus}\text{(C—H)} + 2\Delta H_d^{\ominus}\text{(C—C)} + 5\Delta H_d^{\ominus}\text{(O=O)}$
$\qquad\qquad - 6\Delta H_d^{\ominus}\text{(C=O)} - 8\Delta H_d^{\ominus}\text{(H—O)}$

From data tables, the mean bond enthalpies are:

- C—H $\quad= 413\,\text{kJ}\,\text{mol}^{-1}$
- C—C $\quad= 347\,\text{kJ}\,\text{mol}^{-1}$
- O=O $\quad= 498\,\text{kJ}\,\text{mol}^{-1}$
- C=O $\quad= 805\,\text{kJ}\,\text{mol}^{-1}$
- H—O $\quad= 464\,\text{kJ}\,\text{mol}^{-1}$

$\Delta H_r^{\ominus} = 8 \times 413 + 2 \times 347 + 5 \times 498 - 6 \times 805 - 8 \times 464$
$\qquad = -2054\,\text{kJ}\,\text{mol}^{-1}$

The value for the standard enthalpy of combustion of propane, $\Delta H_c^{\ominus}$, from data tables is $-2219\,\text{kJ}\,\text{mol}^{-1}$. Much of the discrepancy between the above answer and this value is the result of mean bond enthalpies relating to the gaseous state, that is $H_2O(g)$, while enthalpies of combustion relate to the standard state at 298 K and 100 kPa, that is $H_2O(l)$. Adjustment of the calculated value taking this difference of state into account gives $-2226\,\text{kJ}\,\text{mol}^{-1}$, a value quite close to that found in data tables. However, since mean bond enthalpies are used, the answer will not be as accurate as that obtained by direct measurement.

ENERGETICS

1 In an exothermic reaction
 A heat is given out and ΔH is +ve
 B heat is given out and ΔH is −ve
 C heat is taken in and ΔH is +ve
 D heat is taken in and ΔH is −ve

2 Enthalpy changes are measured in
 A kJ **B** $mol\,dm^{-3}$
 C $kJ\,mol^{-1}$ **D** K

3 Butane burns in oxygen to form carbon dioxide and water.

$$C_4H_{10}(g) + 6\tfrac{1}{2}O_2(g) \rightarrow 4CO_2(g) + 5H_2O(g)$$

The standard enthalpy of formation of butane, $\Delta H_f^{\ominus}(C_4H_{10})$ is given by
 A $4\Delta H_c^{\ominus}(C(graphite)) + 5\Delta H_c^{\ominus}(H_2) - \Delta H_c^{\ominus}(C_4H_{10})$
 B $\Delta H_c^{\ominus}(C_4H_{10}) - 4\Delta H_c^{\ominus}(C(graphite)) - 5\Delta H_c^{\ominus}(H_2)$
 C $4\Delta H_c^{\ominus}(C(graphite)) - 5\Delta H_c^{\ominus}(H_2) - \Delta H_c^{\ominus}(C_4H_{10})$
 D $\Delta H_c^{\ominus}(C_4H_{10}) - 4\Delta H_c^{\ominus}(C(graphite)) + 5\Delta H_c^{\ominus}(H_2)$

4 The following equation represents the combustion of methane.

$$CH_4(g) + 2O_2(g) \rightarrow CO_2(g) + 2H_2O(l)$$

Use the following mean bond enthalpies:

- C—H $413\,kJ\,mol^{-1}$
- O=O $498\,kJ\,mol^{-1}$
- C=O $805\,kJ\,mol^{-1}$
- H—O $464\,kJ\,mol^{-1}$

The enthalpy change for this reaction is
 A $-110\,kJ\,mol^{-1}$ **B** $-324\,kJ\,mol^{-1}$
 C $-358\,kJ\,mol^{-1}$ **D** $-818\,kJ\,mol^{-1}$

5 The following equations represent the oxidation of carbon and carbon monoxide to carbon dioxide.

$$C(graphite) + O_2(g) \rightarrow CO_2(g); \quad \Delta H = -393.5\,kJ\,mol^{-1}$$
$$CO(g) + \tfrac{1}{2}O_2(g) \rightarrow CO_2(g); \quad \Delta H = -283.0\,kJ\,mol^{-1}$$

The standard molar enthalpy of formation of carbon monoxide is
 A $-676.5\,kJ\,mol^{-1}$ **B** $-110.5\,kJ\,mol^{-1}$
 C $+110.5\,kJ\,mol^{-1}$ **D** $+676.5\,kJ\,mol^{-1}$

6 Give definitions for the following:
 a standard enthalpy of combustion
 b standard enthalpy of formation
 c standard enthalpy of atomisation
 d Hess's law

7 Cyclohexene reacts with hydrogen to form cyclohexane.

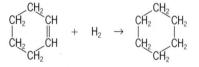

a Use the mean bond enthalpies given to estimate the enthalpy change for this reaction, stating whether it is exothermic or endothermic.
 Mean bond enthalpies:

- C=C $612\,kJ\,mol^{-1}$
- H—H $436\,kJ\,mol^{-1}$
- C—C $347\,kJ\,mol^{-1}$
- C—H $413\,kJ\,mol^{-1}$

b Benzene has three carbon–carbon double bonds.

Use your answer to part **a** to estimate the enthalpy change for the conversion of benzene to cyclohexane.
 c The actual enthalpy change for this reaction is $-208\,kJ\,mol^{-1}$. Explain how the difference between this value and that obtained in part **b** supports the idea that benzene is more stable than might be expected (due to the delocalisation of π electrons from the carbon–carbon double bonds).

8 Copper(II) oxide is reduced to copper by hydrogen.

$$CuO(s) + H_2(g) \rightarrow Cu(s) + H_2O(l)$$

Use the given data to complete the following enthalpy diagram and calculate the enthalpy change for the above reaction.

$\Delta H_f^{\ominus}(CuO) = -157.3\,kJ\,mol^{-1}$
$\Delta H_f^{\ominus}(H_2O) = -285.8\,kJ\,mol^{-1}$

$$Cu(s) + \tfrac{1}{2}O_2(g) + H_2(g)$$

_____

$$CuO(s) + H_2(g)$$

_____

$$Cu(s) + H_2O(l)$$

_____

9 a Give the equation for the complete combustion of ethanol under standard conditions.
 b Calculate the standard molar enthalpy of formation of ethanol from the following data.

$\Delta H_f^{\ominus}(CO_2) = -393.5\,kJ\,mol^{-1}$
$\Delta H_f^{\ominus}(H_2O) = -285.8\,kJ\,mol^{-1}$
$\Delta H_c^{\ominus}(C_2H_5OH) = -1367.3\,kJ\,mol^{-1}$

c Draw an enthalpy diagram for the combustion of ethanol.

10 Draw a suitable enthalpy diagram and use the following data to find the carbon–carbon bond enthalpy in ethane.

$\Delta H_a^{\ominus}(C(graphite)) = 717\,kJ\,mol^{-1}$
$\Delta H_a^{\ominus}(H_2(g)) = 218\,kJ\,mol^{-1}$
$\Delta H_f^{\ominus}(C_2H_6(g)) = -85\,kJ\,mol^{-1}$
mean bond enthalpy, $E(C-H) = 413\,kJ\,mol^{-1}$

1 B

2 C

3 A

4 D

5 B

6 a The enthalpy change when 1 mole of a compound is completely burned in oxygen under standard conditions, all reactants and products being in their standard states.

b The enthalpy change when 1 mole of a compound is formed from its elements under standard conditions, all reactants and products being in their standard states.

c The enthalpy change when 1 mole of its atoms, in the gaseous state, is formed from an element under standard conditions.

d The enthalpy change of a reaction depends only on the initial and final states of the reaction and is independent of the route taken or the number of steps involved.

7 a Bonds broken
$1 \times C{=}C$ $\quad +612\,kJ\,mol^{-1}$
$1 \times H{-}H$ $\quad +436\,kJ\,mol^{-1}$
Total $\quad\quad +1048\,kJ\,mol^{-1}$

Bonds made
$1 \times C{-}C$ $\quad\quad\quad\quad\quad -347\,kJ\,mol^{-1}$
$2 \times C{-}H$ $\quad 2 \times 413\,kJ\,mol^{-1}$ $\quad -826\,kJ\,mol^{-1}$
Total $\quad\quad\quad\quad\quad\quad\quad -1173\,kJ\,mol^{-1}$

Enthalpy change $= 1048 + (-1173) = -125\,kJ\,mol^{-1}$
b Estimated enthalpy change is $3 \times -125 = -375\,kJ\,mol^{-1}$
c The difference between the estimated and the actual enthalpy change is $-167\,kJ\,mol^{-1}$. This indicates that benzene is at a lower energy level than might be expected on the basis of the three carbon–carbon double bonds.

8

Cu(s) + ½O$_2$(g) + H$_2$(g)

$\Delta H_f^{\ominus}$ (CuO)
$= -157.3\,kJ\,mol^{-1}$

$\Delta H_f^{\ominus}$ (H$_2$O)
$= -285.8\,kJ\,mol^{-1}$

CuO(s) + H$_2$(g)

$\Delta H_r^{\ominus}$

Cu(s) + H$_2$O(l)

$\Delta H_f^{\ominus}(H_2O) = \Delta H_f^{\ominus}(CuO) + \Delta H_r^{\ominus}$
$\Delta H_r^{\ominus} = \Delta H_f^{\ominus}(H_2O) - \Delta H_f^{\ominus}(CuO)$
$\Delta H_r^{\ominus} = -285.8 - (-157.3)$
$\quad\quad = -128.5\,kJ\,mol^{-1}$

9 a $C_2H_5OH(l) + 3O_2(g) \rightarrow 2CO_2(g) + 3H_2O(l)$
b $\Delta H_c^{\ominus}(C_2H_5OH) = 2\Delta H_f^{\ominus}(CO_2) + 3\Delta H_f^{\ominus}(H_2O) - \Delta H_f^{\ominus}(C_2H_5OH)$

Rearranging the equation:
$\Delta H_f^{\ominus}(C_2H_5OH)$
$= 2\Delta H_f^{\ominus}(CO_2) + 3\Delta H_f^{\ominus}(H_2O) - \Delta H_c^{\ominus}(C_2H_5OH)$
$= 2(-393.5\,kJ\,mol^{-1}) + 3(-285.8\,kJ\,mol^{-1})$
$\quad -(-1367.\,3\,kJ\,mol^{-1})$
$= -277.1\,kJ\,mol^{-1}$

c

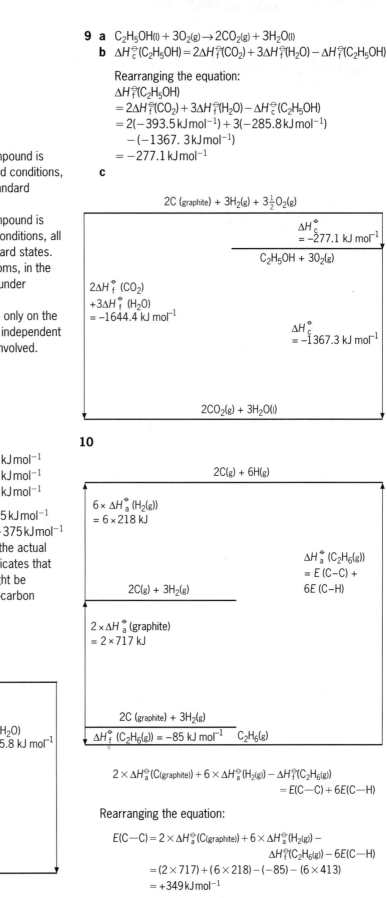

2C (graphite) + 3H$_2$(g) + 3½O$_2$(g)

$\Delta H_c^{\ominus}$
$= -277.1\ kJ\ mol^{-1}$

C$_2$H$_5$OH + 3O$_2$(g)

$2\Delta H_f^{\ominus}$ (CO$_2$)
$+3\Delta H_f^{\ominus}$ (H$_2$O)
$= -1644.4\ kJ\ mol^{-1}$

$\Delta H_c^{\ominus}$
$= -1367.3\ kJ\ mol^{-1}$

2CO$_2$(g) + 3H$_2$O(l)

10

2C(g) + 6H(g)

$6 \times \Delta H_a^{\ominus}$ (H$_2$(g))
$= 6 \times 218\ kJ$

$\Delta H_a^{\ominus}$ (C$_2$H$_6$(g))
$= E\ (C{-}C) +$
$6E\ (C{-}H)$

2C(g) + 3H$_2$(g)

$2 \times \Delta H_a^{\ominus}$ (graphite)
$= 2 \times 717\ kJ$

2C (graphite) + 3H$_2$(g)

$\Delta H_f^{\ominus}$ (C$_2$H$_6$(g)) $= -85\ kJ\ mol^{-1}$ $\quad$ C$_2$H$_6$(g)

$2 \times \Delta H_a^{\ominus}(C_{(graphite)}) + 6 \times \Delta H_a^{\ominus}(H_{2(g)}) - \Delta H_f^{\ominus}(C_2H_{6(g)})$
$\quad\quad\quad\quad\quad\quad\quad = E(C{-}C) + 6E(C{-}H)$

Rearranging the equation:

$E(C{-}C) = 2 \times \Delta H_a^{\ominus}(C_{(graphite)}) + 6 \times \Delta H_a^{\ominus}(H_{2(g)}) -$
$\quad\quad\quad\quad\quad\quad\quad \Delta H_f^{\ominus}(C_2H_{6(g)}) - 6E(C{-}H)$
$= (2 \times 717) + (6 \times 218) - (-85) - (6 \times 413)$
$= +349\,kJ\,mol^{-1}$

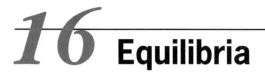

16 Equilibria

Reversible reactions

$$2Na + Cl_2 \rightarrow 2NaCl$$

The above reaction takes place in one direction only. It will carry on until one of the reactants is used up and then it will stop. Under normal circumstances none of the sodium chloride produced will dissociate back into sodium and chlorine.

Similarly, the neutralisation reaction between aqueous solutions of sodium hydroxide and hydrochloric acid carries on until one of the reactants is used up.

$$NaOH(aq) + HCl(aq) \rightarrow NaCl(aq) + H_2O(l)$$

These reactions are said to go to completion.

However, many of the reactions do not go to completion and are **reversible**. For example, the formation of ethyl ethanoate from ethanol and ethanoic acid is a reversible reaction. In one direction ethanol and ethanoic acid react to give ethyl ethanoate and water.

$$\begin{array}{cccc} C_2H_5OH(l) & + & CH_3COOH(l) & \rightarrow & CH_3COOC_2H_5(l) & + & H_2O(l) \\ \text{ethanol} & & \text{ethanoic} & & \text{ethyl} & & \text{water} \\ & & \text{acid} & & \text{ethanoate} \end{array}$$

However, ethyl ethanoate and water also react to give ethanol and ethanoic acid.

$$\begin{array}{cccc} CH_3COOC_2H_5(l) & + & H_2O(l) & \rightarrow & C_2H_5OH(l) & + & CH_3COOH(l) \\ \text{ethyl} & & \text{water} & & \text{ethanol} & & \text{ethanoic} \\ \text{ethanoate} & & & & & & \text{acid} \end{array}$$

The reaction is reversible because it can take place in both directions. Overall, the reaction is best represented by the equation

$$\begin{array}{cccc} C_2H_5OH(l) & + & CH_3COOH(l) & \rightleftharpoons & CH_3COOC_2H_5(l) & + & H_2O(l) \\ \text{ethanol} & & \text{ethanoic} & & \text{ethyl} & & \text{water} \\ & & \text{acid} & & \text{ethanoate} \end{array}$$

The symbol $\rightleftharpoons$ indicates that the reaction is reversible. By convention, the reaction from left to right is called the **forward reaction** and the reaction from right to left is called the **backward** or **reverse reaction**.

Equilibrium

If the components of a reversible reaction are kept together in a sealed container, there eventually comes a time when the rate of the forward reaction is equal to the rate of the backward reaction. Although both reactions are still taking place, the concentrations of the chemicals remain constant. At this point the reaction is said to have reached **equilibrium**. This is described as a **dynamic equilibrium** since the system is not static; it is simply that each component is being formed at exactly the same rate as it is being used up.

The equilibrium constants K_c and K_p

The following equation is for a typical reversible reaction:

$$aA + bB \rightleftharpoons cC + dD$$

where A, B, C and D represent the reactants and products, and a, b, c and d are the numbers of moles of each, respectively.

The rate of the forward reaction is given by the expression

$$\text{rate} = k_1[A]^a \times [B]^b$$

where k_1 is some constant value, and the square brackets [ ] are used to denote **concentration** in $mol\,dm^{-3}$.

Similarly, the rate of the backward reaction is given by the expression

$$\text{rate} = k_2[C]^c \times [D]^d$$

At equilibrium the rates of the forward and backward reactions are equal, so

$$k_1[A]^a \times [B]^b = k_2[C]^c \times [D]^d$$

Rearranging this equation gives

$$\frac{k_2}{k_1} = \frac{[C]^c \times [D]^d}{[A]^a \times [B]^b}$$

$$K_c = \frac{[C]^c[D]^d}{[A]^a[B]^b}$$

K_c is called the **equilibrium constant**. If K_c is large then the equilibrium position in the reaction lies to the right and the concentration of the products of the forward reaction will be high, and concentration of the reactants will be low. Conversely, if K_c is small then the equilibrium position in the reaction lies to the left and the concentration of the products of the forward reaction will be low, and concentration of the reactants will be high.

The value of the equilibrium constant is not changed by a change in concentration or pressure of any species involved in the reaction. Similarly, it is not changed by the use of a catalyst. However, it does vary with temperature and therefore, when a value for the equilibrium constant is stated, the temperature should always be given.

The equilibrium constant does not have a set unit. The units for any particular equilibrium constant will depend upon how many molecules are involved in the reaction. In the reaction between ethanol and ethanoic acid,

$$K_c = \frac{[CH_3COOC_2H_5(l)]^1[H_2O(l)]^1}{[C_2H_5OH(l)]^1[CH_3COOH(l)]^1}$$

The units of concentration are

$$\frac{mol\,dm^{-3} \times mol\,dm^{-3}}{mol\,dm^{-3} \times mol\,dm^{-3}}$$

In this reaction the units cancel out and the K_c has no units, but this is not always the case.

The following equation represents the reversible reaction between hydrogen and iodine to form hydrogen iodide.

$$H_2(g) + I_2(g) \rightleftharpoons 2HI(g)$$

At equilibrium:

- $[H_2(g)] = 1.03 \times 10^{-2}\,mol\,dm^{-3}$
- $[I_2(g)] = 0.16 \times 10^{-2}\,mol\,dm^{-3}$
- $[HI(g)] = 2.74 \times 10^{-2}\,mol\,dm^{-3}$.

What is the equilibrium constant at this particular temperature and what are its units?

$$K_c = \frac{[HI(g)]^2}{[H_2(g)][I_2(g)]}$$

$$= \frac{(2.74 \times 10^{-2})^2}{(1.03 \times 10^{-2})(0.16 \times 10^{-2})} = 45.6$$

$$units = \frac{(mol\,dm^{-3})^2}{(mol\,dm^{-3})(mol\,dm^{-3})} = no\ units$$

For reactions in which all of the reactants and products are gases it is sometimes more convenient to base the equilibrium constant on the **partial pressures** of the gases. Since the partial pressure of any gas in a mixture is proportional to its **mole fraction**, the equilibrium constant, K_p, becomes

$$K_p = \frac{p(C)^c \times p(D)^d}{p(A)^a \times p(B)^b}$$

where $p(C)$ denotes partial pressure of C, for example, expressed in pascals.

Systems in which all of the reactants and products are in the same state, for example aqueous solutions or gases, give rise to **homogeneous equilibria**. Where some of the reactants or products are in different states the result is a **heterogenous equilibrium**.

For example, the thermal decomposition of calcium carbonate involves a heterogenous equilibrium:

$$CaCO_3(s) \rightleftharpoons CaO(s) + CO_2(g)$$

From the equation we could write

$$K_c = \frac{[CaO(s)][CO_2(g)]}{[CaCO_3(s)]}$$

However, $[CaCO_3(s)]$ and $[CaO(s)]$ are both effectively constant, as they are solids, so the expression for the equilibrium constant becomes:

$$K_c = [CO_2(g)]$$

Or, in this case, since carbon dioxide is a gas we could write

$$K_p = p(CO_2)$$

Where a heterogenous equilibrium involves a solid, the solid is not included in the expression for the equilibrium constant since the concentration of the solid is effectively constant. Similarly, water is not included in the expression for the equilibrium constant for an aqueous system.

Le Chatelier's principle

If any change is made to the external conditions (such as temperature, concentration and pressure) of a system at equilibrium, the equilibrium position will alter so as to oppose the change.

The effect of increasing the temperature on the equilibrium position of a reaction depends on whether a reaction is exothermic or endothermic.

reactants $\xrightarrow{exothermic}$ products $\xleftarrow{endothermic}$

If the forward reaction is exothermic (as above) an increase in temperature will cause the equilibrium to move to the left, producing less products. This is because the backward reaction is endothermic so a move to the left results in the added heat energy being absorbed, thus opposing the change in conditions.

reactants $\xrightarrow{endothermic}$ products $\xleftarrow{exothermic}$

Conversely, if the forward reaction is endothermic (as above) a rise in temperature has the opposite effect, causing the equilibrium to move to the right, producing more of the products.

Changes in pressure will only have a significant effect on reactions taking place in the gas phase, and in which the number of moles of product is different from the number of moles of reactant. If the pressure is increased the equilibrium position will move in the direction in which there is the smaller number of moles of gas.

A catalyst has an equal effect on both the forward and the backward reactions. It changes the time taken for the system to reach equilibrium but does not alter the proportions of reactants and products in the equilibrium mixture.

Industrial applications of Le Chatelier's principle

Le Chatelier's principle applies to any reversible reaction, but its importance in the industrial preparation of ammonia (the **Haber process**) and sulphur trioxide (the **Contact process**) merits further discussion.

The Haber process

In the Haber process nitrogen, from the air, and hydrogen, from the steam reforming of natural gas, react over an iron catalyst to form ammonia. The reaction is exothermic.

$$N_2(g) + 3H_2(g) \rightleftharpoons 2NH_3(g); \qquad \Delta H^\ominus = -92\,kJ\,mol^{-1}$$

The forward reaction is exothermic and there is a decrease from 4 moles to 2 moles of gas.

Since the forward reaction is exothermic, a low temperature would produce the highest yield of ammonia in the equilibrium mixture. However, a low temperature would also reduce the rate of reaction so the reaction would take longer to reach equilibrium. This problem is partly solved by using a catalyst which speeds up the reaction rate. In practice a compromise temperature of 370–450 °C is used.

Since the forward reaction involves a decrease in moles of gas a high pressure would produce the highest yield of ammonia. A high pressure also increases the rate at which the ammonia is produced. It would seem that as high a pressure as possible should be used; however, high-pressure equipment is expensive and difficult to maintain. In practice a pressure of 80–110 atmospheres is used.

In the actual process, the reaction mixture is not left to reach equilibrium. A single pass through the converter produces about 15% ammonia. This is removed by cooling to condense the product, and the unreacted nitrogen and hydrogen are recycled.

The Contact process
The Contact process is a stage in the production of sulphuric acid. Sulphur dioxide, made from burning elemental sulphur in dry air, is oxidised to sulphur trioxide, which is dissolved in oleum (fuming sulphuric acid) and diluted to give sulphuric acid.

$$2SO_2(g) + O_2(g) \rightleftharpoons 2SO_3(g); \qquad \Delta H^{\ominus} = -98\,kJ\,mol^{-1}$$

The forward reaction is exothermic and there is a decrease from 3 moles to 2 moles of gas.

The forward reaction is exothermic so a low temperature would produce the highest yield of sulphur dioxide. However, a low temperature would also reduce the rate of reaction, making the process uneconomic. In practice a temperature of 430–630 °C and a vanadium pentoxide catalyst are used. This gives the best balance between increased yield of sulphur trioxide and reasonable reaction rate.

High pressure would produce the highest yield of sulphur dioxide and the rate at which it is produced. However, high pressure is expensive and difficult to maintain, and in this reaction it does not significantly increase the yield of sulphur dioxide so the reaction is carried out at atmospheric pressure.

1 Nitrogen and oxygen react to form nitrogen monoxide according to the following equation.

$$N_2(g) + O_2(g) \rightleftharpoons 2NO(g); \qquad \Delta H = +180\,kJ\,mol^{-1}$$

Which of the following changes in reaction conditions would increase the proportion of NO in the equilibrium mixture?
A higher pressure
B lower pressure
C higher temperature
D lower temperature

2 Iron(III) ions form a distinctive deep red complex with thiocyanate ions.

$$Fe^{3+}(aq) + NCS^-(aq) \rightleftharpoons Fe(NCS)^{2+}(aq)$$

The units of the equilibrium constant, K_c, for this reaction are
A $mol\,dm^{-3}$
B no unit
C $mol^{-1}\,dm^3$
D $mol^{-2}\,dm^6$

3 The following equation represents a reversible reaction between silver ions (Ag^+) and iron(II) ions (Fe^{2+}).

$$Ag^+(aq) + Fe^{2+}(aq) \rightleftharpoons Ag(s) + Fe^{3+}(aq)$$

Which of the following gives the equilibrium constant, K_c, for this reaction at equilibrium?

A $\dfrac{[Ag^+(aq)][Fe^{2+}(aq)]}{[Fe^{3+}(aq)]}$ B $\dfrac{[Ag^+(aq)][Fe^{3+}(aq)]}{[Fe^{2+}(aq)]}$

C $\dfrac{[Fe^{2+}(aq)]}{[Ag^+(aq)][Fe^{3+}(aq)]}$ D $\dfrac{[Fe^{3+}(aq)]}{[Ag^+(aq)][Fe^{2+}(aq)]}$

4 Dinitrogen tetroxide dissociates on heating to give nitrogen dioxide.

$$N_2O_4(g) \rightleftharpoons 2NO_2(g)$$

The equilibrium constant in terms of partial pressures, K_p, for this reaction at equilibrium is

A $\dfrac{p(N_2O_4(g))^2}{p(NO_2(g))}$ B $\dfrac{p(NO_2(g))^2}{p(N_2O_4(g))}$

C $\dfrac{p(N_2O_4(g))}{p(NO_2(g))^2}$ D $\dfrac{p(NO_2(g))}{p(N_2O_4(g))^2}$

5 During the industrial production of nitric acid, ammonia is oxidised to nitrogen monoxide.

$$4NH_3(g) + 5O_2(g) \rightleftharpoons 4NO(g) + 6H_2O(g);$$
$$\Delta H = -909\,kJ\,mol^{-1}$$

Which of the following conditions favours the highest proportion of NO in the equilibrium mixture?
A high pressure and high temperature
B high pressure and low temperature
C low pressure and high temperature
D low pressure and low temperature

6 The industrial preparation of methanol involves the reaction of carbon dioxide and hydrogen.

$$CO_2(g) + 3H_2(g) \rightleftharpoons CH_3OH(g) + H_2O(g);$$
$$\Delta H = -49\,kJ\,mol^{-1}$$

a What conditions favour a high yield of methanol in the equilibrium mixture?
b What would be the effect of increasing the partial pressure of hydrogen on the equilibrium position?
c What would be the effect of increasing the partial pressure of steam on the equilibrium position?
d The original process has now been replaced by a new process.

	Old process	New process
operating temperature	625–675K	475–575K
operating pressure	300 atm	40–100 atm
catalyst	chromium oxide/zinc oxide	copper-based

Discuss the advantages of the new process over the old.

7 In the Contact process sulphur dioxide reacts with oxygen to form sulphur trioxide.

$$2SO_2(g) + O_2(g) \rightleftharpoons 2SO_3(g); \qquad \Delta H^\ominus = -98\,kJ\,mol^{-1}$$

a What is meant by dynamic equilibrium?
b State Le Chatelier's principle.
c Give and explain the effect on the equilibrium position of:
i increasing the pressure at constant temperature
ii increasing the temperature at constant pressure
iii using a catalyst.
d Explain why, industrially, this reaction is carried out at 700–900K, at a pressure of 100–200kPa, over a vanadium pentoxide catalyst.

continues

8 Hydrogen for the Haber process is made by the steam reformation of methane.

$$CH_4(g) + H_2O(g) \rightleftharpoons CO(g) + 3H_2(g);$$
$$\Delta H = +210\,kJ\,mol^{-1}$$

a Give an expression for the equilibrium constant, K_c, for this reaction.
b Discuss what effect changing the reaction temperature would have on the equilibrium position of this reaction.
c Discuss what effect changing the reaction pressure would have on the equilibrium position of this reaction.

9 In the Haber process nitrogen reacts with hydrogen to form ammonia.

$$N_2(g) + 3H_2(g) \rightleftharpoons 2NH_3(g)$$

At equilibrium, the partial pressures of the gases in the equilibrium mixture were nitrogen $2 \times 10^6\,Pa$; hydrogen $6 \times 10^6\,Pa$; ammonia $1.2 \times 10^7\,Pa$.

a Give an expression for the equilibrium constant, K_p, for this reaction.
b Calculate the value of K_p and state its units.

10 Silver(I) ions (Ag^+) form a soluble complex with ammonia.

$$Ag^+(aq) + 2NH_3(aq) \rightleftharpoons Ag(NH_3)_2^+(aq)$$

a Give an expression for the equilibrium constant, K_c, for this reaction.
b In an equilibrium mixture at 298 K, the concentrations of reactants and product are found to be:

- $[Ag^+(aq)] = 2.0 \times 10^{-3}\,mol\,dm^{-3}$
- $[NH_3(aq)] = 2.9 \times 10^{-3}\,mol\,dm^{-3}$
- $[Ag(NH_3)_2^+] = 1.98 \times 10^{-1}\,mol\,dm^{-3}$

Calculate the equilibrium constant for this reaction at 298 K and state its units.
c What is the significance of the magnitude of the equilibrium constant in this reaction?

1 C

2 C

3 D

4 B

5 D

6 a High pressure and low temperature.
b The equilibrium would be shifted to the right, giving more methanol in the equilibrium mixture.
c The equilibrium would be shifted to the left, giving less methanol in the equilibrium mixture.
d The new process is more efficient since it operates at lower temperatures and lower pressures.

7 a The situation in a reversible reaction when the rate of the forward reaction is equal to the rate of the backward reaction and the concentrations of the reactants and products remain constant.
b In a system at equilibrium, the equilibrium position will alter so as to oppose any change in external conditions.
c i The forward reaction involves a decrease from 3 moles to 2 moles of gas. An increase in pressure will shift the equilibrium position to the right in order to oppose the increase in pressure.
ii The forward reaction is exothermic. An increase in temperature will shift the equilibrium position to the left as the backward reaction is endothermic and opposes the increase in temperature.
iii A catalyst will have no effect on the equilibrium position but will allow the reaction to reach the equilibrium position more quickly.
d The temperature is a compromise between a high yield, favoured by a low temperature, and a high reaction rate, favoured by a high temperature. The catalyst also increases the rate of reaction. A higher pressure would improve the yield of product at equilibrium but high-pressure equipment is expensive and the improvement would not justify the extra costs.

8 a $K_c = \dfrac{[CO_{(g)}][H_{2(g)}]^3}{[CH_{4(g)}][H_2O_{(g)}]}$

b The forward reaction is endothermic so an increase in temperature would shift the equilibrium position to the right, producing a greater proportion of carbon monoxide and hydrogen in the equilibrium mixture. A decrease in temperature would have the opposite effect.
c The forward reaction involves an increase in the number of moles of gas from 2 to 4. Increasing the pressure would shift the equilibrium position to the left, producing a greater proportion of methane and steam in the equilibrium mixture. A decrease in pressure would have the opposite effect.

9 a $K_p = \dfrac{p(NH_{3\,(g)})^2}{p(N_{2\,(g)})p(H_{2\,(g)})^3}$

b $K_p = \dfrac{(1.2\times10^7)^2}{(2\times10^6)(6\times10^6)^3}$

$= 3.3\times10^{-13}$

units $= \dfrac{Pa^2}{Pa\times Pa^3}$

$= Pa^{-2}$

10 a $K_c = \dfrac{[Ag(NH_3)_2{}^+{}_{(aq)}]}{[Ag^+{}_{(aq)}][NH_{3\,(aq)}]^2}$

b $K_c = \dfrac{1.98\times10^{-1}}{2.0\times10^{-3}\times(2.9\times10^{-3})^2}$

$= 1.18\times10^7$

units $= \dfrac{mol\,dm^{-3}}{mol\,dm^{-3}\times(mol\,dm^{-3})^2}$

$= mol^{-2}\,dm^6$

c The magnitude of the equilibrium constant gives an indication of the equilibrium position. In this case the value is high, suggesting the equilibrium lies to the right.

The transition metals are the **d block elements** at the centre of the Periodic Table. There are three series of transition metals corresponding to the filling of the 3d, 4d and 5d sub-shells. The first series of transition metals is of particular interest (Table 17.1).

Table 17.1 Electron configuration of the first series of d block elements

Element	Electron configuration	
	3d	4s
scandium	1	2
titanium	2	2
vanadium	3	2
chromium	5	1
manganese	5	2
iron	6	2
cobalt	7	2
nickel	8	2
copper	10	1
zinc	10	2

The 4s sub-shell, although marginally further from the nucleus, is at a lower energy level than the 3d sub-shell and fills with electrons first. Thus at the start of Period 4, the electron configurations of the first two elements are K [Ar]$4s^1$ and Ca [Ar]$4s^2$.

The electron configurations for chromium and copper are not what might be expected when compared with those of the other elements. The electron configuration of chromium might be expected to be $3d^44s^2$ rather than $3d^54s^1$. The reason for the $3d^54s^1$ structure is that it allows the six electrons to exist unpaired in separate sub-shells. This configuration has a lower energy than $3d^44s^2$.

Similarly in copper, a configuration in which all of the d sub-shell is filled leaving a single electron in the 4s sub-shell, $3d^{10}4s^1$, is more stable than the expected $3d^94s^2$ electron configuration.

General properties

The transition metals are harder than s block metals. In general, they have higher melting points and are less reactive. The characteristic properties of the transition metals arise from the partially filled d sub-shell within the atoms or their ions. The elements at each end of the d block, scandium and zinc, are not true transition metals since they only form ions that do not have a partially filled d sub-shell (Table 17.2).

Table 17.2 Electron configurations of scandium and zinc

Element	Electron configuration	Common ion	Electron configuration
scandium	[Ar]$3d^14s^2$	Sc^{3+}	[Ar]
zinc	[Ar]$3d^{10}4s^2$	Zn^{2+}	[Ar]$3d^{10}$

The main characteristics of the transition metals are that they:
- form compounds in which they exhibit different oxidation states
- form coloured ions (topic 18)
- form complex ions (topic 18)
- act as catalysts (both the metals and their compounds) (topic 18).

Variable oxidation states

The ability of transition metals to exhibit different oxidation states in different compounds is central to the behaviour of these elements.

The oxidation state of simple ions is given by the charge they carry. For example:
- Na^+ has an oxidation state of $+1$
- Mg^{2+} has an oxidation state of $+2$
- Cl^- has an oxidation state of -1
- O^{2-} has an oxidation state of -2.

The situation is more complicated in a complex ion. The oxidation state of the central atom in a complex ion is the charge that the ion would have if it was a simple ion. This is found by adding the oxidation states of the various components in the complex ion. These components are assigned oxidation states as shown in Table 17.3.

Table 17.3 Rules for assigning oxidation states

Component	Oxidation state
uncombined elements	0
Group 2 metals in compounds	+2
Group 1 metals in compounds	+1
combined hydrogen, except in metal hydrides	+1
combined hydrogen in metal hydrides	−1
combined halogens	−1
combined oxygen	−2

Example 1

What is the oxidation state of copper in the complex ion $CuCl_4^{2-}$?

- Number of chlorine atoms $= 4$
- Total oxidation number due to chlorine $= 4 \times -1$
$$= -4$$
- Charge on the ion overall $= -2$
- Oxidation state of the central copper atom
$$= -2 - (-4)$$
$$= +2$$

$CuCl_4^{2-}$ is called the tetrachlorocuprate(II) ion.

Example 2

What is the oxidation state of manganese in the complex ion MnO_4^-?

- Number of oxygen atoms $= 4$
- Total oxidation number due to oxygen $= 4 \times -2$
  $$= -8$$
- Charge on the ion $= -1$
- Oxidation state of the central manganese atom
  $$= -1 - (-8)$$
  $$= +7$$

MnO_4^- is called the manganate(VII) ion.

Halogens also exhibit variable oxidation states. In complex ions containing halogens and oxygen, the oxidation number due to oxygen always takes precedence. For example:

- in ClO^-, chlorate(I), the oxidation state of chlorine is $+1$
- in ClO_3^-, chlorate(V), the oxidation state of chlorine is $+5$
- in ClO_4^-, chlorate(VII), the oxidation state of chlorine is $+7$.

Redox reactions

Reactions that involve a reduction must also involve an oxidation; if one reactant is reduced another must be oxidised. Such reactions are described as **redox** reactions.

Historically, the terms oxidation and reduction were applied to reactions involving either the addition or the removal of oxygen.

$$2Cu(s) + O_2(g) \rightarrow 2CuO(s)$$
copper is oxidised

$$Fe_2O_3(s) + 3CO(g) \rightarrow 2Fe(s) + 3CO_2(g)$$
iron oxide is reduced

However, these terms are now used more widely to describe changes in oxidation state. Such a definition covers all of those reactions involving the gain or loss of oxygen and other reactions which do not involve oxygen.

Oxidation is the process of electron loss.
Reduction is the process of electron gain.

Not all reactions are redox reactions. Consider the following examples.

$$Mg(s) + 2HCl(aq) \rightarrow MgCl_2(aq) + H_2(g)$$

- The oxidation state of magnesium in elemental form is 0 while in magnesium chloride it is $+2$, thus the magnesium has been oxidised.
- The oxidation state of hydrogen in hydrochloric acid is $+1$ while as hydrogen gas it is 0 so the hydrogen has been reduced.

This is an example of a redox reaction.
Compare it with the following reaction.

$$CuO(s) + 2HCl(aq) \rightarrow CuCl_2(aq) + H_2O(l)$$

The copper oxide has lost oxygen and thus it may appear to have been reduced; however, consider its oxidation state. Copper has an oxidation state of $+2$ in both copper(II) oxide and copper(II) chloride, thus it has neither been reduced nor oxidised. Similarly, the oxidation state of hydrogen in hydrochloric acid and in water is $+1$. This is not a redox reaction.

It is often easier to consider redox reactions as separate oxidation and reduction reactions, each represented by a **half-equation**. Such equations should only contain the species being oxidised or reduced. In constructing half-equations:

- only one component in each equation changes its oxidation state
- the equation must balance for atoms
- the equation must balance for charge.

The half-equations for the formation of magnesium chloride from magnesium and hydrochloric acid are:

$$Mg(s) \rightarrow Mg^{2+}(aq) + 2e^-$$
$$2H^+(aq) + 2e^- \rightarrow H_2(g)$$

It should be possible to combine half-equations to obtain an overall equation for a reaction. There is no need to include state symbols in this calculation.

Mg	$\rightarrow Mg^{2+} + 2e^-$
$2H^+ + 2e^-$	$\rightarrow H_2$

$$Mg + 2H^+ + 2e^- \rightarrow Mg^{2+} + 2e^- + H_2$$
$$Mg + 2H^+ \rightarrow Mg^{2+} + H_2$$

Notice that **spectator ions**, species that take no part in the reaction, are not included. In the above reaction Cl^- ions are not involved in the reaction.

This approach can be used to derive equations for complicated reactions.

Example 3

Under acidic conditions dichromate(VI) ions, $Cr_2O_7^{2-}$, are reduced to chromium(III) ions, Cr^{3+}, by sulphite ions, SO_3^{2-}, which are themselves oxidised to sulphate ions, SO_4^{2-}. What is the overall equation?

In deriving an overall equation for this reaction, first find the separate half-equations representing the reduction and oxidation reactions.

Reduction: $Cr_2O_7^{2-}$ is reduced to Cr^{3+}. The oxidation state of chromium is reduced from $+6$ to $+3$.

$$Cr_2O_7^{2-} \rightarrow 2Cr^{3+} \qquad \text{(not balanced for O or charge)}$$

Acidic conditions are required. $14H^+$ are required to combine with the oxide ions, forming $7H_2O$.

$$Cr_2O_7^{2-} + 14H^+ \rightarrow 2Cr^{3+} + 7H_2O$$
$$\text{(not balanced for charge)}$$

The half-equation is still not complete. $6e^-$ are also needed. The first half-equation is:

$$Cr_2O_7^{2-}(aq) + 14H^+(aq) + 6e^- \rightarrow 2Cr^{3+}(aq) + 7H_2O(l)$$
$$\text{(equation 1)}$$

Example continues

Oxidation: SO_3^{2-} is oxidised to SO_4^{2-}. Additional oxygen is provided by water, and a surplus of electrons is also produced.

$$SO_3^{2-}(aq) + H_2O(l) \rightarrow SO_4^{2-}(aq) + 2H^+(aq) + 2e^-$$

(equation 2)

Oxidation and reduction involve the transfer of electrons. In combining the two half-equations we use the number of electrons to balance them. In this example, three lots of equation 2 are needed to balance equation 1. The overall equation is:

$$Cr_2O_7^{2-} + 14H^+ + 6e^- \rightarrow 2Cr^{3+} + 7H_2O$$
$$3SO_3^{2-} + 3H_2O \rightarrow 3SO_4^{2-} + 6H^+ + 6e^-$$
$$\overline{}$$
$$Cr_2O_7^{2-} + 14H^+ + 3SO_3^{2-} + 3H_2O$$
$$\rightarrow 2Cr^{3+} + 7H_2O + 3SO_4^{2-} + 6H^+$$

Finally, the equation is simplified by cancelling those water and hydrogen ions that appear on both sides.

$$Cr_2O_7^{2-}(aq) + 8H^+(aq) + 3SO_3^{2-}(aq)$$
$$\rightarrow 2Cr^{3+}(aq) + 4H_2O(l) + 3SO_4^{2-}(aq)$$

Redox titrations

Redox titrations may be used in quantitative volumetric analysis. The end-point may be determined by the use of indicators; however, some redox reactions involve colour changes that make indicators unnecessary.

One oxidising agent in common use is acidified potassium manganate(VII) solution. This is purple while the reduction product, Mn^{2+}, is essentially colourless. Loss of the purple colour indicates when the end-point of a reaction is reached.

$$MnO_4^-(aq) + 8H^+(aq) + 5e^- \rightarrow Mn^{2+}(aq) + 4H_2O(l)$$

The choice of acid for this reaction is important if accurate results are to be obtained. The acid must:

- be strong as a high concentration of hydrogen ions is necessary; for this reason, ethanoic acid cannot be used
- not be an oxidising agent since this would react independently with the species being oxidised; for this reason nitric acid and concentrated sulphuric acid cannot be used
- not be a reducing agent since this would react with the acidified manganate(VII) solution; for this reason hydrochloric acid cannot be used since chloride ions can be oxidised to chlorine.

Potassium manganate(VII) is usually acidified with dilute sulphuric acid.

1 What is the electron configuration of the ion Mn^{2+}?
   A $3d^34s^2$ **B** $3d^44s^1$
   C $3d^54s^0$ **D** $3d^54s^2$

2 Which of the following d block elements is not a true transition metal?
   A titanium   **B** manganese
   C nickel   **D** zinc

3 What is the oxidation state of S in the ion SO_4^{2-}?
   A $+2$ **B** $+4$
   C $+6$ **D** $+8$

4 What is the oxidation state of chromium in the ion $Cr_2O_7^{2-}$?
   A $+4$ **B** $+6$
   C $+8$ **D** $+12$

5 Which of the following equations represents a redox reaction?
   A $2Na(s) + 2H_2O(l) \rightarrow 2NaOH(aq) + H_2(g)$
   B $CuCO_3(s) + 2HCl(aq) \rightarrow CuCl_2(aq) + CO_2(g) + H_2O(l)$
   C $FeO(s) + H_2SO_4(aq) \rightarrow FeSO_4(aq) + H_2O(l)$
   D $AgNO_3(aq) + KCl(aq) \rightarrow AgCl(s) + KNO_3(aq)$

6 **a** State four characteristics of transition metals.
   b Explain why scandium is not a true transition metal although it is a d block metal.
   c Give the electron configuration of chromium and explain why it is different to what might be expected looking at the pattern of electron configuration of the other d block metals.

7 Explain why one of the following is a redox reaction while the other is not. Give details of the oxidation and reduction that take place in the redox reaction.

   $Zn(s) + 2HCl(aq) \rightarrow ZnCl_2(aq) + H_2(g)$ (reaction 1)
   $ZnO(s) + 2HCl(aq) \rightarrow ZnCl_2(aq) + H_2O(l)$ (reaction 2)

8 Copper is oxidised to Cu^{2+} by concentrated nitric acid, which is itself reduced to nitrogen(IV) oxide, NO_2.
   a Write half-equations for the oxidation and reduction reactions.
   b Combine the half-equations to obtain an overall equation for this reaction.

9 When iron(II) is titrated with manganese(VII) under acidic conditions the following changes take place.

   iron(II)→iron(III)   and   manganate(VII)→manganese(II)

   a Write half-equations for the oxidation and reduction reactions.
   b Combine the half-equations to obtain an overall equation for this reaction.

10 A sample of hydrated iron(III) oxide has the formula $Fe_2O_3.xH_2O$. 3.0 g of the sample was dissolved in hydrochloric acid and the solution made up to 250 cm³. 25.0 cm³ of this solution was reduced to the iron(II) state. Subsequently, it was oxidised by 14.6 cm³ of 0.0167 mol dm⁻³ $K_2Cr_2O_7$. Calculate the percentage of Fe_2O_3 in the sample.

1 C **2** D **3** C **4** B **5** A

6 a • They form compounds in which the elements exhibit different oxidation states.
  • They form coloured ions.
  • They form complex ions.
  • They act as catalysts both as metals and as compounds.

  b The electron configuration of the common ion of scandium is $1s^2 2s^2 2p^6 3s^2 3p^6$. It has no other ions that contain a partially filled d sub-shell so it does not exhibit the characteristics of a transition metal.

  c Looking at the pattern of electron configuration of the other transition metals, chromium might be expected to have the electron configuration $3d^4 4s^2$. However, its actual electron configuration, $3d^5 4s^1$, is more stable since it allows the six electrons to exist unpaired in separate sub-shells.

7 Reaction 1 is a redox reaction. Zinc is oxidised; its oxidation state changes from 0, as the metal, to $+2$ in zinc chloride. Hydrogen is reduced; its oxidation state changes from $+1$, in hydrochloric acid, to 0 as hydrogen gas.
  Reaction 2 is not a redox reaction. Zinc has an oxidation state of $+2$ in both zinc oxide and zinc chloride. Hydrogen has an oxidation state of $+1$ in both hydrochloric acid and water.

8 a $Cu_{(s)} \rightarrow Cu^{2+}_{(aq)} + 2e^-$
  $HNO_{3(aq)} + H^+_{(aq)} + e^- \rightarrow NO_{2(g)} + H_2O_{(l)}$
  (Need one H^+ ion to make H_2O and thus one e^- to balance the charges)

  b $Cu \qquad\qquad\qquad\qquad \rightarrow Cu^{2+} + 2e^-$
  $2HNO_3 + 2H^+ + 2e^- \qquad \rightarrow 2NO_2 + 2H_2O$
  _____
  $Cu_{(s)} + 2HNO_{3(aq)} + 2H^+_{(aq)} \rightarrow Cu^{2+}_{(aq)} + 2NO_{2(g)} + 2H_2O_{(l)}$

9 a $Fe^{2+}_{(aq)} \rightarrow Fe^{3+}_{(aq)} + e^-$
  $MnO_4^-{}_{(aq)} + 8H^+_{(aq)} + 5e^- \rightarrow Mn^{2+}_{(aq)} + 4H_2O_{(l)}$
  (Four oxygen atoms in MnO_4^- so this will form $4H_2O$. Need $8H^+$ to make $4H_2O$ and thus $5e^-$ to balance the charges)

  b $5Fe^{2+} \qquad\qquad\qquad\qquad \rightarrow 5Fe^{3+} + 5e^-$
  $MnO_4^- + 8H^+ + 5e^- \qquad \rightarrow Mn^{2+} + 4H_2O$
  _____
  $5Fe^{2+}_{(aq)} + MnO_4^-{}_{(aq)} + 8H^+_{(aq)}$
  $\qquad\qquad \rightarrow 5Fe^{3+}_{(aq)} + Mn^{2+}_{(aq)} + 4H_2O_{(l)}$

10 $6Fe^{2+} \qquad\qquad\qquad\qquad \rightarrow 6Fe^{3+} + 6e^-$
  $Cr_2O_7^{2-} + 14H^+ + 6e^- \qquad \rightarrow 2Cr^{3+} + 7H_2O$
  _____
  $Cr_2O_7^{2-}{}_{(aq)} + 14H^+_{(aq)} + 6Fe^{2+}_{(aq)}$
  $\qquad\qquad \rightarrow 2Cr^{3+}_{(aq)} + 7H_2O_{(l)} + 6Fe^{3+}_{(aq)}$

  1 mole of $Cr_2O_7^{2-}$ oxidises 6 moles of Fe^{2+}, so it is equivalent to 3 moles of Fe_2O_3.

  Amount of $Cr_2O_7^{2-}$ used

  $= \dfrac{14.6}{1000} \, dm^3 \times 0.0167 \, mol \, dm^{-3}$

  $= 2.43 \times 10^{-4} \, mol$

  Amount of Fe_2O_3 present in sample
  $= 2.43 \times 10^{-4} \, mol \times 3 \times 10$
  $= 0.00729 \, mol$

  If $x = 0$, the amount of Fe_2O_3 present in the sample would be $\dfrac{3.0}{160} = 0.0188 \, mol$

  Therefore the percentage of Fe_2O_3 present in the sample is:

  $\dfrac{0.00729 \times 100}{0.0188} = 38.8\%$

18 Transition metals 2

Coloured ions

Solutions of transition metal compounds are frequently coloured, unlike solutions of the compounds formed by the s block metals. The colour depends on:

- the transition metal involved
- the oxidation state of the transition metal
- the number of groups (ligands) surrounding the transition metal ion
- the nature of the groups (ligands) surrounding the transition metal ion.

The colour arises from transitions of electrons in d orbitals between the ground state and an excited state. The frequency, ν, corresponding to the colour is related to the energy difference, ΔE, between the two states by the equation

$$\Delta E = h\nu$$

where h is a universal constant known as **Planck's constant**.

The colour is often a useful aid to identifying the transition metal in a compound and may be useful in qualitative analysis. However, it should be borne in mind that the compounds of a transition metal may exhibit more than one colour, depending on their structure.

- $[Cr(H_2O)_6]^{2+}$ is blue.
- $[Cr(H_2O)_6]^{3+}$ is violet.
- $[CrO_4]^{2-}$ is yellow.

Table 18.1 gives typical colours of the more common ions.

Table 18.1 Transition ion colours

Ion	Colour
$[Ti(H_2O)_6]^{3+}$	purple
$[V(H_2O)_6]^{3+}$	green
$[Cr(H_2O)_6]^{3+}$	violet
$[Mn(H_2O)_6]^{2+}$	pink
$[Mn(H_2O)_6]^{3+}$	violet
$[Fe(H_2O)_6]^{2+}$	pale green
$[Fe(H_2O)_6]^{3+}$	yellow
$[Co(H_2O)_6]^{2+}$	pink
$[Ni(H_2O)_6]^{2+}$	green
$[Cu(H_2O)_6]^{2+}$	blue

When the reaction of a transition metal involves a colour change, this is the result of a change in one or more of the following:

- Oxidation number, for example

$$[Fe(H_2O)_6]^{2+} \rightarrow [Fe(H_2O)_6]^{3+}$$
green   pale violet

Note that $[Fe(H_2O)_6]^{3+}$ is pale violet but readily hydrolyses in solution producing $[Fe(H_2O)_5OH]^{2+}$, which is responsible for the characteristic yellow colour of iron(III) compounds in solution.

- Ligand, for example

$$[Cr(H_2O)_6]^{3+} \rightarrow [Cr(NH_3)_6]^{3+}$$
violet   purple

- Coordination number or number of ligands (usually also involves a change of ligand), for example

$$[Cu(H_2O)_6]^{2+} \rightarrow [CuCl_4]^{2-}$$
blue   green

In some reactions all three of the above changes may occur, for example

$$[CrO_4]^{2-} \rightarrow [Cr(H_2O)_4Cl_2]^{3+}$$
yellow   green

Formation of complexes

A complex compound, or **complex**, consists of a central metal ion, or atom, surrounded by ions or molecules called **ligands**. When a complex forms each ligand donates an electron pair to the central ion, forming a coordinate bond. The number of ligands bonded to the central ion is called the **coordination number**.

For $[Cu(H_2O)_6]^{2+}$, the oxidation number is +2 and the coordination number is 6.

In the complex shown above, hexaaquacopper(II), the oxygen atom of each water molecule donates an electron pair to the copper(II) ion. Ligands that donate a single pair of electrons are called **unidentate**. Examples of unidentate ligands include H_2O, NH_3, OH^-, Cl^- and CN^-.

Some ligands contain two donor atoms and are thus able to form two coordinate bonds with the central metal ion. These are called **bidentate** and include 1,2-diaminoethane, often referred to as 'en', and the ethanedioate ion.

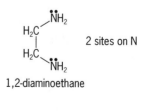

1,2-diaminoethane

ethanedioate ion

Some ligands contain multiple donor atoms and are described collectively as **multidendate**. The ligand known as edta (ethylenediaminetetraacetic acid) or bis[di(carboxymethyl)amino]ethane has six donor atoms, four sites on O and two on N.

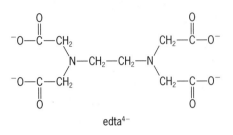

edta^{4-}

$$[Cu(H_2O)_6]^{2+} + edta^{4-} \rightarrow [Cu(edta)]^{2-} + 6H_2O$$

Haem is an iron(II) complex with a multidentate ligand.

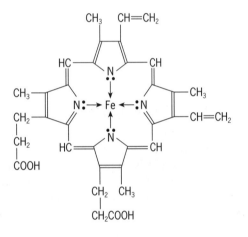

This is similar in structure to haemoglobin, the complex responsible for the red colour of blood.

Haemoglobin combines loosely with oxygen in the lungs, forming oxyhaemoglobin, and it is in this form that oxygen is transported around the body. Unfortunately, haemoglobin also readily combines with carbon monoxide, forming carboxyhaemoglobin. The affinity of haemoglobin for carbon monoxide is 500 times greater than for oxygen. Thus, in an atmosphere containing carbon monoxide, insufficient oxygen may be transported in the blood, leading to unconsciousness or even death.

Complex ions may be cations or anions. The oxidation state of the central metal ion is the number of positive charges that need to be given to the metal ion to balance the overall charge and the charges on the ligands.

- $[Mn(H_2O)_6]^{2+}$: H_2O has no charge and the overall charge is $+2$, therefore the oxidation state of Mn must be $+2$.
- $[CoCl_4]^{2-}$: each Cl carries a charge of -1 and the overall charge is -2, so the oxidation state of Co must be $-2-(-4)=+2$.

The shape of a complex is determined to a large extent by the coordination number. The shape adopted keeps the ligands as far apart from each other as possible. In general:

- 2-coordinate complexes are linear, for example

  dichlorocuprate(I)

- 4-coordinate complexes are tetrahedral or occasionally square planar, for example

  tetrachlorocobaltate(II)

  tetracyanonickelate(II)

- 6-coordinate complexes are octahedral, for example

  hexaamminechromium(III)

Naming complex ions

The name of a complex ion has three parts.

- The number of ligands bonded – the following prefixes are used:

2	di
3	tri
4	tetra
5	penta
6	hexa

- The name of the ligand – here is a list of common ligands:

NH_3	ammine
H_2O	aqua
CO	carbonyl
Cl^-	chloro
CN^-	cyano
OH^-	hydroxo

In oxoanions, such as ClO^-, the presence of oxygen is denoted by 'oxo'. However, this is often not included in the common name. For example, $[CrO_4]^{2-}$ is correctly called tetraoxochromate(VI) but this is often shortened to simply chromate(VI).

- The central atom – this is written followed by its oxidation state. Where the complex is a cation the name of the element is used; where the complex is an anion a modified form using the ending '-ate' is used (Table 18.2).

Table 18.2 Naming the central atom in a complex ion

In a cation	In an anion
aluminium	aluminate
chromium	chromate
copper	cuprate
iron	ferrate
manganese	manganate
nickel	nickelate
platinum	platinate
silver	argentate
vanadium	vanadate

Here are some examples.

- $[Ag(NH_3)_2]^+$ is called diamminesilver(I) – the complex used in Tollens' reagent (topic 29).
- $[Ag(CN)_2]^-$ is called dicyanosilver(I) – this complex is used in electroplating.
- $[Al(OH)_4]^-$ is called tetrahydroxoaluminate(III).
- $[Fe(H_2O)_6]^{2+}$ is called hexaaquairon(II).
- $[Cu(NH_3)_4(H_2O)_2]^{2+}$ is called diaquatetraamminecopper(II).

Where **isomerism** is possible further modification of the name is needed to identify the compound.

Dichlorodiammineplatinum(II) is square planar and thus there are two structures possible; one in which similar ligands are adjacent, and one where they are opposite. These are identified as *cis* and *trans*.

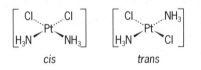

cis   *trans*

The compound *cis*-dichlorodiammineplatinum(II) is more commonly known as cisplatin and is an important drug in the treatment of certain types of cancer.

Catalysis

A **catalyst** is a substance that alters the rate of a chemical reaction but remains unchanged at the end of it. Those catalysts that increase the rate of attainment of equilibrium are of great importance to industry since they increase the speed with which new materials can be produced.

Catalaysts may be classified as **homogenous**, in the same phase (solid, liquid or gas) as the reactants, or **heterogeneous**, in a different phase from the reactants.

The ability to exist in different oxidation states makes transition metals and their salts particularly useful as catalysts.

- In the **Contact process**, vanadium(v) oxide is used to catalyse the oxidation of sulphur dioxide to sulphur trioxide.

$$2SO_2(g) + O_2(g) \xrightleftharpoons{V_2O_{5(s)}} 2SO_3(g)$$

This reaction involves the reduction and subsequent reoxidation of the catalyst.

$SO_2 + V_2O_5 \rightarrow SO_3 + V_2O_4$ (equation 1)
$2V_2O_4 + O_2 \rightarrow 2V_2O_5$ (equation 2)

The oxidation state of vanadium changes from +5 to +4 in equation 1, and back from +4 to +5 in equation 2.

- The **Haber process** for the manufacture of ammonia is another important industrial process involving a transition metal catalyst. In this case it is iron.

$$N_2(g) + 3H_2(g) \xrightleftharpoons{Fe(s)} 2NH_3(g)$$

- A large number of reactions are catalysed on the surface of solid catalysts. The surface provides active sites where reactions can occur, thus an increase in the surface area will increase the effect of the catalyst.

1 How many electron pairs does a bidentate ligand donate to the central metal ion in a complex?
A 1 **B** 2 **C** 3 **D** 4

2 What is the oxidation state of Cr in the complex $[Cr(H_2O)_4Cl_2]^+$?
A +1
B +2
C +3
D +4

3 What is the likely shape of the complex $[Fe(H_2O)_6]^{2+}$?
A linear
B octahedral
C square planar
D tetrahedral

4 What is the likely colour of the complex $[Cu(H_2O)_2(NH_3)_4]^{2+}$?
A blue
B green
C red
D yellow

5 The name of the complex ion $[CoCl_4]^{2-}$ is
A tetrachlorocobalt(II)
B tetrachlorocobaltate(II)
C tetrachlorocobalt(III)
D tetrachlorocobaltate(III)

6 Calculate the oxidation states of vanadium in the following series of complexes.
a $[VO_4]^{3-}$
b $[VO(H_2O)_5]^{2+}$
c $[V(H_2O)_6]^{2+}$
d $[V(H_2O)_6]^{3+}$

7 Name the following complex ions.
a $[V(H_2O)_6]^{2+}$
b $[Ni(CN)_4]^{2-}$
c $[MnO_4]^{2-}$
d $[CuCl_4]^{2-}$
e $[Fe(CN)_6]^{3-}$
f $[Al(H_2O)_2(OH)_4]^-$

8 a Discuss the factors that determine the colour of a complex ion.
b Explain why, in each of the following pairs, the complex ions are a different colour.
i $[Co(H_2O)_6]^{2+}$ (pink) and $[Fe(H_2O)_6]^{2+}$ (green)
ii $[Cu(H_2O)_6]^{2+}$ (blue) and $[Cu(NH_3)_4(H_2O)_2]^{2+}$ (blue-violet)
iii $[Co(H_2O)_6]^{2+}$ (pink) and $[CoCl_4]^{2-}$ (blue)
iv $[Cr(H_2O)_6]^{2+}$ (blue) and $[Cr(H_2O)_6]^{3+}$ (violet)

9 a Draw likely structures for each of the following complexes and name them.
i $[Al(H_2O)_6]^{3+}$
ii $[Pt(NH_3)_4]^{2+}$ (this complex ion is not tetrahedral)
iii $[Ag(CN)_2]^-$
iv $[Ni(CO)_4]$
b Each of the following complex ions might be expected to exhibit isomerism. In each case explain why and draw the two possible isomers.
i $[Pt(Cl)_2(NH_3)_2]$ (this complex is square planar)
ii $[Cr(H_2O)_4(Cl)_2]^+$ (this complex is octahedral)

10 a Explain the terms bidentate and multidentate.
b Draw the structure of the following complex ions. In each case clearly show which are the donor atoms on the ligand.
i $[Ni(en)_3]^{2+}$
ii $[Cu(edta)]^{2-}$

1 B **2** C **3** B **4** A **5** B

6 a +5 **b** +4 **c** +2 **d** +3

7 a hexaaquavanadium(II)
b tetracyanonickelate(II)
c tetraoxomanganate(VI) or manganate(VI)
d tetrachlorocuprate(II)
e hexacyanoferrate(III)
f tetrahydroxodiaquaaluminate(III)

8 a • The transition metal involved: each has its own characteristic colours.
  • The oxidation state of the transition metal.
  • The number of ligands that surround the transition metal ion.
  • The nature of the ligands that surround the transition metal ion.
b i The complex ions contain different transition metals.
ii The complex ions contain different ligands.
iii The complex ions contain different numbers of different ligands.
iv The oxidation state of the metal ion is different.

9 a i Octahedral – hexaaquaaluminium(III)

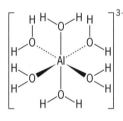

ii Square planar – tetraammineplatinum(II)

iii Linear – dicyanoargentate(I)

$$[NC-Ag-CN]^-$$

iv Tetrahedral – tetracarbonylnickel(0)

b i The two Cl (or NH_3) ligands could be placed next to each other (*cis*) or opposite each other (*trans*) around the platinum ion.

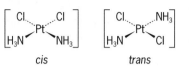

cis   *trans*

ii The two Cl ligands could be placed next to each other (*cis*) or opposite each other in the octahedral structure.

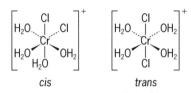

cis   *trans*

10 a Bidentate – ligands containing two donor atoms which are thus able to form two coordinate bonds with the central metal ion.
Multidentate – ligands containing multiple donor atoms which are thus able to form multiple coordinate bonds with the central metal ion.
b Draw the donor atoms in first around the metal ion and then complete the ligands. The donor atoms are shown in bold in the following structures.
i

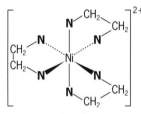

ii

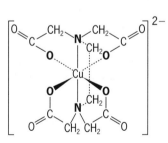

19 Acids and bases

Acids have the following general properties:

- they do not conduct electricity when pure but readily dissolve in water to form electrolytes
- they change the colour of indicators
- they react with some metals to produce salts and hydrogen
- they react with metal oxides to produce salts and water
- they react with metal carbonates to produce salts, carbon dioxide and water.

Any theory of acids and bases needed to account for these properties. There have been several major theories of acids and bases over the years.

Arrhenius theory

In 1884, the Swedish chemist Arrhenius defined acids as substances which dissociate to give hydrogen ions (H^+) when dissolved in water. He also suggested that bases were substances which dissociate in water to produce hydroxide ions (OH^-).

This definition is limited to chemicals in aqueous solution. It did not, for example, account for the properties of organic substances such as amines. These show basic properties but do not contain the OH group.

Brønsted–Lowry theory

In 1923, Brønsted and Lowry proposed a more general picture of acids and bases. In this theory an acid is defined as a **proton donor**, a proton being a hydrogen ion, H^+. Conversely, a base is defined as a **proton acceptor**. Using these definitions, it is possible to explain reactions other than those that take place in aqueous solution. For example, consider the reaction between hydrogen chloride gas and ammonia gas.

$$HCl(g) + NH_3(g) \rightleftharpoons NH_4^+(s) + Cl^-(s)$$

In the forward reaction HCl acts as an acid since it donates a proton and NH_3 acts as a base since it accepts a proton. Also, consider the back reaction; in this case NH_4^+ acts as an acid or proton donor and Cl^- as a base or proton acceptor.

In any acid–base reaction there are always two acids and two bases, one on each side of the equation. The acid on one side is formed from the base on the other and vice versa.

$$\underset{\text{acid 1}}{HA(aq)} + \underset{\text{base 1}}{H_2O(l)} \rightarrow \underset{\text{acid 2}}{H_3O^+(aq)} + \underset{\text{base 2}}{A^-(aq)}$$

HA is an acid and A^- is its conjugate base; together they form a **conjugate pair**. Similarly, H_3O^+ is an acid and H_2O is its conjugate base; they also form a conjugate pair.

Strong acids dissociate completely in solution. They are acids according to the Brønsted–Lowry definition since they donate protons.

$$H_2SO_4(l) + 2H_2O(l) \rightarrow 2H_3O^+(aq) + SO_4^{2-}(aq)$$

In the case of **weak acids**, dissociation is only partial but they still donate protons.

$$CH_3COOH(aq) + H_2O(l) \rightleftharpoons H_3O^+(aq) + CH_3COO^-(aq)$$

Acids that dissociate to produce one proton per molecule, like hydrochloric acid, nitric acid and ethanoic acid, are called **monoprotic** acids. Sulphuric acid dissociates to produce two protons per molecule and is thus described as **diprotic**.

Strong bases are fully ionised in aqueous solution.

$$NaOH(aq) \rightarrow Na^+(aq) + OH^-(aq)$$

Ammonia dissolves in water to give an alkaline solution.

$$NH_3(g) + H_2O(l) \rightleftharpoons NH_4^+(aq) + OH^-(aq)$$

However, aqueous ammonia is only a **weak base** since this reaction only goes to partial completion.

Bases act as proton acceptors. The following reaction is common to all **neutralisation** reactions.

$$OH^-(aq) + H_3O^+(aq) \rightarrow 2H_2O(l)$$

Notice that in aqueous solution, hydrogen ions exist in their hydrated form, H_3O^+ oxonium ions, although, for simplicity, they are often written in equations simply as protons, H^+.

Under the Brønsted–Lowry theory water is able to behave as both an acid and a base, depending upon the reaction:

- as an acid

$$NH_3(g) + H_2O(l) \rightleftharpoons NH_4^+(aq) + OH^-(aq)$$

- as a base

$$CH_3COOH(aq) + H_2O(l) \rightleftharpoons H_3O^+(aq) + CH_3COO^-(aq)$$

The Brønsted–Lowry theory provides a more comprehensive view of acids and bases than the Arrhenius theory; however, it is still limited since it only applies to reactions involving proton transfer.

The Lewis theory

In 1938, the American chemist G N Lewis produced a theory that extended the concept of acids and bases beyond proton transfer. Lewis defined an acid as any species that can accept a pair of electrons from a base to form a covalent bond. A Lewis base is thus any species that can donate a pair of electrons in the formation of a covalent bond.

The Lewis theory is consistent with the Brønsted–Lowry picture of acids and bases; for example:

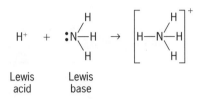

<table>
<tr><td>Lewis acid</td><td>Lewis base</td></tr>
</table>

However, the strength of the Lewis theory is that it also includes other reactions like the following, which would not be considered acid–base reactions under other theories.

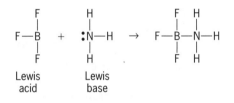

<table>
<tr><td>Lewis acid</td><td>Lewis base</td></tr>
</table>

Acid–base indicators

Acid–base indicators are usually weak acids that dissociate to give an ion that is a different colour to the acid.

$$\text{HIn}_{(aq)} + \text{H}_2\text{O}_{(l)} \rightleftharpoons \text{H}_3\text{O}^+_{(aq)} + \text{In}^-_{(aq)}$$
colour A   colour B

A change in pH will change the position of the equilibrium and therefore the colour of the solution. For example, the addition of H^+ ions causes the equilibrium to shift to the left and the solution will be colour A. The addition of OH^- will remove H^+ ions, shifting the equilibrium to the right and causing the solution to appear colour B.

Phenolphthalein is an example of an acid–base indicator.

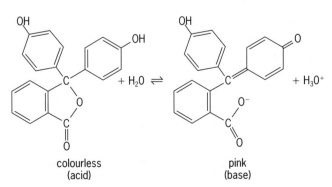

colourless (acid)   pink (base)

Indicators and titrations

The ideal indicator for any acid–base titration is one that gives an easily detectable colour change at the end-point of the reaction. However, not all indicators are suitable for all acid–base titrations. Different indicators change colour over different pH ranges (Table 19.1).

Table 19.1 Acid–base indicators

Indicator	Colour		pH range over which colour change occurs
	Acid	**Base**	
bromocresol green	yellow	blue	3.8–5.4
bromothymol blue	yellow	blue	6.0–7.6
methyl orange	red	yellow	3.2–4.4
methyl red	yellow	red	4.8–6.0
phenolphthalein	colourless	pink	8.2–10.0
phenol red	yellow	red	6.8–8.4

Notice that most of these indicators do not change colour at exactly pH 7. This means that the **end-point** of the titration, the point at which the indicator undergoes the maximum colour change, occurs at a different time to the **equivalence point** of the titration, the point at which there are equivalent amounts of acid and base. However, the difference between the two is not significant provided a suitable choice of indicator is made.

The indicator to be used in a titration depends upon the strength of the acid and base being used and on whether the acid is being added to the base or vice versa.

Strong acid with strong base

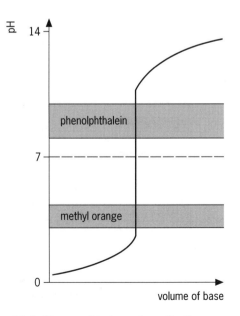

Figure 19.1 Strong acid–strong base titration

At the end-point of a strong acid–strong base titration (Figure 19.1) the pH changes by 5–6 pH units when 1 drop of acid or alkali is added. Many indicators could be used to detect the end-point of this reaction. It is easier to detect the appearance of a red colour rather than its disappearance. If acid is to be added to base then methyl orange (yellow to red) is a good choice. If base is to be added to acid then phenolphthalein (colourless to pink) is a good choice.

Strong acid with weak base

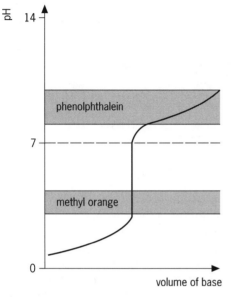

Figure 19.2 Strong acid–weak base titration

The pH change is still significantly great to give a colour change for the addition of 1 drop of base. The pH at equivalence is less than pH 7 (Figure 19.2). The indicator which is used will need to change colour below pH 7. Methyl orange would be a good choice whereas phenolphthalein would be a bad choice.

Strong base with weak acid

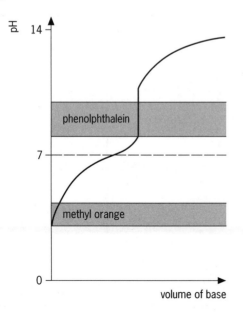

Figure 19.3 Weak acid–strong base titration

Again the pH change is significantly great to give a colour change for the addition of 1 drop of acid. The pH at equivalence is greater than pH 7 (Figure 19.3). The indicator which is used will need to change colour at pH values above 7. Phenolphthalein would be a good choice whereas methyl orange would be a bad choice.

Weak acid with weak base

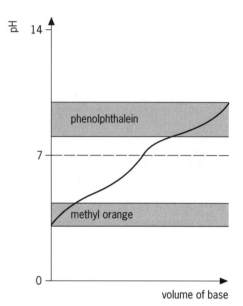

Figure 19.4 Weak acid–weak base titration

The pH changes too slowly around the equivalence point to give a colour change with the addition of 1 drop of acid (Figure 19.4). It is not usual to titrate weak acid with weak alkali but if it must be done then a pH meter is necessary to find the equivalence point accurately. There is no indicator suitable for this type of titration.

Standard enthalpy of neutralisation

Neutralisation occurs when an acid reacts with an alkali to form a salt and water. Neutralisation reactions are exothermic.

The standard enthalpy of neutralisation is the enthalpy change per mole of water formed when an acid and an alkali react at 298 K and 100 kPa. When the reaction involves a strong acid and a strong alkali the value is invariably around $-57\,\text{kJ}\,\text{mol}^{-1}$ since the following reaction is common to all such reactions.

$$H^+(aq) + OH^-(aq) \rightarrow H_2O(l)$$

Where the reaction involves a weak acid and/or alkali the value is less (Table 19.2).

Table 19.2 Enthalpy of neutralisation

Reaction	$\Delta H^\ominus/\text{kJ}\,\text{mol}^{-1}$
$HCl(aq) + NaOH(aq) \rightarrow NaCl(aq) + H_2O(l)$	-57.9
$CH_3COOH(aq) + NaOH(aq) \rightarrow CH_3COONa(aq) + H_2O(l)$	-56.1
$HCl(aq) + NH_4OH(aq) \rightarrow NH_4Cl(aq) + H_2O(l)$	-53.4
$CH_3COOH(aq) + NH_4OH(aq) \rightarrow CH_3COONH_4(aq) + H_2O(l)$	-50.4

1 The following reaction occurs when ethanoic acid is dissolved in water.

$$CH_3COOH(aq) + H_2O(l) \rightleftharpoons H_3O^+(aq) + CH_3COO^-(aq)$$

Under the Brønsted–Lowry theory of acids and bases, which of the following are acting as acids in the equilibrium mixture?
A only CH_3COOH
B H_2O and CH_3COO^-
C CH_3COOH and H_3O^+
D CH_3COO^- and H_3O^+

2 Under the Brønsted–Lowry theory water behaves
A neither as an acid or a base
B as an acid only
C as a base only
D as an acid or a base depending upon the reaction

3 Which of the following graphs shows how the pH changes during a titration in which a weak base is added to a strong acid?

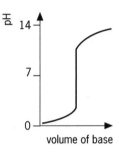

A

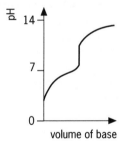

B

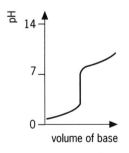

C

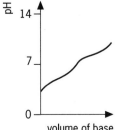

D

4 The following reaction occurs when ammonia dissolves in water.

$$NH_3(g) + H_2O(l) \rightleftharpoons NH_4^+(aq) + OH^-(aq)$$

NH_4^+ is the
A conjugate base of $NH_3(g)$
B conjugate acid of $NH_3(g)$
C conjugate base of $H_2O(l)$
D conjugate acid of $H_2O(l)$

5 In a titration sodium hydroxide solution was added dropwise to dilute hydrochloric acid containing methyl orange indicator. The colour change at the end-point is
A colourless to pink
B red to yellow
C yellow to blue
D yellow to red

6 a Define a Lewis acid and a Lewis base.
  b Identify the Lewis acid and the Lewis base in each of the following reactions.
    i $BF_3 + F^- \rightarrow BF_4^-$
    ii $NH_3 + H^+ \rightarrow NH_4^+$
    iii $Cu^{2+} + 4NH_3 \rightarrow Cu(NH_3)_4^{2+}$
    iv $Ni + 4CO \rightarrow Ni(CO)_4$

7 a Explain what is meant by a conjugate acid–base pair in the Brønsted–Lowry theory of acids and bases.
  b Give one example of a conjugate acid–base pair in each of the following equations.
    i $HCN + H_2O \rightleftharpoons CN^- + H_3O^+$
    ii $NH_3 + Na^+Cl^- \rightleftharpoons Na^+NH_2^- + HCl$
  c Give an example of water acting as a Brønsted–Lowry acid and an example of it acting as a Brønsted–Lowry base.

8 With the help of suitable diagrams, explain why methyl orange would be a suitable indicator for use in a titration involving a weak base with a strong acid, but not for a titration involving a strong base with a weak acid.

9 a What is an acid–base indicator?
  b The following equation represents the equilibrium of a weak acid indicator (HIn) with its conjugate base (In⁻).

$$HIn(aq) + H_2O(l) \rightleftharpoons H_3O^+(aq) + In^-(aq)$$

Apply the equilibrium law to the above to obtain an expression for K_{In}, the indicator dissociation constant, and show that at the end-point in a titration $K_{In} = [H_3O^+]$.

10 The following sequence shows a progression in the theory of acids and bases.

Arrhenius → Brønsted–Lowry → Lewis

Give brief details of each theory and explain how the latter two theories provided a broader picture of acids and bases.

1 C

2 D

3 C

4 B

5 B

6 a Lewis acid: any species that can accept a pair of electrons from a base to form a covalent bond.
Lewis base: any species that can donate a pair of electrons in the formation of a covalent bond.

b

	Lewis acid	Lewis base
i	BF_3	F^-
ii	H^+	NH_3
iii	Cu^{2+}	NH_3
iv	Ni	CO

7 a In any acid–base reaction there are always two acids and two bases, one on each side of the equation. The acid on one side is formed from the base on the other and vice versa. An acid and its conjugate base together form a conjugate pair.

b i HCN and CN^- or H_3O^+ and H_2O
ii NH_3 and NH_2^- or HCl and Cl^-

c As an acid, a proton donor

$$NH_{3(g)} + H_2O_{(l)} \rightleftharpoons NH_4^+{}_{(aq)} + OH^-{}_{(aq)}$$

As a base, a proton acceptor

$$HCl_{(aq)} + H_2O_{(l)} \rightarrow H_3O^+{}_{(aq)} + Cl^-{}_{(aq)}$$

8 Methyl orange changes colour in the pH range 3.2 to 4.4. In a titration curve for a base being added to an acid, there is a sharp increase in pH at the end-point (apart from when a weak base is added to a weak acid) giving a vertical portion to the graph. In order for an indicator to give an accurate indication of the end-point the range within which it changes colour must be within the vertical portion of the graph.

In a strong acid–weak base titration the pH range over which methyl orange changes colour is contained within the vertical portion of the graph so it gives an accurate indication of the equivalence point.

strong acid–weak base

In a weak acid–strong base titration the pH range over which methyl orange changes colour is well outside the vertical portion of the graph. The indicator would change colour before the equivalence point.

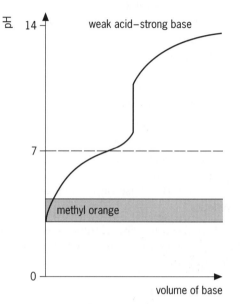

weak acid–strong base

9 a An acid–base indicator is a substance that changes colour with pH.

b $K_{In} = \dfrac{[H_3O^+{}_{(aq)}][In^-{}_{(aq)}]}{[HIn_{(aq)}]}$

At the end-point of the titration $[HIn_{(aq)}] = [In^-{}_{(aq)}]$, thus $K_{In} = [H_3O^+{}_{(aq)}]$.

10 The Arrhenius theory defined acids as substances that dissociate in water to produce hydrogen ions, and bases as substances that dissociate in water to produce hydroxide ions. These definitions were restricted to aqueous chemistry.

The Brønsted–Lowry theory defined acids as proton donors and bases as proton acceptors. This included Arrhenius acids and bases but also included reactions in non-aqueous systems, as is often the case in organic chemistry.

The Lewis theory defined acids as species that can accept a pair of electrons from a base to form a covalent bond. This theory is consistent with both the Arrhenius and Brønsted–Lowry theories but has the added advantage that it can be applied to reactions that do not involve proton transfer.

20 Kinetics

Reactions occur when molecules collide with sufficient force to provide the activation energy needed to start the reaction. Not every collision between molecules gives rise to a reaction, but every set of molecules which do react will have collided together.

Factors which influence the rate of a reaction will have an effect on the rate at which molecules collide, or the average energy of the molecules, or both.

Maxwell–Boltzmann distribution

As all the molecules of a particular chemical, element or compound have the same mass, the energy of the particles is directly related to their speed. In any mixture of moving molecules, the speed at which each molecule is moving will vary considerably. Some will be moving very quickly while others will be moving very slowly. Most of the molecules will be moving with speeds close to an average value.

The Maxwell–Boltzmann graph shows how the number of molecules in a sample are distributed at different energies at a particular temperature (Figure 20.1).

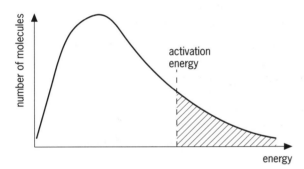

Figure 20.1 Maxwell–Boltzmann distribution

Notice that there are no molecules at zero energy. Also, there are relatively few molecules at very high energy; however, there is no maximum energy value.

In order to react, molecules need to have a minimum amount of energy, called the **activation energy**. In Figure 20.1 only those molecules in the shaded area have sufficient energy to react should they collide with molecules of another reactant.

Factors that affect the rate of a reaction

1 Temperature

When temperature increases the average speed of the molecules will increase. Figure 20.2 shows the Maxwell–Boltzmann distribution at two temperatures; T_2 is greater than T_1.

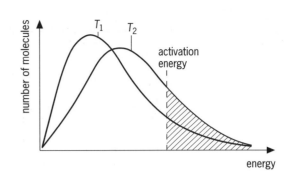

Figure 20.2 Maxwell–Boltzmann distribution at increased temperature

The number of molecules is constant so the areas under the two curves are the same. However, the average energy of the molecules at T_2 is greater, thus there are more molecules with enough energy to react. Also, if the particles are moving more quickly within the same space they will collide more often. The consequence of this is that as temperature increases so the rate of reaction will increase.

2 Concentration and pressure

An increase in the concentration of a chemical, or the pressure of a gas, means that there will be more molecules within a given space. This means that they will collide more often; thus as concentration or pressure increases so the rate of reaction increases.

3 Physical state

In order for a reaction to take place the reactants must come into contact with each other. If all the reactants are gases or liquids it is easy for them to mix, giving the maximum opportunity for the molecules to collide. If one of the reactants is a solid then the reaction can only take place at the surface of the solid. The smaller the size of the particles the greater the surface area at which the reaction can take place.

Consider a cube with sides 2 cm in length, as shown in Figure 20.3 (overleaf):

- surface area of each face $= 2 \times 2$
$$= 4\,cm^2$$
- therefore the total surface area $= 6 \times 4$
$$= 24\,cm^2$$

However, consider the cube to be divided into 8 cubes with sides 1 cm in length:

- surface area of each face $= 1 \times 1$
$$= 1\,cm^2$$
- total surface area $= 6 \times 1 \times 8$
$$= 48\,cm^2$$

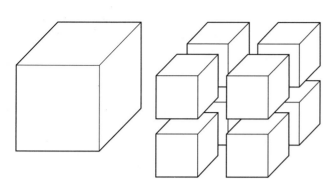

Figure 20.3 Increasing surface area

This is why a finely divided powder reacts more quickly than large lumps of the same chemical. The larger the surface area the faster the rate of reaction.

If one of the reactants is a liquid and one is a gas, or if the two reactants are immiscible liquids, then the reaction can only take place at the interface. The larger the surface area of the interface the faster the reaction will take place.

4 Catalyst

A catalyst is a substance which alters the rate of reaction (most frequently increasing it) without being used up or permanently changed chemically. Catalysts work by lowering the activation energy needed to start the reaction (Figure 20.4).

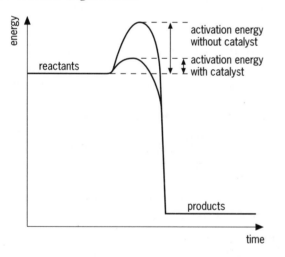

Figure 20.4 Reaction pathways for catalysed and uncatalysed reactions

The catalyst provides an alternative pathway for the reaction, along which the activation energy is less.

Since a catalyst lowers the activation energy this means that, at any given temperature, there will be more molecules with sufficient energy to react (Figure 20.5).

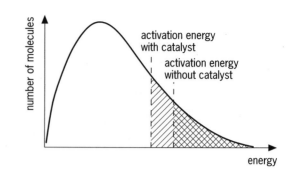

Figure 20.5 Maxwell–Boltzmann distribution for a catalysed reaction

Measuring the rate of reaction

The **rate** of any reaction is the speed at which the reactants are converted to products. This can be quantified as the change in the concentration of reactants or products with time.

$$\text{reaction rate} = \frac{\text{change in concentration}}{\text{time}}$$

Whether the change in concentration to be measured is of reactants or of products will depend upon which procedure is the easier.

The change in concentration can be measured by:

- appearance or disappearance of colour in products or reactants
- volume of gas evolved
- changes in pH
- heat produced
- changes in pressure (for reactions involving gases and a change in the number of molecules).

Order of reaction

Consider a typical reaction in which two reactants, A and B, react to form products.

$$A + B \rightarrow \text{products}$$

What effect would altering the concentrations of the reactants, A and B, have on the rate of the reaction?

We can write a mathematical equation, the **rate equation**, that describes how the rate of a reaction varies with the concentration of reactants at a given temperature. The order of reaction is the power to which the concentrations of the reactants are raised in this equation.

The concentrations of the reactants in a reaction are conveniently represented using square brackets, thus [A] means the concentration of A and [B] means the concentration of B.

To work out the order of a reaction with respect to any reactant in a reaction we can double the concentration of that particular reactant (while keeping the concentrations of the other reactants and all the reaction conditions constant) and see what effect this has on the rate of the reaction.

- If doubling [A] has no effect on the rate then the reaction is said to be **zero order** with respect to A (Figure 20.6).

Figure 20.6 A zero-order reaction

- If doubling [A] doubles the rate then the reaction is said to be **first order** with respect to A (Figure 20.7).

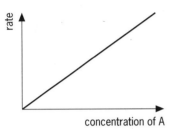

Figure 20.7 A first-order reaction

- If doubling [A] increases the rate fourfold then the reaction is said to be **second order** with respect to A (Figure 20.8).

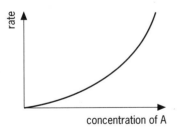

Figure 20.8 A second-order reaction

It is unusual for reactions to be more than second order with respect to any reactant.

Rate equation

A rate equation expresses the relationship between the concentrations of the reactants and the rate of the reaction. The rate equation gives important information that can be used to help determine the **mechanism** of the reaction. The orders for the rate equation need to be found experimentally. It is not possible to work out the rate equation from the reaction equation.

$$A + B \rightarrow products$$

The rate for this reaction will be proportional to $[A]^m$ and $[B]^n$ where m and n are the orders of reaction with respect to A and B. The values of m and n will usually be integers or zero.

$$rate \propto [A]^m[B]^n$$

This expression is modified by the introduction of a constant to give the rate equation:

$$rate = k[A]^m[B]^n$$

where k is a constant (the **rate constant**).

$$k = \frac{rate}{[A]^m[B]^n}$$

The overall order of the reaction is $m + n$. Care needs to be taken in determining the units of the rate constant since these depend on the overall order of a reaction.

The units for the rate of a reaction will always be $mol\,dm^{-3}\,s^{-1}$; however, the units of concentration in the denominator of the equation will depend on the overall order of the reaction.

Consider a reaction in which the rate is zero order with respect to the only reactant. The rate equation for this reaction is

$$rate = k[A]^0$$

From this

$$k = \frac{rate}{[A]^0} \quad therefore \quad k = rate$$

The units of a zero-order rate constant are those of the rate, which are $mol\,dm^{-3}\,s^{-1}$.

For a first-order reaction:

$$rate = k[A]^1$$

From this

$$k = \frac{rate}{[A]^1} \quad or \quad k = \frac{rate}{[A]}$$

The units of concentration will be $mol\,dm^{-3}$ so the units of the rate constant will be:

$$\frac{mol\,dm^{-3}\,s^{-1}}{mol\,dm^{-3}} = s^{-1}$$

Similarly, for a second-order reaction:

$$rate = k[A]^2 \quad or \quad rate = k[A]^1[B]^1$$

From this

$$k = \frac{rate}{[A]^2} \quad or \quad k = \frac{rate}{[A][B]}$$

The units of (concentration)^2 will be $(mol\,dm^{-3})^2$ so the units of the rate constant will be:

$$\frac{mol\,dm^{-3}\,s^{-1}}{(mol\,dm^{-3})(mol\,dm^{-3})} = mol^{-1}\,dm^3\,s^{-1}$$

Half-life for first-order reactions

For any first-order reaction the time taken for the concentration of reactant to fall by half is constant (Figure 20.9). This time is called the **half-life**.

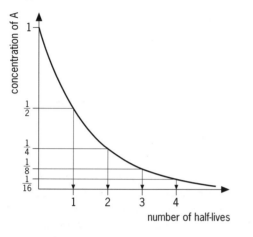

Figure 20.9 First-order reaction and half-life

Half-lives can vary from fractions of a second for fast reactions to millions of years for a very slow reaction.

Radioactive decay follows a similar reaction profile to a first-order reaction. Carbon-dating is worked out using radioactive carbon-14, which has a half-life of 5730 years.

Reaction mechanisms

The rate equation for a reaction is often very different from the reaction equation.

Iodine reacts with propanone in acid solution according to the following equation:

$$I_2(aq) + CH_3COCH_3(aq) \rightarrow CH_3COCH_2I(aq) + H^+(aq) + I^-(aq)$$

Many reactions do not take place in a single step, but rather as a series of steps. The overall rate of the reaction will be determined by the rate of the slowest step and this step is called the **rate-determining step**.

In the above reaction the rate-determining step involves propanone and hydrogen ions, but not iodine. The rate equation for this reaction is

$$\text{rate} = k[CH_3COCH_3(aq)][H^+(aq)]$$

- Step 1 – this is the rate-determining step and must therefore be the slowest stage of the reaction.

- Step 2 – fast

- Step 3 – fast

- Step 4 – fast

Knowledge of the rate-determining step is often useful in working out the mechanism for a reaction.

1 The following graph shows the
Maxwell–Boltzmann distribution of molecules at
temperatures T_1 and T_2.

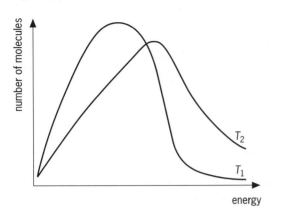

Which of the following statements is true?
A The total energy of the molecules is the same at
T_1 and T_2
B The area under graph T_1 is greater than the area
under graph T_2
C T_1 is greater than T_2
D The average energy of the molecules is greater
at T_2 than T_1

2 When a catalyst is introduced into a mixture of
reactant molecules it
A increases the average energy of the molecules
B lowers the energy needed for the molecules to
react
C increases the proportion of molecules at higher
energies
D lowers the number of molecules at the average
energy value

3 A second-order rate constant has the units
A $mol^{-1}\,dm^3\,s^{-1}$
B s^{-1}
C $mol\,dm^{-3}\,s^{-1}$
D $mol^{-2}\,dm^6\,s^{-1}$

4 The reaction between bromine and methanoic acid
is first order with respect to bromine. In a reaction
it takes 200 s for the concentration of bromine to
fall to half the initial value. How long from the start
of the reaction will it take for the concentration to
fall to one-eighth the initial value?
A 400 s
B 600 s
C 800 s
D 1000 s

5 A reaction is second order with respect to reactant
X. Which of the following graphs shows how the
rate of reaction varies with [**X**]?

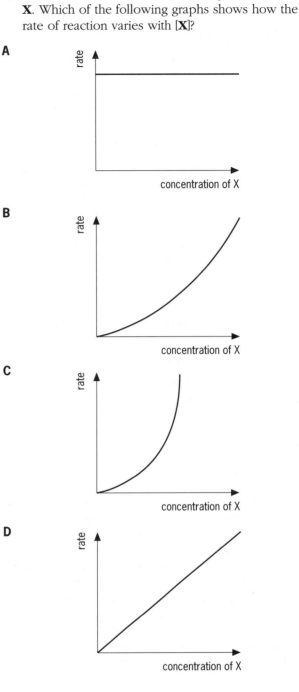

6 a Sketch a graph to show the distribution of
molecules at different energies in a sample of a
substance and comment on the proportion of
molecules at each end of the energy scale.
b With reference to your graph explain the effect
of a catalyst in a chemical reaction.

continues

7 The decomposition of hydrogen peroxide is first order.

$$2H_2O_2(aq) \rightarrow 2H_2O(l) + O_2(g)$$

a Give the rate equation for the decomposition of hydrogen peroxide.

b The rate constant for this reaction at room temperature is $8.25 \times 10^{-4} s^{-1}$. Calculate the concentration of hydrogen peroxide needed to give an initial rate of decomposition of $5 \times 10^{-3} mol\,dm^{-3}\,s^{-1}$.

c Sketch a graph to show how the concentration of hydrogen peroxide changes with time and indicate how you would find the half-life for this reaction.

8 The following table gives some information about a reaction in which sucrose ($C_{11}H_{22}O_{11}$), of initial concentration $0.100\,mol\,dm^{-3}$, is hydrolysed by $1.0\,mol\,dm^{-3}$ hydrochloric acid.

Time/s	Sucrose concentration/mol dm^{-3}
0	0.100
100	0.076
200	0.058
300	0.047
400	0.037
500	0.028
600	0.020

a Plot a graph of sucrose concentration against time.

b Determine two half-life values from your graph and from them deduce the order of the reaction with respect to sucrose.

c The following table shows what happens to the initial rate of reaction when the sucrose concentration remains constant while the acid concentration is varied.

Concentration of hydrochloric acid/mol dm^{-3}	Initial rate of the reaction /mol dm^{-3} s^{-1}
0.20	1.24×10^{-4}
0.40	2.45×10^{-4}

Deduce the order of reaction with respect to hydrochloric acid and give the rate equation for this reaction.

9 The following equation represents the reaction between 2-bromo-2-methylpropane and sodium hydroxide.

$$(CH_3)_3CBr + OH^- \rightarrow (CH_3)_3COH + Br^-$$

The following table contains information about three experiments in which the concentrations of the reactants were varied but the temperature remained constant.

Experiment	Initial concentration of (CH$_3$)$_3$CBr/ mol dm^{-3}	Initial concentration of OH$^-$ / mol dm^{-3}	Initial rate of the reaction/ mol dm^{-3} s^{-1}
1	1.0×10^{-3}	1.0×10^{-1}	0.5×10^{-3}
2	2.0×10^{-3}	1.0×10^{-1}	1.0×10^{-3}
3	2.0×10^{-3}	0.5×10^{-1}	1.0×10^{-3}

a Find the order of reaction with respect to the two reactants. Explain your reasoning.

b What do the orders of reaction of the two reactants indicate about the mechanism of this reaction?

c Write the rate equation for this reaction and find the value of the rate constant, and its units, at this temperature.

d Calculate the initial rate of reaction when the initial concentration of (CH$_3$)$_3$CBr is $5.0 \times 10^{-3} mol\,dm^{-3}$ and the initial concentration of OH$^-$ is $2.0 \times 10^{-3} mol\,dm^{-1}$.

10 Nitrogen oxide and hydrogen react at high temperatures to form nitrogen and steam.

$$2NO(g) + 2H_2(g) \rightarrow N_2(g) + 2H_2O(g)$$

The following table contains information about six experiments involving this reaction.

Experiment	Initial concentration of NO/ mol dm^{-3}	Initial concentration of H$_2$/ mol dm^{-3}	Initial rate of N$_2$ production /mol dm^{-3} s^{-1}
A	2×10^{-3}	5×10^{-3}	0.5×10^{-3}
B	4×10^{-3}	5×10^{-3}	2.0×10^{-3}
C	6×10^{-3}	5×10^{-3}	4.5×10^{-3}
D	5×10^{-3}	2×10^{-3}	1.25×10^{-3}
E	5×10^{-3}	4×10^{-3}	2.50×10^{-3}
F	5×10^{-3}	6×10^{-3}	3.75×10^{-3}

a Find the orders of reaction with respect to the two reactants and the overall order of reaction. Explain your reasoning.

b Write the rate equation for this reaction and find the value of the rate constant, and its units, at this temperature.

1 D

2 B

3 A

4 B

5 B

6 a

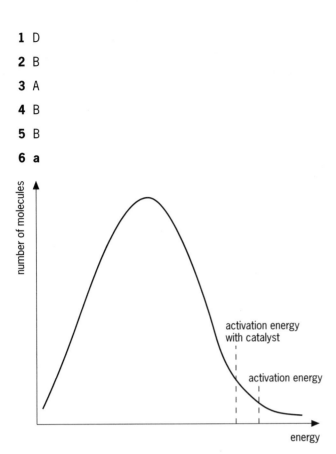

There are no molecules at zero energy. The proportion of molecules at very high energy is small but there is no upper limit to the energy a molecule may possess.

b A molecule can only react if it attains the activation energy. A catalyst lowers the activation energy so that a greater proportion of molecules will have sufficient energy to react.

7 a rate $= k[H_2O_{2(aq)}]$

b $[H_2O_{2(aq)}] = \dfrac{rate}{k}$

$= \dfrac{5 \times 10^{-3}\,mol\,dm^{-3}\,s^{-1}}{8.25 \times 10^{-4}\,s^{-1}}$

$= 6.06\,mol\,dm^{-3}$

c

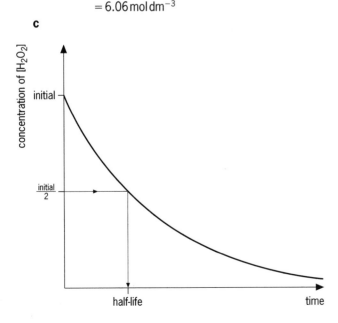

8 a

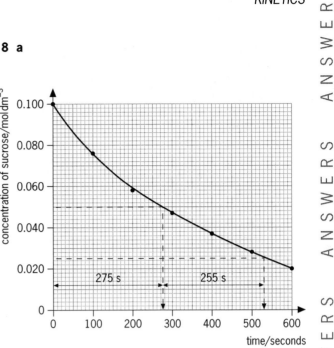

b Half-lives around 260–270 s. Allowing for a small degree of experimental error the half-life is constant and indicates the reaction is first order with respect to sucrose.

c When the concentration doubles, the initial rate doubles, indicating that the reaction is first order with respect to hydrochloric acid. The rate equation for this reaction is:

rate $= k$[sucrose][hydrochloric acid]

9 a In experiments 1 and 2 the concentration of OH^- remains constant but the concentration of $(CH_3)_3CBr$ is doubled. This doubles the initial rate of reaction thus the reaction is first order with respect to $(CH_3)_3CBr$.

In experiments 2 and 3 the concentration of $(CH_3)_3CBr$ remains constant but the concentration of OH^- is halved. This has no effect on the initial rate of reaction thus the reaction is zero order with respect to OH^-.

b Since the rate depends only on the concentration of $(CH_3)_3CBr$ the rate-determining step must involve the breaking of the C—Br bond.

c rate $= k[(CH_3)_3CBr]$ therefore $k = \dfrac{rate}{[(CH_3)_3CBr]}$

Substituting these values from any of the experiments will produce the same answer.

$k = \dfrac{0.5 \times 10^{-3}}{1.0 \times 10^{-3}}$

$= 5.0 \times 10^{-1}$

units $= \dfrac{mol\,dm^{-3}\,s^{-1}}{mol\,dm^{-3}} = s^{-1}$

d rate $= k[(CH_3)_3CBr]$
$= 5.0 \times 10^{-1}\,s^{-1} \times 5.0 \times 10^{-3}\,mol\,dm^{-3}$
$= 2.5 \times 10^{-3}\,mol\,dm^{-3}\,s^{-1}$

10 a In experiments A–C the concentration of H_2 remains constant while the concentration of NO is doubled and tripled. When the concentration of NO is doubled the initial rate of N_2 increases fourfold; when the concentration of NO is tripled the initial rate of N_2 increases ninefold. Thus, the reaction is second order with respect to NO.

In experiments D–F the concentration of NO remains constant while the concentration of H_2 is doubled and tripled. When the concentration of H_2 is doubled the initial rate of N_2 increases twofold; when the concentration of H_2 is tripled the initial rate of N_2 increases threefold. Thus, the reaction is first order with respect to H_2.

The overall order of the reaction is $2 + 1 = 3$.

b rate $= k[NO_{(g)}]^2[H_{2(g)}]$, where k is the rate constant. Rearranging the equation:

$$k = \frac{\text{rate}}{[NO_{(g)}]^2[H_{2(g)}]}$$

Substituting the values from any of the experiments A–F will give the value of k. For example, from experiment A:

$$k = \frac{0.5 \times 10^{-3}}{(2 \times 10^{-3})^2(5 \times 10^{-3})}$$

$$= 2.5 \times 10^4$$

$$\text{units} = \frac{\text{mol dm}^{-3}\,\text{s}^{-1}}{(\text{mol dm}^{-3})^2(\text{mol dm}^{-3})}$$

$$= \text{mol}^{-2}\,\text{dm}^6\,\text{s}^{-1}$$

21 Organic series of compounds 1

Haloalkanes

The haloalkanes are a homologous series of compounds of general formula $C_nH_{2n+1}X$, where X is F, Cl, Br or I.

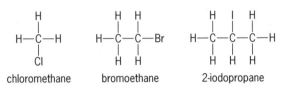

chloromethane   bromoethane   2-iodopropane

Halogens are very electronegative so carbon–halogen bonds are polar (Table 21.1). The electrons in the carbon–halogen bond are unevenly distributed, leaving the carbon atom electron-deficient ($\delta+$) and the halogen atom electron rich ($\delta-$).

Table 21.1 Haloalkanes and bond strength

Element	Electronegativity	C—X bond strength/kJ mol^{-1}
F	4.0	467
Cl	3.0	346
Br	2.8	290
I	2.5	228

The electron-deficient carbon atom in a carbon–halogen bond is liable to attack by **nucleophiles**. These are ions or molecules with lone pairs of electrons. When nucleophilic attack occurs, the nucleophile replaces the halogen atom. The carbon–halogen bond breaks and a halide ion is released.

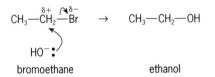

The rate of such reactions depends on the strength of the carbon–halogen bond. Fluoroalkanes are very unreactive because of the great strength of the carbon–fluorine bond. Chloroalkanes are slow to react for the same reason, but bromoalkanes react at a reasonable rate and are often used as starting materials in a sequence of reactions.

When haloalkanes are warmed with aqueous sodium hydroxide or potassium hydroxide, **alcohols** are formed. The hydroxide ion acts as a nucleophile.

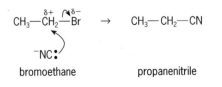

bromoethane   ethanol

When haloalkanes are warmed with an aqueous or alcoholic solution of potassium cyanide, **nitriles** are formed. The cyanide ion acts as a nucleophile.

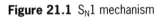

bromoethane   propanenitrile

This reaction is particularly useful since it provides a means of adding a carbon atom to the carbon chain. The nitrile may be reduced by hydrogen in the presence of a nickel catalyst, or by lithium tetrahydridoaluminate(III) (LiAlH$_4$) in dry ethoxyethane to give an **amine**. This amine has one more carbon atom in the chain than the original haloalkane.

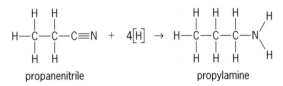

propanenitrile   propylamine

Nucleophiles are not always anions. When haloalkanes are warmed with ammonia, primary amines are formed. The acid, HBr, immediately reacts with the basic amine to produce a salt.

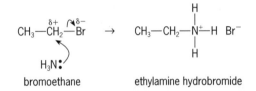

bromoethane   ethylamine hydrobromide

S$_N$1 and S$_N$2 mechanisms

Nucleophilic substitution reactions may go by two distinct reaction mechanisms, which are widely referred to as **S$_N$1** and **S$_N$2** reactions.

The alkaline hydrolysis of 2-bromo-2-methylpropane occurs via an S$_N$1-type mechanism (Figure 21.1). The first step involves the formation of a **carbocation**. This ion then reacts with a hydroxide ion to form the product. Methyl groups are electron releasing so the intermediate carbocation is stabilised.

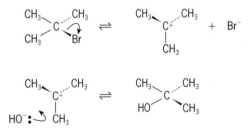

Figure 21.1 S$_N$1 mechanism

The first step is slower than the second and is thus the rate-determining step for this reaction. The rate of the reaction is first order with respect to 2-bromo-2-methylpropane but zero order with respect to hydroxide ions. In 'S$_N$1', S = substitution, N = nucleophilic, 1 = first order overall.

The alkaline hydrolysis of bromoethane, however, cannot proceed in the same way because the carbocation which would form by this mechanism would not be stable. Only one electron-releasing methyl group is attached to the carbon atom that would carry the charge. Instead, the alkaline hydrolysis of bromoethane occurs via an S_N2-type mechanism (Figure 21.2). This involves the formation of an **intermediate** in which both the nucleophile and the leaving group are partially bonded to the carbon atom.

Figure 21.2 S_N2 mechanism

The intermediate involves both bromoethane and the hydroxide ion, thus the reaction is first order with respect to bromoethane and first order with respect to hydroxide ions. The 2 in 'S_N2' refers to the reaction being second order overall.

Elimination reactions

The alkaline hydrolysis of a haloalkane may also result in an **elimination** reaction, in which an **alkene** is formed. Hydrogen bromide is eliminated from bromoethane producing ethene.

In practice, the alkaline hydrolysis of a haloalkane produces a mixture of products since both substitution and elimination occur. The substitution reaction is favoured by warming with an aqueous solution of sodium hydroxide or potassium hydroxide, while the elimination reaction is favoured by warming with a solution of potassium hydroxide in ethanol. Potassium hydroxide is preferred to sodium hydroxide for the latter reaction since it is more soluble in ethanol.

Alcohols

The alcohols form a homologous series of general formula $C_nH_{2n+1}OH$. They all contain the functional group —OH, which is called the **hydroxyl group**.

Ethanol is the commonest alcohol, and is the alcohol found in 'alcoholic' drinks. There are two main industrial processes for the production of ethanol.

- In **fermentation**, live yeast cells convert sugars into ethanol and carbon dioxide.

$$C_6H_{12}O_6(aq) \xrightarrow{yeast} 2C_2H_5OH(aq) + 2CO_2(g)$$

This process is normally carried out at around 35 °C. At lower temperatures the yeast is less active whereas at higher temperatures the yeast is killed. Fermentation only produces an aqueous solution of up to 15% ethanol. At concentrations above this the yeast is killed. Fermentation is widely used in the production of beers, wines and spirits, but is of little importance in producing ethanol for industrial use.

- The preferred industrial method of producing ethanol is by the **hydration** of ethene using steam. This reaction is carried out at 300 °C and 6–7 MPa (60–70 atmospheres) in the presence of a phosphoric acid catalyst. The reaction mechanism is thought to involve the addition of a proton to form a carbocation which then reacts with water.

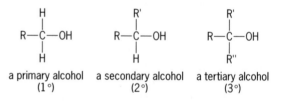

This process produces pure ethanol and is a much faster reaction than fermentation.

Alcohols are classified as **primary**, **secondary** or **tertiary**, depending on the position of the carbon atom to which the hydroxyl group is attached.

Some reactions are common to all alcohols while in others, primary, secondary and tertiary alcohols behave differently.

continues

Formation of alkenes from alcohols

All alcohols can be converted to alkenes by the loss of water. **Dehydration** is brought about by heating the alcohol with concentrated sulphuric acid or concentrated phosphoric acid.

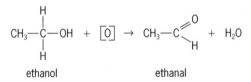

The mechanism involves an acid-catalysed elimination in which the alcohol molecule initially gains a proton, which is subsequently lost. Some alcohols may form more than one alkene but, owing to the stability of the intermediate carbocation, one tends to be the major product.

Oxidation of alcohols

Primary, secondary and tertiary alcohols behave differently in their reactions with oxidising agents such as acidified potassium dichromate(VI). In oxidation reactions the symbol [O] is often used to represent the oxidant in order to simplify the equation.

- Primary alcohols may initially be oxidised to **aldehydes**. Ethanol is oxidised to ethanal.

ethanol   ethanal

An aldehyde still has one hydrogen atom attached to the carbon of the carbonyl group. Aldehydes may be oxidised further to **carboxylic acids**. Ethanal is oxidised to ethanoic acid.

ethanal   ethanoic acid

In practice, when a primary alcohol is oxidised by dripping it into warm acidified potassium dichromate(VI) solution the aldehyde formed immediately distils off since it has a lower boiling point than the alcohol. If the primary alcohol is to be oxidised through to the carboxylic acid the reaction mixture must be heated under **reflux** in order to prevent the loss of the intermediate aldehyde.

- Secondary alcohols are oxidised to **ketones**. Propan-2-ol is oxidised to propanone.

propan-2-ol   propanone

A ketone has no hydrogen atoms attached to the carbon of the carbonyl group and cannot easily be oxidised further.

- Tertiary alcohols are not easily oxidised.

Reduction to form alcohols

Aldehydes and ketones may be reduced to alcohols using hydrogen in the presence of a nickel or platinum catalyst.

However, this reagent will also hydrogenate carbon–carbon double bonds. Sodium tetrahydridoborate(III) in methanol also reduces aldehydes and ketones to alcohols but has no effect on carbon–carbon double bonds. The reason for this is that the reaction proceeds via nucleophilic attack on the carbonyl group. The nucleophile is attracted by the slightly positive carbon atom in the carbonyl group but repelled by the electron cloud of the carbon–carbon double bond.

1 Which of the following types of reaction is typical of a haloalkane?

A electrophilic addition

B electrophilic substitution

C nucleophilic addition

D nucleophilic substitution

2 Which of the following compounds is a tertiary alcohol?

A

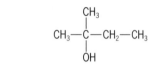

$$CH_3-\overset{\overset{\displaystyle CH_2OH}{|}}{CH}-CH_2-CH_3$$

B

$$CH_3-\overset{\overset{\displaystyle CH_3}{|}}{CH}-CH_2-CH_2OH$$

C

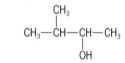

$$CH_3-\overset{\overset{\displaystyle CH_3}{|}}{\underset{\underset{\displaystyle OH}{|}}{C}}-CH_2-CH_3$$

D

$$CH_3-\overset{\overset{\displaystyle CH_3}{|}}{CH}-\underset{\underset{\displaystyle OH}{|}}{CH}-CH_3$$

3 Which of the following groups of compounds act as Brønsted–Lowry bases?

A R—Cl

B R—H

C R—NH$_2$

D R—CH=CH$_2$

4 Which of the following could **not** be made directly from a primary alcohol?

A an aldehyde

B an alkene

C a carboxylic acid

D a ketone

5 The following equation represents the nucleophilic substitution of a bromoalkane which proceeds via an S$_N$2 mechanism.

$$R-Br + N^- \rightarrow R-N + Br^-$$

The rate equation for this reaction is

A $k[N^-]$

B $k[R-Br][N^-]$

C $k[R-Br]$

D $k[R-Br]/[N^-]$

6 This question is about ethanol, ethanal and ethanoic acid.

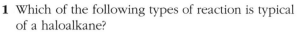

ethanol   ethanal   ethanoic acid

a Briefly describe how you could bring about reactions A, B and C.

b The boiling points of these compounds are as follows: ethanol 78 °C, ethanal 23 °C, ethanoic acid 118 °C. Explain the differences in boiling points and predict the solubility of each compound in water in terms of their structure.

7 a Describe the two main industrial processes for making ethanol, giving details of the starting material and the reaction conditions required. Explain which is the preferred method of making ethanol for use as a solvent.

b Give one example each of a primary, a secondary and a tertiary alcohol and explain how they behave differently with an oxidising agent.

8 Describe the reaction(s) involved in converting bromoethane (CH$_3$CH$_2$Br) into:

a ethylamine (CH$_3$CH$_2$NH$_2$)

b propylamine (CH$_3$CH$_2$CH$_2$NH$_2$)

c ethanal (CH$_3$CHO).

9 The following table shows the boiling points of some organic compounds.

Compound	Structure	Boiling point/K
ethane	CH$_3$CH$_3$	185
chloroethane	CH$_3$CH$_2$Cl	285
ethylamine	CH$_3$CH$_2$NH$_2$	290
ethanol	CH$_3$CH$_2$OH	351

With the help of suitable diagrams, explain why the boiling points of these compounds are so different, in terms of the forces of attraction between molecules.

10 Warming 2-bromopropane with aqueous sodium hydroxide produces propan-2-ol.

$$CH_3CHBrCH_3 + NaOH \rightarrow CH_3CHOHCH_3 + NaBr$$

a What type of reaction is this?

b Show how this reaction might proceed via an S$_N$1 mechanism, and via an S$_N$2 mechanism.

c Explain how a study of the reaction rate would indicate the actual mechanism of this reaction.

1 D

2 C

3 C

4 D

5 B

6 a Reaction A – warm with acidified aqueous potassium dichromate(VI) and allow product to distil off.
Reaction B – heat with acidified aqueous potassium dichromate(VI) under reflux.
Reaction C – react with hydrogen in the presence of a nickel or platinum catalyst or react with sodium tetrahydridoborate(III) in methanol.

b Ethanal has no —OH group and does not contain hydrogen bonding. The forces of attraction between molecules are relatively weak so the boiling point is low. Both ethanol and ethanoic acid contain an —OH group and contain hydrogen bonding. The forces of attraction between molecules are relatively strong so the boiling points are higher than that of ethanal. The ethanoic acid molecule has a greater density of electrons surrounding it than the ethanol molecule. The result is stronger van der Waal's forces between molecules and thus a higher boiling point. Ethanol, ethanal and ethanoic acid are all able to hydrogen bond with water molecules so they would all be expected to be soluble.

7 a Fermentation converts sugars, such as glucose, into ethanol and carbon dioxide.

$$C_6H_{12}O_6 \rightarrow 2C_2H_5OH + 2CO_2$$

The reaction is carried out at around 35 °C and normal atmospheric pressure in the presence of yeast.
Ethanol may also be obtained from the hydration of ethene using steam.

$$CH_2{=}CH_2 + H{-}OH \rightarrow C_2H_5OH$$

The reaction is carried out at 300 °C and 6–7 MPa. Fermentation gives an aqueous solution of about 15% ethanol. The ethanol has to be separated by distillation and dried before it can be used as a solvent. Hydration of ethene produces a pure product and, for this reason, is the preferred method of making ethanol.

b Primary alcohol, for example ethanol: a primary alcohol can be oxidised to an aldehyde (ethanal) and then to a carboxylic acid (ethanoic acid).
Secondary alcohol, for example propan-2-ol: a secondary alcohol can be oxidised to a ketone (propanone).
Tertiary alcohol, for example 2-methylpropan-2-ol: a tertiary alcohol cannot readily be oxidised.

8 a This conversion can be achieved in one step by a nucleophilic substitution reaction using ammonia.

$$CH_3CH_2Br + NH_3 \rightarrow CH_3CH_2NH_2 + HBr$$

b This conversion is carried out in two steps. The first step is a nucleophilic substitution reaction using cyanide ions to form ethanenitrile.

$$CH_3CH_2Br + CN^- \rightarrow CH_3CH_2CN + Br^-$$

The ethanenitrile is then reduced using nickel/hydrogen or lithium tetrahydridoaluminate(III).

$$CH_3CH_2CN + 2H_2 \rightarrow CH_3CH_2CH_2NH_2$$

c This conversion is carried out in two steps. The first is a nucleophilic substitution reaction using hydroxide ions to form ethanol.

$$CH_3CH_2Br + OH^- \rightarrow CH_3CH_2OH + Br^-$$

The ethanol is then oxidised by adding warm acidified potassium dichromate(VI) and collecting the ethanal as it distils off.

$$CH_3CH_2OH + [O] \rightarrow CH_3CHO + H_2O$$

9 In the ethane molecule there is no overall dipole. Attraction between molecules is limited to temporary, induced dipole–dipole attractions, or van der Waals forces. These are the weakest of intermolecular forces and are easily overcome, so ethane has a very low boiling point.

$$\overset{\delta+\;\;\;\delta-}{H_3C{-}CH_3}$$

$$\overset{\delta-\;\;\;\delta+}{H_3C{-}CH_3}$$

In the chloroethane molecule there is a permanent dipole due to the high electronegativity of the chlorine atom. The chlorine atom has a greater share of the electrons in the carbon–chlorine bond leaving it δ− and the carbon atom δ+. Oppositely charged parts of the molecule attract each other by a greater force than is the case between molecules of ethane.

$$CH_3{-}\overset{\delta+}{CH_2}{-}\overset{\delta-}{Cl}$$

$$\overset{\delta-}{Cl}{-}\overset{\delta+}{CH_2}{-}CH_3$$

In the amino group of ethylamine the electronegative nitrogen atom has a greater share of the electrons in the nitrogen–hydrogen bonds leaving it δ− and the hydrogen atoms δ+. The hydrogen atoms are attracted to the lone pair of electrons on the nitrogen atom of another molecule of ethylamine. This gives rise to hydrogen bonding, which is the strongest of the intermolecular forces.

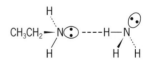

In ethanol, molecules are also attracted by hydrogen bonding. The hydrogen bonding in alcohols is stronger than in analogous amines because oxygen has a higher electronegativity than nitrogen so the O—H bond is more strongly polarised than the N—H bond. For this reason, the alcohols have higher boiling points.

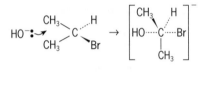

10 a Nucleophilic substitution.

  b An S_N1 mechanism involves the formation of a carbocation.

  An S_N2 mechanism involves the formation of an intermediate.

  c If this reaction proceeds via an S_N1 mechanism we would expect it to be first order with respect to 2-bromopropane and zero order with respect to sodium hydroxide. However, if it proceeds via an S_N2 mechanism we would expect it to be first order with respect to both 2-bromopropane and sodium hydroxide, and so second order overall.

22 Organic series of compounds 2

Amines

Amines are derivatives of ammonia. Alkyl or aryl groups replace one (primary), two (secondary) or three (tertiary) of the hydrogen atoms.

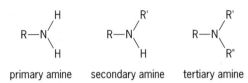

primary amine   secondary amine   tertiary amine

Amines have a characteristic fishy smell. Their boiling points are less than those of analogous alcohols because the hydrogen bonding in amines is weaker. This is because oxygen is more electronegative than nitrogen so the O—H bond is more strongly polarised than the N—H bond. Primary amines such as methylamine (CH_3NH_2) and phenylamine ($C_6H_5NH_2$) are of most interest.

Like ammonia, amines act as Brønsted–Lowry bases since the lone pair of electrons on the nitrogen atom of the molecule is available to form a bond with a proton.

$$RNH_2 + H^+ \rightleftharpoons RNH_3^+$$

Alkyl amines are stronger bases than ammonia because the alkyl group is electron releasing, thus making the lone pair of electrons on the nitrogen atom more available for bonding. Conversely, aryl amines are weaker bases than ammonia because the nitrogen lone pair becomes involved in the aromatic delocalisation of electrons.

Alkyl amines may be conveniently prepared in a two-step synthesis from haloalkanes. In the first step, the haloalkane undergoes nucleophilic substitution to form the nitrile.

$$RBr + CN^- \rightarrow R—C\equiv N + Br^-$$

The nitrile is subsequently reduced using hydrogen and nickel or lithium tetrahydridoaluminate(III) ($LiAlH_4$, a powerful reducing agent).

$$R—C\equiv N + 2H_2 \rightarrow RCH_2NH_2$$

Aryl amines are usually prepared by the reduction of aryl nitro compounds using tin and dilute hydrochloric acid.

$$ArNO_2 + 3H_2 \rightarrow ArNH_2 + H_2O$$

Both ammonia and amines act as Lewis bases, or nucleophiles, and take part in nucleophilic substitution reactions. The formation of primary amines by the reaction of ammonia with a haloalkane has already been described in topic 21. Depending on reaction conditions, primary amines may react further with haloalkanes, eventually culminating in the formation of quaternary salts.

$$NH_3 + CH_3Br \rightarrow \quad CH_3NH_2 \quad + HBr$$
primary amine

$$CH_3NH_2 + CH_3Br \rightarrow \quad (CH_3)_2NH \quad + HBr$$
secondary amine

$$(CH_3)_2NH + CH_3Br \rightarrow \quad (CH_3)_3N \quad + HBr$$
tertiary amine

$$(CH_3)_3N + CH_3Br \rightarrow (CH_3)_4N^+Br^-$$
quaternary salt

Carbonyl compounds

Aldehydes and ketones are carbonyl compounds. They each contain the carbonyl function group and have the general formula $C_nH_{2n}O$.

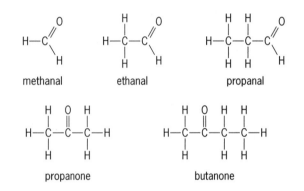

methanal   ethanal   propanal

propanone   butanone

The carbonyl group is planar: the carbon atom, the oxygen atom and the two other atoms attached to the carbon atom are all in the same plane.

Reduction of carbonyl compounds

Aldehydes are reduced to primary alcohols and ketones to secondary alcohols. There are two convenient ways in which this can be achieved.

- Reaction of the carbonyl compound with hydrogen in the presence of a nickel or platinum catalyst. This reagent has the disadvantage that it will also saturate any carbon–carbon double bonds present in the molecule.

$$CH_3COCH_3 + H_2 \rightarrow CH_3CHOHCH_3$$
propanone   propan-2-ol

$$CH_2{=}CHCHO + 2H_2 \rightarrow CH_3CH_2CH_2OH$$
prop-2-enal   propan-1-ol

- Reaction of the carbonyl compound with sodium tetrahydridoborate(III) ($NaBH_4$) in methanol. This reagent has no effect on carbon–carbon double bonds and can be used for selective reduction of aldehydes and ketones.

The reason for this selectivity is that the reagent adds hydrogen via nucleophilic attack on the carbonyl group. This is discussed in some detail in the following section.

Nucleophilic addition to carbonyl compounds

The carbon–oxygen bond in a carbonyl compound is unsaturated in the same way as the carbon–carbon bond in an alkene and is thus subject to addition reactions. However, unlike the carbon–carbon double bond in an alkene, the electron cloud in a carbonyl group is not evenly distributed between the two atoms.

$$\underset{\delta+}{}C=\underset{\delta-}{}O$$

The oxygen atom is more electronegative and has a greater share of the electron density. The electron-deficient ($\delta+$) carbon atom is attractive to nucleophiles, ions or molecules that are looking to donate electrons in order to form bonds. A typical example of **nucleophilic addition** to the carbonyl group is the formation of hydroxynitriles by the addition of HCN.

The mechanism involves nucleophilic attack by the cyanide ion on the electron-deficient carbon atom. The resulting anion then gains a proton.

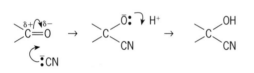

The reaction is not usually carried out using hydrogen cyanide since this is highly poisonous. The gas is usually generated within the reaction by a carefully measured amount of dilute sulphuric acid and potassium cyanide.

A negative charge indicates the presence of one extra electron. In order to form a bond two electrons are needed, so a lone pair of electrons is shown forming the bond between the nucleophile and the carbon atom, and between the oxygen atom and the proton.

Notice that all aldehydes (except methanal) and all unsymmetrical ketones react with HCN to produce compounds which contain an asymmetric carbon atom. However, the product obtained is an optically inactive racemic mixture (see topic 12). The reason for this is that the cyanide nucleophile may attack from either side of the planar carbonyl group with equal probability.

This reaction provides a convenient way of increasing the carbon chain length of a compound. The nitrile group can be hydrolysed to form a carboxylic acid group.

Carboxylic acids

Carboxylic acids contain the functional group —COOH. At first glance this might appear to be a combination of a carbonyl group (as in aldehydes and ketones) and a hydroxyl group (as in alcohols), and therefore it may be expected that carboxylic acids have the combined properties of these groups; however, this is not the case. In fact the two groups have interacted to the extent that carboxylic acids have mainly different properties.

Acidity of carboxylic acids

The —COOH group in carboxylic acids will hydrogen bond with water. Low molecular mass carboxylic acids are very soluble in water; however, as the alkyl or aryl group component of the molecule increases the solubility in water decreases. Even those carboxylic acids which are very soluble in water are only weak acids, because of the small degree of dissociation that occurs.

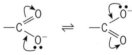

ethanoic acid   ethanoate ion

The dissociation constant, K_a, for ethanoic acid is $1.75 \times 10^{-5}\,mol^2\,dm^{-6}$. This means that only about four molecules of ethanoic acid in every 1000 are ionised at any one time.

Notwithstanding the low degree of dissociation, carboxylic acids are much more acidic than alcohols. When a proton is lost from the —OH group of a carboxylic acid the remaining negative charge is delocalised between the two oxygen atoms.

$$-C\overset{O}{\underset{O^-}{\lessgtr}} \rightleftharpoons -C\overset{O^-}{\underset{O}{\lessgtr}}$$

This means that the charge on each oxygen atom is effectively halved, thus giving a more stable anion.

The anion is further stabilised, leading to greater dissociation, if substituents attached to the carbon atom adjacent to the —COOH group are electron-withdrawing, like chlorine for instance (Table 22.1).

Table 22.1 Acidity of carboxylic acids

Carboxylic acid	Formula	pK$_a$ value*	Strength of acid
ethanoic acid	CH$_3$COOH	4.76	
monochloroethanoic acid	CH$_2$ClCOOH	2.86	
dichloroethanoic acid	CHCl$_2$COOH	1.29	increasing strength
trichloroethanoic acid	CCl$_3$COOH	0.65	

*See topic 26 for explanation of pK$_a$ values

Trichloroethanoic acid is a much stronger acid than ethanoic acid because a greater proportion of its molecules are ionised.

Reaction of carboxylic acids to form salts

Carboxylic acids react with metals, bases, alkalis and carbonates to form **salts**; however, these reactions are much less vigorous than similar reactions with strong mineral acids.

$$2CH_3COOH + Mg \rightarrow (CH_3COO^-)_2Mg^{2+} + H_2$$
ethanoic acid   metal   magnesium ethanoate

$$2HCOOH + CaO \rightarrow (HCOO^-)_2Ca^{2+} + H_2O$$
methanoic acid   base   calcium methanoate

$$C_6H_5COOH + NaOH \rightarrow C_6H_5COO^-Na^+ + H_2O$$
benzoic acid   alkali   sodium benzoate

$$2CH_3CH_2COOH + Na_2CO_3 \rightarrow$$
propanoic acid   carbonate
$$2CH_3CH_2COO^-Na^+ + CO_2 + H_2O$$
sodium propanoate

Esters

Esters contain the functional group —COOR, where R may be an alkyl or an aryl group. They are formed by the reaction of carboxylic acids with alcohols in the presence of a strong acid catalyst such as concentrated sulphuric acid.

$$RCOOH + R'OH \rightarrow RCOOR' + H_2O$$
carboxylic acid   alcohol   ester

Ester formation involves the loss of water. There are two possible ways in which this may happen, involving the loss of an —OH group from the carboxylic acid or from the alcohol.

$$R-\overset{O}{\overset{\|}{C}}-O\!\mid\!H + H-O\!\mid\!R' \rightarrow R-\overset{O}{\overset{\|}{C}}-O-R' + H_2O$$

$$R-\overset{O}{\overset{\|}{C}}\!\mid\!O-H + H\!\mid\!O-R' \rightarrow R-\overset{O}{\overset{\|}{C}}-O-R' + H_2O$$

Isotopic labelling using radioactive oxygen-18 shows the -OH group is lost from the carboxylic acid and the oxygen atom from the alcohol is retained in the ester.

$$R-\overset{O}{\overset{\|}{C}}-O-H + H-\overset{*}{O}-R' \rightarrow R-\overset{O}{\overset{\|}{C}}-\overset{*}{O}-R' + H_2O$$

Naming esters frequently causes confusion, perhaps because the molecular *formulae* of esters are often given with that part of the ester derived from the carboxylic acid being written first. Naming esters is simple if these guidelines are followed.

- The alcohol part of an ester is written at the beginning of the ester name.

   CH_3- methyl
   CH_3CH_2- ethyl

- The acid part of an ester is named as if it was the ionic carboxylate group in a salt of a carboxylic acid, and it comes at the end of the ester name, for example:

   CH_3COO^- ethanoate
   $CH_3CH_2COO^-$ propanoate

- Combining these two parts:

   $CH_3COOCH_2CH_3$ ethyl ethanoate
   $CH_3CH_2COOCH_3$ methyl propanoate

Esters have no —OH group so they cannot hydrogen bond like carboxylic acids or alcohols. Consequently, they are more volatile than carboxylic acids and are insoluble in water.

Esters are widely used as solvents for other organic substances. Many have a fruity smell and are used for food flavourings and in cosmetics. This fruity smell is often given as a means of identifying the presence of esters.

Hydrolysis of esters

When esters are heated with an alkali, such as sodium hydroxide, they are readily hydrolysed to form an alcohol and a carboxylate salt. This reaction may also be described as a **saponification** reaction.

$$CH_3CH_2COOCH_3 + NaOH \rightarrow CH_3CH_2COONa + CH_3OH$$
   methyl   sodium   methanol
   propanoate   propanoate

This reaction is important in the industrial production of soaps from fats and oils.

Most naturally occurring fats and oils are esters of propane-1,2,3-triol (glycerine) with three long-chain carboxylic acids. These acids may contain from 7 to 21 carbon atoms and are often called **fatty acids** since they are obtained from fats (Table 22.2).

When the fats are boiled with sodium hydroxide, propane-1,2,3-triol and a mixture of the sodium salts of the three carboxylic acids are formed. These salts are what we call **soaps**.

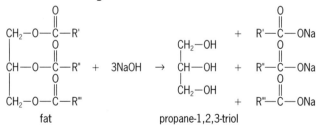

Table 22.2 Common fatty acids

Name	Formula	Found in
palmitic acid	$CH_3(CH_2)_{14}COOH$	animal and vegetable fats
stearic acid	$CH_3(CH_2)_{16}COOH$	animal and vegetable fats
oleic acid	$CH_3(CH_2)_7CH=CH(CH_2)_7COOH$	most fats and oils
linoleic acid	$CH_3(CH_2)_4CH=CHCH_2CH=CH(CH_2)_7COOH$	soya-bean oil and nut oil

Amino acids

Amino acids are compounds that contain both an amine group and a carboxylic acid functional group. When the amine and carboxylic acid groups are attached to the same carbon atom, they are termed **α-amino acids**, with the general formula $H_2NCHRCOOH$, where R may be hydrogen or another organic group. Proteins found in living tissue are made from α-amino acids.

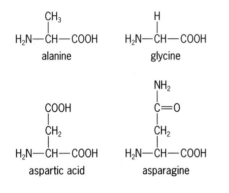

Amino acids are soluble in water. By the loss of a proton from the carboxylic acid group and the gain of a proton on the amine group, amino acids are able to form ions which carry both a positive and a negative charge. Such ions are called **zwitterions**.

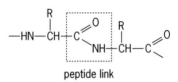

Certain amino acids are essential to our well being and we obtain them from the food we eat. Proteins consist of long chains of amino acids joined together by **peptide** (—CONH—) links.

peptide link

Digestion of proteins involves the hydrolysis of peptide links to produce amino acids for use by the body.

1 Which of the following compounds is likely to be most volatile?

A
$$CH_3—\overset{\overset{\displaystyle O}{\|}}{C}—CH_2—OH$$

B
$$HO—CH_2—CH_2—C\overset{\displaystyle O}{\underset{\displaystyle OH}{<}}$$

C
$$CH_3—CH_2—C\overset{\displaystyle O}{\underset{\displaystyle OH}{<}}$$

D
$$CH_3—C\overset{\displaystyle O}{\underset{\displaystyle O—CH_3}{<}}$$

2 Which of the following types of reaction is typical of the carbonyl group in an aldehyde or ketone?
A electrophilic addition
B electrophilic substitution
C nucleophilic addition
D nucleophilic substitution

3 Which of the following is formed when $CH_2{=}CHCH_2CHO$ is reacted with sodium tetrahydridoborate(III) in methanol?
A $CH_3CH_2CH_2CH_2OH$
B $CH_2{=}CHCH_2CH_2OH$
C $CH_3CH_2CH_2CHO$
D $CH_3CH_2CH_2COOH$

4 Which of the following compounds is an α-amino acid?

A
$$HS—CH_2—\overset{\overset{\displaystyle H}{|}}{\underset{\underset{\displaystyle NH_2}{|}}{C}}—COOH$$

B
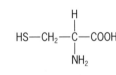

C
$$CH_3—CH_2—C\overset{\displaystyle O}{\underset{\displaystyle NH_2}{<}}$$

D
$$H_2N—CH_2—CH_2—COOH$$

5 An ester has the molecular formula $CH_3CH_2COOCH_2CH_3$. The name of this ester is
A ethyl ethanoate
B ethyl propanoate
C propyl ethanoate
D propyl propanoate

6 Benzoic acid is soluble in both water and benzene. In benzene some of the benzoic acid molecules are dimerised as shown below.

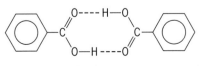

Two benzoic acid molecules become associated together by hydrogen bonding.

a Explain why hydrogen bonds form between benzoic acid molecules.
b Why does benzoic acid dimerise in benzene but not in water?
c In terms of bond angles, explain why the ring formed in a benzoic acid dimer might be expected to be stable.
d A sample of benzoic acid in benzene has an average relative molecular mass of 183. Calculate the fraction of benzoic acid molecules that are dimerised in this sample.

7 The following equation represents an addition reaction.

$$CH_3CHO + HCN \rightarrow CH_3CH(OH)CN$$

a Write a mechanism for this reaction.
b Draw the two enantiomers of the product, 2-hydroxypropanenitrile, and explain why this reaction results in a racemic mixture of the product.
c Draw and name the structure of the product when 2-hydroxypropanenitrile is heated with dilute aqueous acid.

8 Sodium tetrahydridoborate(III) in methanol can be used to reduce the carbonyl group in an aldehyde or ketone. The reaction mechanism involves nucleophilic attack.
a What is a nucleophile?
b Using the symbol 'Nu' to represent a nucleophile, draw a diagram to show nucleophilic attack on a carbonyl group.
c Explain why sodium tetrahydridoborate(III) in methanol will reduce a carbon–oxygen double bond but not a carbon–carbon double bond.

9 a What is K_a? Write an expression for K_a for an acid HA.
b The strengths of three acids are compared below.

hydrochloric acid $>$ trichloroethanoic acid $>$ ethanoic acid

Explain why the acidity of the acids varies in this way.
c Give two examples of reactions which demonstrate that carboxylic acids react as typical acids to form salts.

10 a Name the starting materials and describe the reaction conditions needed for the formation of propyl ethanoate.
b During esterification a molecule of water is lost. Explain how isotopic labelling can be used to determine which reacting species loses the —OH group needed to form water.
c Describe the hydrolysis of fats and oils by heating with sodium hydroxide and state the use of the carboxylic acid salts formed.

1 D

2 C

3 B

4 A

5 B

6 **a** In the carbon–oxygen double bond the oxygen atom is more electronegative and has a greater share of the bonding electrons, leaving it δ−. In the oxygen–hydrogen bond, the hydrogen atom is lower in electronegativity and has a lesser share of the bonding electrons, leaving it δ+. The oppositely charged parts of the molecules are attracted to each other.

b In water the benzoic acid molecules are able to hydrogen bond with water molecules which are present in a much higher concentration than the benzoic acid molecules.

c The ring formed by the dimer is effectively a hexagon with a pair of opposite sides elongated. The O—C—O and C—O—H bond angles are 120°. None of the bond angles need to be distorted in order to form the dimer ring so it is expected to be stable.

d The relative molecular mass of benzoic acid is 122 and its dimer, 244. If a fraction α of the benzoic acid is dimerised then:

$$122 \times (1 - \alpha) + 244 \times \alpha = 183$$
$$\therefore \quad 122 - 122\alpha + 244\alpha = 183$$
$$\therefore \quad \alpha = 0.5$$

7 **a**

b

CH₃⎯C(OH)(H)⎯CN and HO⎯C(CH₃)(H)⎯NC

Ethanal is planar about the carbonyl group. The initial nucleophilic attack by the cyanide ion may come from above or below the plane of the carbonyl group and as each is equally likely a racemic mixture is formed.

c $CH_3CH(OH)COOH$, 2-hydroxypropanoic acid.

8 **a** A nucleophile is an ion or molecule that can donate electrons.

b

$$\overset{\delta+}{C}=\overset{\delta-}{O} \quad \rightarrow \quad C—O^-$$
Nu: — Nu⁺

c In a carbonyl group the electrons between the carbon atom and the oxygen atom are not evenly distributed due to a difference in the electronegativity of these atoms. The oxygen atom has a greater share of the electrons making it charged δ− and leaving the carbon atom charged δ+. In a carbon–carbon double bond the electrons are distributed evenly and the unpolarised electron cloud repels the nucleophile, H^-.

9 **a** K_a is the dissociation constant. For an acid
$$HA \rightleftharpoons H^+ + A^-$$

$$K_a = \frac{[H^+][A^-]}{[HA]}$$

b Hydrochloric acid is the strongest acid because it is fully dissociated in solution whereas the other two acids are only partially dissociated.
   Trichloroethanoic acid is stronger than ethanoic acid because it is able to dissociate to a greater degree. This is because the chlorine atoms on the carbon atom next to the carboxylic acid group are electron withdrawing. They increase the stability of the cation produced by dissociation.

c Examples may include the reaction of a carboxylic acid, such as ethanoic acid, with metals, bases, acids or carbonates.

10 **a** Ethanoic acid and propan-1-ol. A small amount of concentrated sulphuric acid is added and the mixture is heated.

b Propan-1-ol is used in which the oxygen atom is radioactive oxygen-18. In forming water, if the —OH group comes from the alcohol the water will be radioactive but not the ester. Conversely, if the —OH group comes from the ethanoic acid the ester will be radioactive but not the water.

c Fats and oils are esters of propane-1,2,3-triol with three long-chain carboxylic acids. Hydrolysis of fats and oils by heating with sodium hydroxide produces propane-1,2,3-triol and sodium salts of three carboxylic acids. Sodium salts of long-chain carboxylic acids are the basis of soaps.

Modern chemical analysis uses instrumental techniques in conjunction with traditional qualitative and volumetric analysis. The four main techniques used are:

- mass spectrometry
- infrared (IR) spectroscopy
- nuclear magnetic resonance (n.m.r.) spectroscopy
- ultraviolet and visible spectroscopy.

In determining the structure of an organic compound a chemist will often draw on the evidence provided by all of these techniques.

Mass spectrometry

When a sample of an organic compound is introduced into a mass spectrometer, a stream of electrons bombards the molecules. This creates a series of positive ions. Some will be the **molecular ion**, M^+, while others will be **fragment ions** formed by the breaking up of the molecules. The masses of the molecular ion and the fragment ions provide evidence about the structure of the compound.

Figure 23.1 is a mass spectrum of butane. The horizontal axis shows the **mass:charge ratio** (written m/e or m/z). If the charge is 1, then this ratio is equal to the relative molecular mass of the ion.

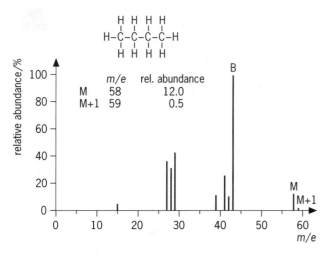

Figure 23.1 Mass spectrum of butane

The spectrum has several interesting features. The molecular ion, M, is present at $m/e = 58$ as expected but there is also a smaller peak at $m/e = 59$, the M + 1 peak. This is due to the presence of carbon-13 in some of the butane molecules. Carbon-13 has a natural

abundance of 1.1%. The ratio M:M + 1 is approximately $100 : n \times 1.1$, where n is the number of carbon atoms in the molecule.

This ratio can be used to estimate the number of carbon atoms in the molecule. In this case the ratio of M:M + 1 is 12:0.5. The approximate number of carbon atoms is given by:

$$\frac{0.5}{12 \times 1.1} \times 100 = 4 \text{ (nearest whole number)}$$

The molecular ion is not the largest peak in the spectrum. This is at $m/e = 43$ and corresponds to the ion $C_3H_7^+$. It is formed by the loss of a methyl group from the molecular ion.

$$C_4H_{10}^+ \rightarrow C_3H_7^+ + CH_3$$

Similarly, we can assign ions to the other major peaks in the spectrum. For example, we might expect butane also to fragment by losing an ethyl group. The resulting ion, $C_2H_5^+$, would give a peak at $m/e = 29$, which can be seen in the spectrum. The way in which a molecule fragments (the **fragmentation pattern**) gives some indication of its structure.

Compounds containing elements that have two (or more) isotopes present in significant proportions, for example chlorine and bromine, show more than one molecular ion.

2-bromopropane has two molecular ions at $m/e = 122$ and 124, corresponding to molecular ions containing bromine-79 and bromine-81, respectively (Figure 23.2). The size of the peaks is similar, since the two isotopes of bromine are present in similar amounts.

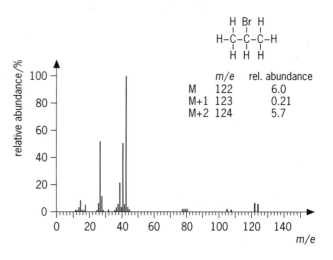

Figure 23.2 Mass spectrum of 2-bromopropane

In chloroalkanes the peak sizes reflect the relative abundance of the chlorine isotopes chlorine-35 (75.77%) and chlorine-37 (25.23%). For chloroethane, the molecular ion peaks appear at $m/e = 64$ and $m/e = 66$ in the approximate ratio of $3:1$, corresponding to the ions $C_2H_5{}^{35}Cl^+$ and $C_2H_5{}^{37}Cl^+$, respectively (Figure 23.3).

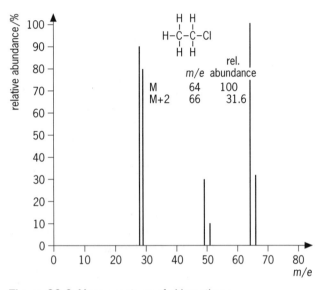

Figure 23.3 Mass spectrum of chloroethane

The situation is even more complicated where a molecule contains two atoms of chlorine or bromine. The pattern of peaks corresponding to molecular ions for dichloroethane, $C_2H_4Cl_2$, is shown in Table 23.1.

Table 23.1 Peak sizes in the mass spectrum of dichloroethane

Ion	m/e value	Ratio
$C_2H_4{}^{35}Cl{}^{35}Cl^+$	98	9
$C_2H_4{}^{35}Cl{}^{37}Cl^+$ $C_2H_4{}^{37}Cl{}^{35}Cl^+$	100	6
$C_2H_4{}^{37}Cl{}^{37}Cl^+$	102	1

Modern high-resolution mass spectrometers are linked to computers. These are able to match precise masses with molecular formulae.

Infrared spectroscopy

The atoms in a molecule are in constant motion relative to each other, as the bonds between them bend and stretch. The energy needed for this motion is in the infrared region of the electromagnetic spectrum. Each bond vibrates at a particular frequency that depends on the strength of the bond and the masses of the atoms bonded together.

The frequency of vibration for a particular bond is similar, but not identical, in all molecules. Thus it is possible to make deductions about the structure of a molecule by observing what frequencies it absorbs in the infrared region.

The infrared region can be conveniently divided into four regions, as shown in Figure 23.4. (Wavenumber is equal to 1/wavelength.)

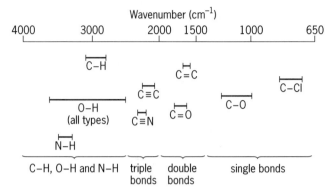

Figure 23.4 Areas of infrared spectra

- The region below $1500\,cm^{-1}$ is called the **fingerprint region** and corresponds to the vibrations of many single bonds. The pattern in this region of the spectrum is usually complex and of little use in identifying individual bonds. However, because of its complexity, each molecule has a unique fingerprint region. Comparing this region of an infrared spectrum with those of known compounds can identify an unknown compound.
- The region between 1500 and $2000\,cm^{-1}$ corresponds to the vibration of double bonds. Those of particular interest in organic molecules are C=C and C=N, which generally appear as small absorptions, and C=O, which is usually a large absorption.
- The region between 2000 and $2500\,cm^{-1}$ corresponds to the vibration of triple bonds. Those found in organic molecules are C≡C and C≡N.
- The region between 2500 and $4000\,cm^{-1}$ corresponds to the vibration of C—H, O—H and N—H bonds. The absorption of the latter two bonds is frequently broad due to the effects of hydrogen bonding.

Important absorptions that identify functional groups are shown in the following infrared spectra.

- Figure 23.5 shows the infrared spectrum of benzoic acid, with the O—H and C=O absorptions marked.

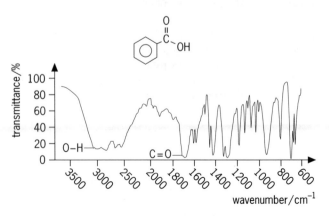

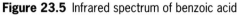

Figure 23.5 Infrared spectrum of benzoic acid

- Figure 23.6 shows the infrared spectrum of benzonitrile, with the C≡N absorption marked.

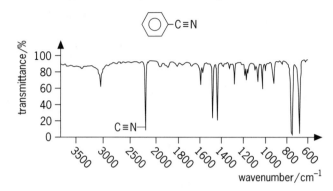

Figure 23.6 Infrared spectrum of benzonitrile

- Figure 23.7 shows the infrared spectrum of propylamine, with the N—H absorption marked.

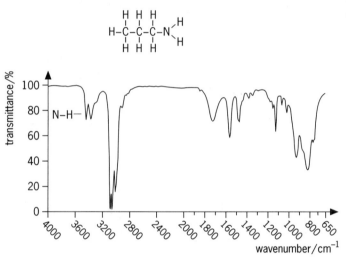

Figure 23.7 Infrared spectrum of propylamine

Nuclear magnetic resonance (n.m.r.) spectroscopy

Nuclei with odd mass numbers have a property of spin. When a nucleus spins it induces a weak magnetic field which can interact with an external magnetic field. When an external magnetic field is applied the magnetic field induced by the nuclear spin can align either in the same direction as the external field or in the opposite direction to the external field. This gives two possible energy levels (Figure 23.8).

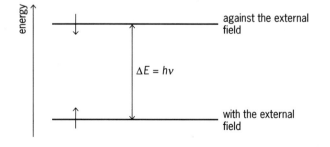

Figure 23.8

If a sample of a compound is irradiated with radiation from the radio region of the electromagnetic spectrum then at some frequency, ν, absorption of energy results in transitions between these two levels.

In reality, spectra are usually measured by keeping the frequency of the radiation constant and varying the strength of the external magnetic field. The scale is not in frequency units, as might be expected, but as a **chemical shift** from a standard compound, tetramethylsilane (TMS), $(CH_3)_4Si$. Its protons absorb therefore at 0.00δ and most protons bonded to carbon absorb at positive values of δ. TMS is chemically inert and miscible with most organic solvents.

The spin due to protons, ^1H, is often used to provide an n.m.r. spectrum of an organic compound. Chemical shifts for hydrogen (δ) are expressed in parts per million (ppm) and typically have values between 0 and 10.

The n.m.r. spectrum gives four important pieces of information about the chemical structure of a compound.

- The number of absorptions indicates the number of different 'types' of protons. These are more correctly described as protons in different chemical environments.
- The position of each absorption gives information about the environment of the protons involved (Table 23.2).
- The intensities of the absorptions, indicated by the areas under the peaks, give the ratio of protons in each environment.
- Different protons on adjacent carbon atoms cause each other's signal to split. The signal of protons on a carbon atom adjacent to another carbon atom having n hydrogen atoms will be split into $n+1$ peaks. This is called **spin–spin coupling** and gives additional information about the compound.

Table 23.2 Proton chemical shifts

Environment of proton	Typical shift from TMS
R—CH$_3$	0.9
R$_2$CH$_2$	1.3
R$_3$CH	2.0
CH$_3$—CO—OR	2.0
R—CO—CH$_3$	2.1
Ar—CH$_3$	2.3
R—C≡C—H	2.6
R—O—CH$_3$	3.8
R—O—H	4.5
RHC=CH$_2$	4.9
RHC=CH$_2$	5.9
Ar—OH	7.0
Ar—H	7.3
R—CO—H	9.7
R—CO—OH	11.5

In propane, the protons are in two different environments, the —CH₃ group and the —CH₂— group. The number of protons is 6 in 2×(—CH₃) and 2 in —CH₂—, a ratio of 6:2 (or 3:1). In the n.m.r. spectrum of propane, the peaks are in ratio of intensity of 3:1 (Figure 23.9).

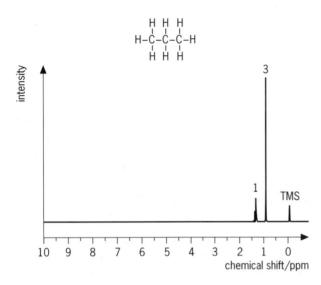

Figure 23.9 n.m.r. spectrum of propane

The n.m.r. spectrum of ethanol shows three peaks, indicating that there are three different 'types' of proton. The absorption at about 1.1δ has an intensity of 3 due to the protons in —CH₃. Similarly the absorption around 3.6δ is due to the protons in —CH₂— and the absorption at 4.5δ is due to the proton in —OH (Figure 23.10).

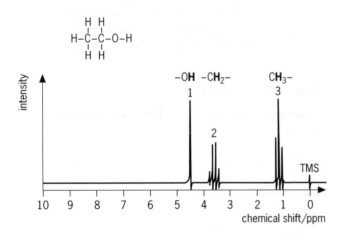

Figure 23.10 n.m.r. spectrum of ethanol

There are two other interesting features of this n.m.r. spectrum.

• The signal at 1.1δ is split into a triplet. This is due to the different possible arrangements of the protons on the adjacent —CH₂— group in relation to the two possible energy levels created by the applied magnetic field (Figure 23.11).

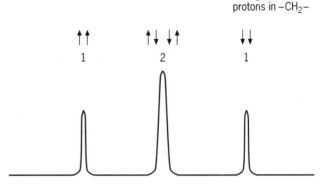

Figure 23.11 Coupling to form a triplet

Similarly, the signal at 3.6δ is split into a quartet due to coupling with the protons on the —CH₃ group (Figure 23.12).

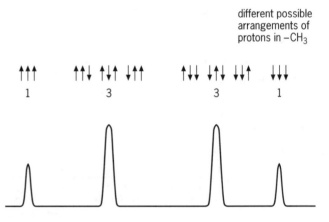

Figure 23.12 Coupling to form a quartet

Spin–spin coupling is often useful to confirm a suspected structure.

• The peak at 4.5δ is lost when the sample is shaken with deuterium oxide, D₂O. Deuterium, ²H, does not have a spin property since it does not have an odd mass number. When ethanol is shaken with D₂O the following occurs:

$$CH_3CH_2—OH + D_2O \rightleftharpoons CH_3CH_2—OD + DHO$$

The proton is exchanged and thus the signal at 4.5δ is lost. This test for the presence or absence of an easily exchangeable (**labile**) proton can often provide useful additional information.

Ultraviolet and visible spectroscopy

Ultraviolet and visible spectroscopy are concerned with the movement of electrons between energy levels as the result of the absorption of energy in the ultraviolet and visible regions of the electromagnetic spectrum.

The energy change for a particular transition corresponds to the absorption of a particular wavelength of light. It is these transitions that explain why some compounds are coloured; however, a lack of colour should not be interpreted as a lack of absorption since many compounds absorb in the ultraviolet region.

Most transition metals form complex ions that are coloured. The colour varies not only between elements, but also between different oxidation states of the same element and, within a series of complexes of the same element, with different ligands.

It should be borne in mind that the energy absorbed in the visible region of the spectrum causes a complex to have the complementary colour to that absorbed. Figure 23.13 shows the absorption spectrum of the complex $[Ti(H_2O)_6]^{3+}$.

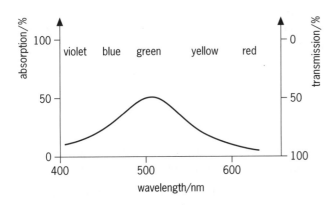

Figure 23.13 Visible spectrum of $[Ti(H_2O)_6]^{3+}$

There is a broad absorption around 500 nm which means that the orange, yellow and green parts of the spectrum have been removed, leaving a mixture of the red and violet components. Aqueous solutions of Ti^{3+} salts appear purple.

A comparison of the absorption spectra of $[Ni(H_2O)_6]^{2+}$ and $[Ni(NH_3)_6]^{2+}$ illustrates the effect of different ligands on the colour of a complex ion (Figure 23.14).

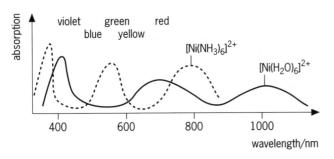

Figure 23.14 Effect of ligands on the colour of a complex ion

The $[Ni(H_2O)_6]^{2+}$ complex is green while the $[Ni(NH_3)_6]^{2+}$ complex is blue.

Organic compounds that absorb in the ultraviolet and visible regions contain double or triple bonds and/or non-bonded pairs of electrons. These structures are called **chromophores** and they are responsible for the absorptions.

Alkanes show no absorption in the ultraviolet or visible regions since they have neither double/triple bonds nor non-bonded electrons; however, alkenes do have absorptions. Alkenes that have a small chromophore absorb in the ultraviolet region (Figure 23.15).

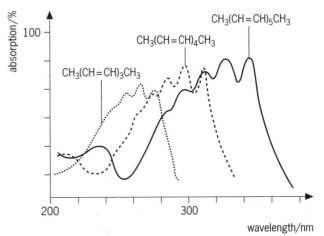

Figure 23.15 Length of alkene chain alters ultraviolet absorption

Notice that extending the chromophore shifts the absorption to longer wavelength and increases its intensity.

In β-carotene, the substance responsible for the colour of carrot roots, the chromophore is extended to eleven carbon–carbon double bonds.

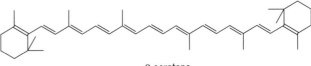

β-carotene

Absorption is shifted to 450 nm, which is in the blue region of the spectrum. This leaves the remaining light to be reflected, giving an orange colour (Figure 23.16).

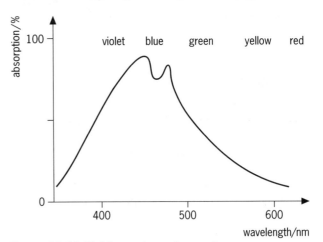

Figure 23.16 Visible spectrum of β-carotene

Determining structure

It is usually the case that the elucidation of structure involves using the techniques described together with whatever other data, such as elemental analysis, is available. The spectra in Figure 23.17a–c were obtained from an unknown organic compound.

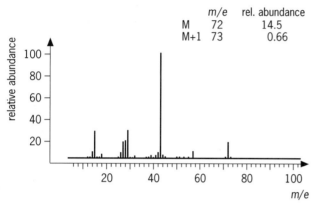

	m/e	rel. abundance
M	72	14.5
M+1	73	0.66

a mass spectrum

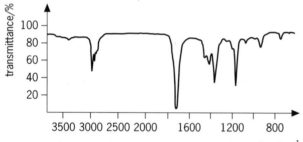

b infrared spectrum

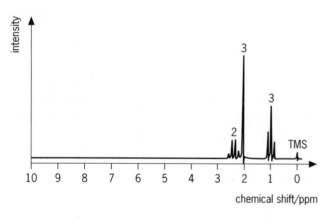

c n.m.r. spectrum

Figure 23.17 Spectra for an unknown compound

- Evidence from the mass spectrum.
  - The molecular ion, M, appears to be at $m/e = 72$. (Sometimes molecular ions are very unstable and are not seen on the mass spectrum.)

- The ratio of M:M+1 is 14.5:0.66. The approximate number of carbon atoms in the compound is given by:

$$\frac{0.66 \times 100}{14.5 \times 1.1} = 4 \text{ (to the nearest whole number)}$$

  - The large peak at $m/e = 43$ may be due to the ion $CH_3C^+=O$.
  - It appears that the compound has a relative molecular mass of 72 and contains 4 carbon atoms.
- Evidence from the infrared spectrum.
  - There is a large absorption around $1710\,cm^{-1}$. This is likely to be due to C=O. (Peaks relating to C=N and C=C would not have such a large absorption.)
  - There is no broad peak in the region $3000–3500\,cm^{-1}$ indicating that there is no —OH group present.
- Evidence from the n.m.r. spectrum.
  - There is a total of 8 hydrogen atoms in three different environments, in the ratio 2:3:3.

The evidence suggests that the compound has the formula C_4H_8O ($4 \times 12 + 8 \times 1 + 1 \times 16 = 72$). The presence of a C=O group and three different 'types' of protons suggests the following possible structure.

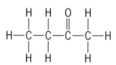

We can look back at the spectra for evidence to confirm or contradict this proposed structure.

- In the mass spectrum there are significant peaks at:

$m/e = 15$	due to CH_3^+
$m/e = 29$	due to $CH_3CH_2^+$
$m/e = 43$	due to CH_3CO^+
$m/e = 57$	due to $CH_3CH_2CO^+$

These fragments are consistent with the proposed structure.
- In the n.m.r. spectrum:

3 protons are not coupled — due to $\mathbf{CH_3}$—C=O
3 protons coupled with 2 protons (triplet) — due to $\mathbf{CH_3}CH_2$—
2 protons coupled with 3 protons (quadruplet) — due to $CH_3\mathbf{CH_2}$—

All of the evidence is consistent with the proposed structure, indicating that the compound is butan-2-one. As an additional check, the fingerprint region of the infrared spectrum of this compound could be compared to that of butan-2-one.

1 The ratio of the $M:M+1$ peaks in the mass spectrum of an organic compound was $31.1:2.1$. The approximate number of carbon atoms in the molecule is

A 4   **B** 5
C 6   **D** 7

2 Which of the following compounds will show a strong absorption in the region $1600–1800\,cm^{-1}$ of the infrared spectrum?

A chloromethane   **B** ethylbenzene
C propanone   **D** butan-1-ol

3 How many peaks (ignoring fine structure) would be seen in the n.m.r. spectrum of ethanal?

A 1   **B** 2
C 3   **D** 4

4 Here is the absorption spectrum of aqueous Cr^{3+}.

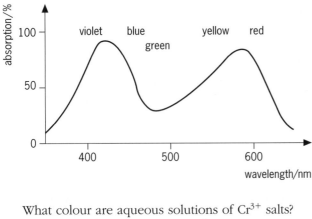

What colour are aqueous solutions of Cr^{3+} salts?

A green   **B** red
C purple   **D** blue

5 With which of the following compounds would shaking with D_2O remove one of the peaks in the n.m.r. spectrum?

A butan-2-one   **B** ethanal
C hexane   **D** propanoic acid

6 The following diagram shows details from the mass spectrum of bromoethane.

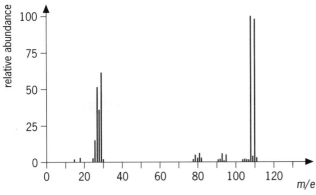

a Account for the peaks at 108 and 110 and comment on the significance of their relative abundances.

b Explain in what ways the pattern of molecular ions would be different in chloroethane.

7 **a** The following infrared spectra are of ethanol, ethanal and ethanoic acid. Identify each spectrum and explain how you reached your conclusions.

b The following diagram shows the fingerprint region of the infrared spectrum of one of the above compounds. Which compound is it?

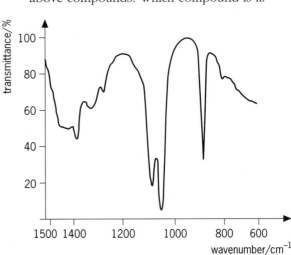

8 a The following diagram shows the n.m.r. spectrum of ethylbenzene.

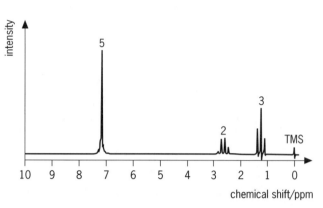

Assign each peak to the protons in this compound and explain the splitting of the peaks at 1.2δ and 2.7δ.

b Explain why shaking with D_2O has different effects on the n.m.r. spectra of benzaldehyde and benzoic acid.

9 The following spectra are of propanoic acid, CH_3CH_2COOH. By considering the structure of this compound, account for the detail shown in the spectra as fully as you can.

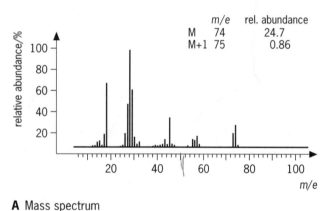

	m/e	rel. abundance
M	74	24.7
M+1	75	0.86

A Mass spectrum

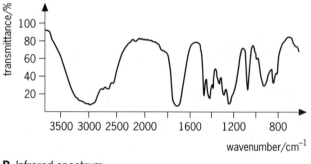

B Infrared spectrum

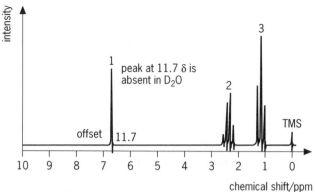

C n.m.r. spectrum

10 Identify the following compound from the spectra given.

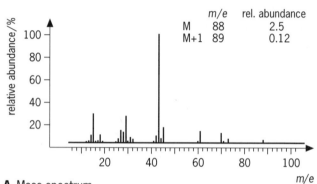

	m/e	rel. abundance
M	88	2.5
M+1	89	0.12

A Mass spectrum

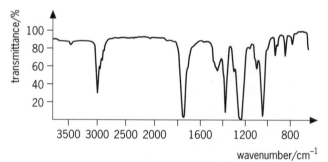

B Infrared spectrum

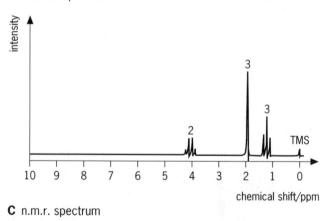

C n.m.r. spectrum

1 C

Approximate number of carbon atoms $= \dfrac{2.1 \times 100}{31.1 \times 1.1} = 6$

2 C **3** B **4** A **5** D

6 a The molecular ion at $m/e = 110$ is the ion $[CH_3CH_2{}^{81}Br]^+$ while the molecular ion at $m/e = 108$ is the ion $[CH_3CH_2{}^{79}Br]^+$.

The relative abundances of $6.0 : 5.7$ suggest that the isotopes of bromine are present in similar proportions.

b If the compound was chloroethane we would expect molecular ions at:

$m/e = 66$ corresponding to the ion $[CH_3CH_2{}^{37}Cl]^+$
$m/e = 64$ corresponding to the ion $[CH_3CH_2{}^{35}Cl]^+$

We would expect the relative abundances of the ions to be in the approximate ratio of $1 : 3$, reflecting the proportions of the isotopes of chlorine.

7 a **A** is ethanoic acid. The broad absorption around $2800–3200\,cm^{-1}$ indicates the presence of an —OH group and the large absorption in the region around $1700\,cm^{-1}$ indicates that a C=O group is present.

B is ethanol. The broad absorption around $3400\,cm^{-1}$ indicates the presence of an —OH group while the absence of a significant absorption in the region between $1600–1800\,cm^{-1}$ indicates that a C=O group is not present.

C is ethanal. The broad absorption around $1750\,cm^{-1}$ indicates the presence of a C=O group while the absence of a significant absorption in the region around $3500\,cm^{-1}$ indicates that an —OH group is not present.

b The fingerprint region corresponds to compound B which is ethanol.

8 a There are three 'types' of proton in ethylbenzene. (The environments of the protons attached to the benzene ring may be taken as the same.)

The peak at $7.1\,\delta$ is due to the five protons attached to the benzene ring.

The peak at $1.2\,\delta$ is due to the protons on the —CH$_3$ group. It is split into a triplet due to coupling with the two protons on the neighbouring —CH$_2$— group. They can align themselves with the external magnetic field in three possible ways:

↑↑   ↑↓ ↓↑   ↓↓

The peak at $2.7\,\delta$ is due to the protons on the —CH$_2$— group. It is split into a quartet due to coupling with the three protons on the neighbouring —CH$_3$ group. They can align themselves with the external magnetic field in four possible ways:

↑↑↑   ↑↑↓ ↑↓↑ ↓↑↑   ↑↓↓ ↓↑↓ ↓↓↑   ↓↓↓

b Benzoic acid has a proton on the —COOH group that can be exchanged. If it is shaken with D_2O this proton will exchange with deuterium and the peak due to it disappears from the n.m.r. spectrum. Shaking benzaldehyde with D_2O will not alter its n.m.r. spectrum since the proton on the —CHO group is not labile.

9 A Mass spectrum:
Molecular ion is at $m/e = 74$
($12 \times 3 + 1 \times 6 + 2 \times 16 = 74$)
The ratio of M : M + 1 is $24.7 : 0.86$; from this the estimated number of carbon atoms in the molecule is:

$\dfrac{0.86 \times 100}{24.7 \times 1.1} = 3.17$
$= 3$ (nearest whole number)

The ions responsible for the major peaks are possibly:

$m/e = 45$	$[COOH]^+$	$m/e = 27$	$[CH_2{=}CH]^+$
$m/e = 29$	$[CH_3{-}CH_2]^{\pm}$	$m/e = 18$	$[H_2O]^+$
$m/e = 28$	$[CO]^+$		

B Infrared spectrum:
Absorption around $1720\,cm^{-1}$ due to C=O group.
Absorption around $3000\,cm^{-1}$ due to —OH group; broad due to hydrogen bonding.

C n.m.r. spectrum:
Peak at $1.1\,\delta$ due to the protons on —CH$_3$; triplet due to coupling with —CH$_2$—.
Peak at $2.4\,\delta$ due to the protons on —CH$_2$—; quartet due to coupling with —CH$_3$.
Peak at $11.7\,\delta$ due to proton on —COOH; exchanges with D_2O.

10 • Considering evidence from the mass spectrum:
The molecular ion is at $m/e = 88$.
The ratio of M : M + 1 is $2.5 : 0.12$; from this the estimated number of carbon atoms in the molecule is:

$\dfrac{0.12 \times 100}{2.5 \times 1.1} = 4$ (nearest whole number)

• Considering evidence from the infrared spectrum:
There is a large absorption around $1750\,cm^{-1}$ that suggests the presence of a C=O group. Absence of a broad absorption between $3000–3500\,cm^{-1}$ indicates that there is no —OH group in the molecule.
• Considering evidence from the n.m.r. spectrum:
Three signals indicate there are three different 'types' of protons in the ratio $2 : 3 : 3$.

Reviewing the evidence:
88 (relative molecular mass) $- 4 \times 12$ (carbons) $= 40$
possibly $2 \times O + 8 \times H$. This would be consistent with both the infrared and n.m.r. spectra.
Therefore, the molecular formula of the compound is $C_4H_8O_2$.

The infrared spectrum shows a C=O group but no —OH group. The compound cannot be an acid or an alcohol. Since it has two oxygen atoms an ester is likely. There are two possibilities:

$$CH_3{-}C\begin{smallmatrix}\diagup O\\ \diagdown O{-}CH_3\end{smallmatrix} \qquad\qquad CH_3{-}CH_2{-}C\begin{smallmatrix}\diagup O\\ \diagdown O{-}CH_3\end{smallmatrix}$$
X   **Y**

Both esters A and B have three 'types' of hydrogen atoms and would exhibit the coupling shown. However, the absorption at $2.0\,\delta$ is consistent with a CH$_3$CO—O—R structure. The large peak at $m/e = 43$ in the mass spectrum is likely to be due to CH$_3$CO, so the evidence suggests the compound is structure X, ethyl ethanoate.

24 Aromatic chemistry

Aromatic chemistry is concerned with the chemistry of benzene, C_6H_6, and compounds containing a benzene ring. The term 'aromatic' was originally used to refer to substances with a pleasant aroma, but this is not the case with many aromatic compounds. Nowadays, the term **arenes** is often used to describe compounds based on a benzene ring structure.

Structure of benzene

Faraday first isolated benzene in 1825. The molecular formula of this compound was found to be C_6H_6 but its structure remained something of a mystery. It wasn't until 1865 that Friedrich August Kekulé suggested a ring structure of alternating single and double bonds.

This structure initially satisfied many of the properties of benzene, but it soon became clear that the problem was more complex than had first appeared. The proposed structure could not explain why there was only one form of 1,2-disubstituted benzene compounds such as 1,2-dibromobenzene, and not two.

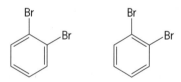

Kekulé subsequently suggested that the benzene molecule existed as two rapidly alternating structures.

The modern view of benzene is of a resonance hybrid of the two Kekulé forms. This can be conveniently considered in terms of its molecular orbitals (Figure 24.1).

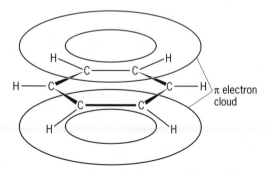

Figure 24.1 π bonding in benzene

Each carbon atom uses three of the four electrons in its outer shell to form σ bonds with two carbon atoms and one hydrogen atom. The fourth electron exists in a p orbital above and below the plane of the molecule. Six electrons, one from each carbon atom in the ring, are **delocalised** within the six overlapping p orbitals, forming a π electron cloud above and below the ring.

In this hexagonal structure all six of the carbon–carbon bonds are identical in length, a length that is intermediate between a carbon–carbon single bond and a carbon–carbon double bond (Table 24.1).

Table 24.1 Carbon–carbon bond lengths

Compound	Carbon–carbon bond length/nm
alkane	0.154
benzene	0.140
alkene	0.134

For ease of drawing, this structure is often represented as:

Evidence of the stability of benzene

The delocalisation of electrons in benzene confers a certain stability to the molecule. This becomes evident when considering the **enthalpy of hydrogenation** of the molecule. This is the enthalpy change when 1 mole of an unsaturated compound reacts completely with hydrogen.

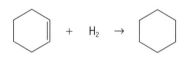

The enthalpy of hydrogenation of cyclohexene is $-120\,\text{kJ mol}^{-1}$.

If benzene was simply cyclohexatriene (an alkene containing three carbon–carbon double bonds) we might reasonably expect the enthalpy of hydrogenation of benzene to be $3 \times -120\,\text{kJ mol}^{-1} = -360\,\text{kJ mol}^{-1}$, since there are three carbon–carbon double bonds in this structure.

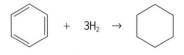

The enthalpy of hydrogenation of benzene is actually $-208\,\text{kJ mol}^{-1}$.

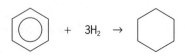

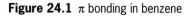

These values indicate that benzene must be lower in energy than cyclohexatriene by $152 \, \text{kJ} \, \text{mol}^{-1}$ (Figure 24.2).

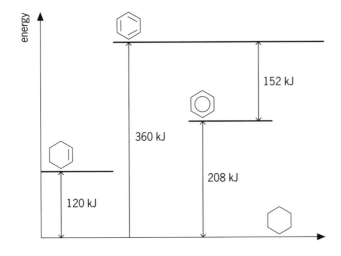

Figure 24.2 Energy diagram for the hydrogenation of benzene

The stability resulting from the delocalisation of electrons prevents benzene from acting as a typical alkene. In order for the addition of bromine across one of the carbon–carbon double bonds to occur, a reaction that occurs readily in alkenes such as ethene, the ring of delocalised electrons would need to be broken and this would require far too much activation energy.

Benzene does undergo addition, but such reactions require vigorous conditions and usually result in a completely saturated product. For example, hydrogenation of benzene requires heating at 150 °C in the presence of a catalyst and the product is cyclohexane.

Electrophilic substitution reactions of benzene

Substitution reactions allow benzene to maintain its π electron cloud structure and the stability associated with it. These delocalised electrons attract positive, electron-deficient species that are known as **electrophiles**. Thus, the most common reactions of benzene, and other aromatic compounds, are **electrophilic substitution** reactions.

Nitration of benzene occurs at 50 °C using a mixture of concentrated sulphuric and nitric acids.

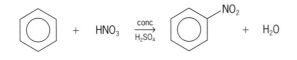

The concentrated sulphuric acid and concentrated nitric acid react to form the nitronium ion, NO_2^+, the electrophile involved in the substitution reaction.

$$HNO_3 + 2H_2SO_4 \rightleftharpoons NO_2^+ + H_3O^+ + 2HSO_4^-$$

The nitronium ion is attracted to the electron-rich ring and a bond is formed between the nitrogen and a carbon in the ring.

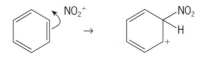

The intermediate cation is unstable since the delocalisation of π electrons in the ring is disrupted. Rapid loss of a proton restores the delocalised ring of electrons.

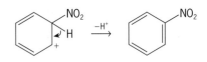

Temperature control of this reaction is important to prevent di- and tri-substitution taking place.

Aromatic nitrates provide an important route to the synthesis of aromatic amines. Reduction of the nitro group is achieved by catalytic hydrogenation, using Ni/H_2 for example, or by using combinations of metal/acid, such as Sn/HCl.

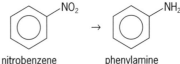

nitrobenzene   phenylamine

The aromatic amines are themselves important intermediates in the manufacture of dyes and other useful products.

Sulphonation of benzene occurs by a similar electrophilic substitution mechanism. Benzene is reacted with fuming sulphuric acid and the attacking species is the sulphur trioxide molecule. The result is benzenesulphonic acid.

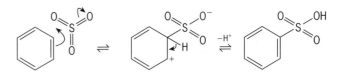

In contrast to nitration, further sulphonation of benzene to obtain benzene-1,3-disulphonic acid is difficult and requires vigorous reaction conditions.

Aryl compounds that contain a sulphonic acid group will often dissolve in water. Solubility in water has significant advantages for chemicals such as dyes.

Sulphonation is also important in the production of sodium salts of long-chain alkylsulphonic acids, which are used as anionic detergents.

Alkylation and acylation of benzene are achieved by **Friedel–Crafts** reactions. These involve the use of a haloalkane (for alkylation) or an alkanoyl (acyl) chloride (for acylation) and a Lewis acid catalyst such as aluminium chloride.

In **alkylation**, the haloalkane forms a coordinate compound with the aluminium chloride; this leads to the generation of an electrophilic species (R^+ in the following equation).

$$R-Cl + AlCl_3 \rightarrow \begin{array}{c} \overset{\delta+}{R} \cdots \cdots Cl \cdots \cdots \overset{\delta-}{AlCl_3} \\ \text{or} \\ R^+ \ AlCl_4^- \end{array}$$

The formation of ions depends on the nature of R. Primary haloalkanes do not produce ions; however, secondary and tertiary haloalkanes show an increasing tendency to form carbocations. Alkylation is similar to nitration in that it involves electrophilic attack on the benzene ring, followed by the loss of a proton.

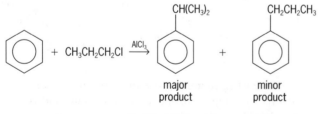

There are two main problems associated with Friedel–Crafts alkylation, which are multiple or polysubstitution and the rearrangement of the carbocation.

Nitration and sulphonation reactions can easily be stopped at the monosubstitution stage since the added groups are electron withdrawing and deactivate the benzene ring. However, an alkyl group is electron releasing so the benzene ring is activated and becomes an even more attractive target for an electrophile. Polysubstitution can, to some extent, be controlled by using an excess of benzene; however, a mixture of products requiring separation is generally obtained from this reaction.

When haloalkanes are used in Friedel–Crafts alkylation reactions the major product is invariably that which results from the most stable carbocation. For example alkylation using 1-chloropropane results in a mixture of two products, the major product being 1-methylethylbenzene.

The reason for this is the rearrangement of the initially formed primary carbocation into a more stable secondary carbocation.

Acylation of benzene using acyl halides results in the formation of ketones.

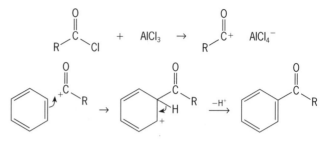

Unlike alkylation, polysubstitution is not a problem in acylation since the acyl group is electron withdrawing and deactivates the ring against subsequent electrophilic attack. However, for the reaction to go to completion the mole ratio of the Lewis acid to reactants must be greater than 1 since the ketone product forms a complex with the Lewis acid.

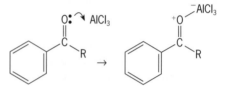

The preparation of alkylbenzenes is often better achieved not by direct alkylation of benzene but by acylation followed by reduction of C=O using zinc amalgam and concentrated hydrochloric acid (**Clemmensen reduction**).

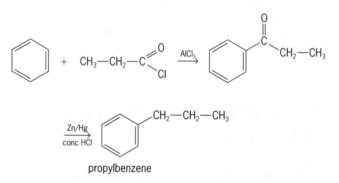

propylbenzene

Chlorobenzene is formed by the reaction of chlorine on benzene in the presence of a Lewis acid catalyst such as aluminium chloride. In this reaction the catalyst is sometimes referred to as a **halogen carrier**.

$$Cl-Cl + AlCl_3 \rightarrow \overset{\delta+}{Cl} \cdots \cdots Cl \cdots \cdots \overset{\delta-}{AlCl_3}$$

major   minor
product   product

1-methylethylbenzene   propylbenzene

$$CH_3-CH-\overset{+}{C}H_2 \rightarrow CH_3-\overset{+}{C}H-CH_3$$
$$\overset{|}{H}$$

primary carbocation   secondary carbocation,
                        more stable

1 The following 'prismatic' structure, in which all carbon–carbon bonds are of equal length, was also considered as a possible structure for benzene.

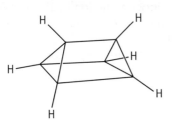

How many different di-substituted isomers, in which the groups are on adjacent carbon atoms, are possible for this structure?

A 1
B 2
C 3
D 4

2 Which of the following is correct about the composition of the delocalised electron cloud above and below the benzene ring?

	Number of electrons	Number of p orbitals
A	6	6
B	6	12
C	12	6
D	12	12

3 Which of the following is the attacking species in the nitration of benzene?

A NO_2^+
B NO_2^-
C NO_2
D NO_3^-

4 How many different isomers of dimethylbenzene are possible?

A 3
B 4
C 5
D 6

5 Which of the following compounds would **not** decolorise bromine water?

A

B   CH=CH₂

C

D

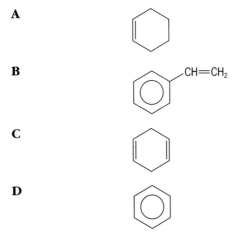

6 At the same time as the Kekulé structure for benzene was considered several other possible structures were proposed including 'Dewar' benzene.

a Explain why the Dewar structure for benzene was unsatisfactory.
b Dewar benzene has been made and is a colourless liquid. Explain whether the chemical properties of Dewar benzene resemble those of benzene or an alkene.

7 Friedel–Crafts alkylation reactions are electrophilic substitution reactions which place an alkyl group on the benzene ring. A haloalkane will react with benzene in the presence of a Lewis acid catalyst, such as aluminium chloride, to produce an alkyl benzene.

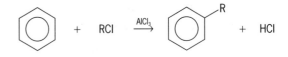

a What is a Lewis acid?
b Explain how chloroethane is activated by coordination with aluminium chloride to generate an electrophilic species.
c Taking the electrophile involved in the reaction to be $CH_3CH_2^+$, draw a mechanism for the ethylation of benzene.
d Suggest one possible drawback to making ethylbenzene by this reaction.

8 Complete hydrogenation of benzene produces cyclohexane.

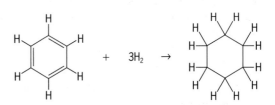

a Use the average bond dissociation energies in the following table to calculate the molar enthalpy of hydrogenation of benzene.

Bond	Average bond dissociation energy/kJ mol⁻¹
C—C	347
C=C	612
C—H	413
H—H	436

b The experimental value for the standard molar enthalpy of hydrogenation of benzene is $-208\,kJ\,mol^{-1}$. Comment on the difference between this value and the value calculated in part **a**.

9 The following equations represent steps in the generation of the nitronium ion for nitration of benzene.

$$H_2SO_4 + HNO_3 \rightleftharpoons H_2NO_3^+ + HSO_4^- \qquad \text{(equation 1)}$$
$$H_2NO_3^+ \rightleftharpoons H_2O + NO_2^+ \qquad \text{(equation 2)}$$
$$H_2SO_4 + H_2O \rightleftharpoons H_3O^+ + HSO_4^- \qquad \text{(equation 3)}$$

a The rate-determining step in this reaction involves the breakdown of the protonated nitric acid. Which of the above equations represents this step?

b Combine the above equations to give an overall equation for the generation of the nitronium ion.

c Copy and complete the following by showing the movement of electrons during the breakdown of the protonated nitric acid molecule.

d The following diagram shows one possible resonance hybrid of the intermediate cation formed during the nitration of benzene.

Show how the positive charge on the intermediate may be delocalised onto two other carbon atoms in the benzene ring by moving pairs of electrons.

10 It might be expected that any cyclic compound containing alternate carbon–carbon single and double bonds would be aromatic (have similar properties to benzene). However, it has been shown that compounds are aromatic only if the number of delocalised electrons is equal to $4n + 2$ where n is a whole number.

When $n = 1$ the number of delocalised electrons is $4 \times 1 + 2 = 6$ as in benzene.

a Use this rule to predict whether each of the following is aromatic in its properties.

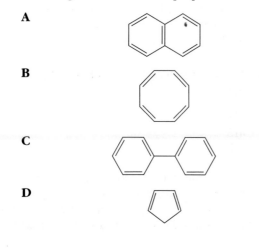

A

B

C

D

b Some cyclic compounds containing atoms other than carbon also show aromatic properties. Where an atom has a lone pair of electrons, such as nitrogen, the lone pair can take the place of the two electrons in a carbon–carbon double bond.

i Show that the following compound contains six delocalised electrons.

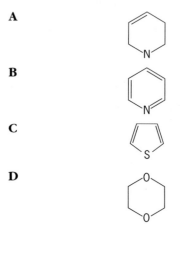

ii Predict whether each of the following is aromatic or not.

A

B

C

D

1 B

2 A

3 A

4 A

5 D

6 a The objections to the Dewar structure are similar to those to the structure first proposed by Kekulé. The structure shows single and double bonds but all of the carbon–carbon bonds were known to be the same length. Also, in Dewar benzene, three different di-substituted compounds in which the substituents are on adjacent carbon atoms are possible but only one form was known.

b In the Dewar structure it is not possible to gain stability by the delocalisation of π electrons in the carbon–carbon double bonds. Dewar benzene acts like a typical alkene, readily undergoing addition reactions.

7 a An electron acceptor.

b The carbon–chlorine bond in chloroethane is polarised, leaving the chlorine atom $\delta-$. Coordination of chloroethane with aluminium chloride leads to the formation of $CH_3CH_2^{\delta+} \ldots Cl \ldots^{\delta-}AlCl_3$.

c

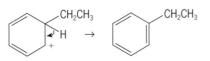

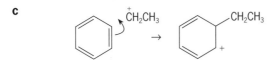

d Di- and tri-substitution may also take place, leading to a mixture of products.

8 a Bonds broken:

$3 \times C=C \quad 3 \times 612 = 1836$
$3 \times H-H \quad 3 \times 436 = 1308$

total energy needed $= 3144 \, kJ \, mol^{-1}$

Bonds formed:

$3 \times C-C \quad 3 \times 347 \quad = 1041$
$6 \times C-H \quad 6 \times 413 \quad = 2478$

total energy released $= 3519 \, kJ \, mol^{-1}$

Molar enthalpy of hydrogenation of benzene
$= $ energy needed $-$ energy released
$= 3144 - 3519$
$= -375 \, kJ \, mol^{-1}$

b The experimental value for the molar enthalpy of hydrogenation of benzene is $167 \, kJ \, mol^{-1}$ less negative than that calculated. This shows that benzene must be at a lower energy level than might be expected if it is considered as a triene. This is proof of the stability of the delocalised electrons in benzene.

9 a Equation 2

b $2H_2SO_4 + HNO_3 \rightleftharpoons NO_2^+ + H_3O^+ + 2HSO_4^-$

c

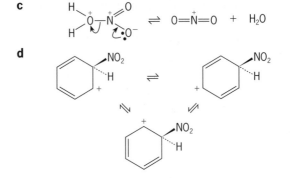

d

10 a A and C are aromatic in nature; B and D are not.

b i In addition to the four electrons from the carbon–carbon double bonds, the lone pair of electrons on the nitrogen atom is also available for delocalisation in the ring, giving a total of six electrons.

ii B and C are aromatic in nature; A and D are not. In B the lone pair of electrons in the N atom is not involved in the π bond, thus there are six electrons in the ring and not eight.

25 Thermodynamics

Standard enthalpy change

The enthalpy change, ΔH, in a system is defined as the amount of heat given out or taken in when the system undergoes a physical or chemical change in which it remains at constant pressure (this was introduced in topic 15).

Enthalpy changes are usually given in a standard state, at a pressure of 1 bar (100 kPa) and a temperature of 298 K. Standard enthalpy changes are denoted by the symbol $\Delta H_{298}^{\ominus}$. The following are examples of standard enthalpy changes that will be used later in discussing the Born–Haber cycle.

- Electron affinity, $\Delta H_{ea}^{\ominus}$, the standard enthalpy change that occurs when 1 mole of an atom, ion or molecule gains an electron to form an anion in the gas phase.

$$Cl(g) + e^- \rightarrow Cl^-(g); \qquad \Delta H_{ea}^{\ominus} = -349\,kJ\,mol^{-1}$$

- Enthalpy of atomisation, $\Delta H_{at}^{\ominus}$, the standard enthalpy change that occurs when 1 mole of gaseous atoms is formed from an element under standard conditions.

$$Na(s) \rightarrow Na(g); \qquad \Delta H_{at}^{\ominus} = +107\,kJ\,mol^{-1}$$

- Enthalpy of hydration, $\Delta H_{hyd}^{\ominus}$, the standard enthalpy change that occurs when 1 mole of a gaseous anion or cation is hydrated.

$$X^+(g) \xrightarrow{\text{water}} X^+(aq)$$

or

$$X^-(g) \xrightarrow{\text{water}} X^-(aq)$$

In reality it is, of course, impossible to measure the enthalpy of hydration of a single ion in isolation since its co-ion will always be present. For this reason what is determined is, in fact, the sum of the hydration enthalpies of the two ions.

$$H^+(g) + Cl^-(g) \xrightarrow{\text{water}} H^+(aq) + Cl^-(aq);$$
$$\Delta H_{hyd}^{\ominus}(H^+) + \Delta H_{hyd}^{\ominus}(Cl^-) = -1454\,kJ\,mol^{-1}$$

To obtain a value for a single species of ion it is necessary to use a 'defined value' for $H^+(g) \rightarrow H^+(aq)$. This is defined to be $-1090\,kJ\,mol^{-1}$. Using this the value for $Cl^-(g) \rightarrow Cl^-(aq)$ can be calculated.

$$Cl^-(g) \xrightarrow{\text{water}} Cl^-(aq);$$
$$\Delta H_{hyd}^{\ominus}(Cl^-) = -1454 - (-1090) = -364\,kJ\,mol^{-1}$$

- Enthalpy of solution, $\Delta H_{sol}^{\ominus}$, the enthalpy change when 1 mole of an ionic solid dissolves in an excess of water. There needs to be sufficient water to ensure that the ions are separated and do not interact with each other.

$$NaCl(s) \xrightarrow{\text{water}} Na^+(aq) + Cl^-(aq);$$
$$\Delta H_{sol}^{\ominus} = 4\,kJ\,mol^{-1}$$

- Ionisation energy, $\Delta H_i^{\ominus}$, the standard enthalpy change for the removal of an electron from 1 mole of an atom, ion or molecule in the gas phase to form a cation in the gas phase.

$$Na(g) \rightarrow Na^+(g) + e^-; \qquad \Delta H_i^{\ominus} = +496\,kJ\,mol^{-1}$$

- Lattice enthalpy, $\Delta H_l^{\ominus}$, the enthalpy change when 1 mole of a solid ionic lattice is formed from separate ions in the gas phase.

$$Na^+(g) + Cl^-(g) \rightarrow NaCl(s); \qquad \Delta H_l^{\ominus} = -787\,kJ\,mol^{-1}$$

Notice that all lattice enthalpies are negative since energy is released when oppositely charged ions form an ionic lattice.

Born–Haber cycle

The enthalpy change of formation, $\Delta H_f^{\ominus}$, of a solid ionic compound can be broken up into a sequence of discernible steps. These can be represented as a cycle, known as a Born–Haber cycle (Figure 25.1).

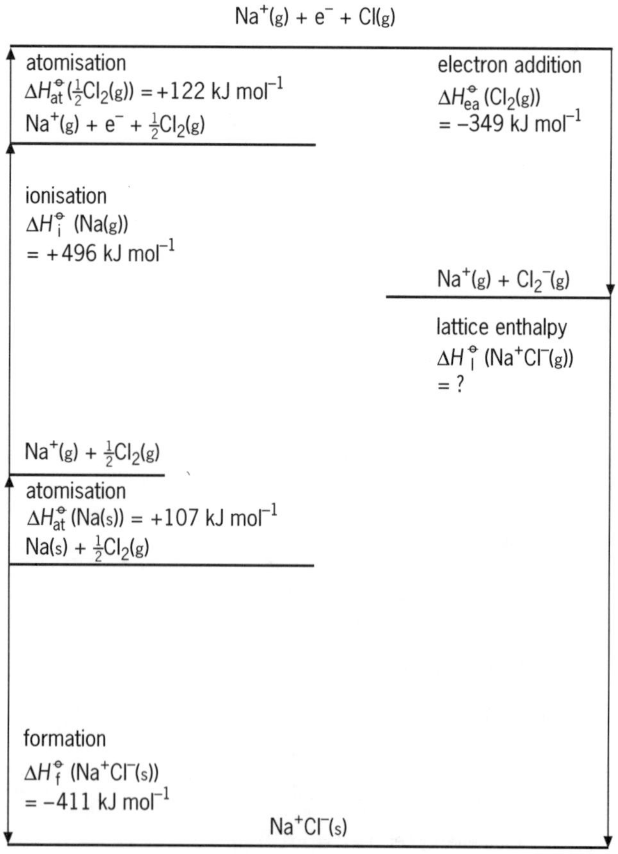

Figure 25.1 Born–Haber cycle for NaCl

The sum of the energy changes round the cycle is zero, that is $\Sigma_{cycle}\Delta H_{step}^{\ominus} = 0$.

$$-\Delta H_f^{\ominus}(Na^+Cl^-(s)) + \Delta H_{at}^{\ominus}(Na(s)) + \Delta H_i^{\ominus}(Na(g))$$
$$+ \Delta H_{at}^{\ominus}(\tfrac{1}{2}Cl_2(g)) + \Delta H_{ea}^{\ominus}(Cl) + \Delta H_l^{\ominus}(Na^+Cl^-(s)) = 0$$

Thus the value of any unknown enthalpy change in the cycle can be found provided all of the others are known. Lattice energies cannot be determined directly, but values can be obtained indirectly by using such a cycle as the one shown above.

$$\Delta H_{l}^{\ominus}(Na^{+}Cl^{-}(s)) = +\Delta H_{f}^{\ominus}(Na^{+}Cl^{-}(s)) - \Delta H_{at}^{\ominus}(Na(s))$$
$$- \Delta H_{i}^{\ominus}(Na(g)) - \Delta H_{at}^{\ominus}(\tfrac{1}{2}Cl_{2}(g)) - \Delta H_{ea}^{\ominus}(Cl)$$

$$= (-411) - (+107) - (+496) - (+122)$$
$$- (-349)$$

$$= -787\,kJ\,mol^{-1}$$

Entropy change, ΔS

Many of the changes with which we are familiar occur spontaneously. When a cup of hot tea is left on a table it cools spontaneously; energy is lost to the surroundings.

Many chemical reactions are **spontaneous**. Once activation energy is reached there is a loss of energy to the surroundings. Exothermic reactions are spontaneous. The reactants are at a higher enthalpy than the products. Energy is lost to the surroundings thus the change in enthalpy, ΔH, is less than 0.

$$CH_{4}(g) + 2O_{2}(g) \rightarrow CO_{2}(g) + 2H_{2}O(g); \qquad \Delta H < 0$$

The combustion of methane releases a large amount of energy. However, the release of energy is not the only factor that determines whether a reaction is spontaneous. Endothermic reactions may also be spontaneous.

$$N_{2}(g) + O_{2} \rightarrow 2NO(g); \qquad \Delta H > 0$$

Clearly there is another factor to be considered and that factor is **entropy**, S.

Entropy is most easily understood as the degree of disorder of a system and it increases as disorder increases. Unlike enthalpy, it is possible to assign an absolute entropy value to every substance under standard conditions. Table 25.1 gives some examples.

Table 25.1 Values of standard entropy

Substance	Standard entropy at 298K, $S_{298}^{\ominus}/J\,mol^{-1}K^{-1}$
C(diamond)	2.4
C(graphite)	5.7
SiO$_2$(s)	41.8
NaCl(s)	72.1
H$_2$O(l)	69.9
H$_2$O(g)	188.7
CH$_3$OH(l)	127.0
CH$_3$OH(g)	239.7
O$_2$(g)	205
CO(g)	197.6
CO$_2$(g)	213.6

Two trends are evident from the standard entropy values given in Table 25.1.

- Entropy increases with the complexity of the molecule. Simple molecules have lower entropies than complicated ones.
- Entropy increases on going from solid to liquid to gas. Gases have the highest entropies.

Figure 25.2 shows how the entropy of a substance changes with increasing temperature.

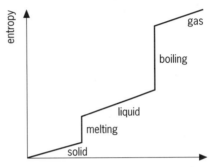

Figure 25.2 Entropy changes with temperature

From the graph it is clear that:

- entropy increases with temperature
- a change of state results in a rapid increase in entropy
- entropy increases more during boiling than during melting.

The **second law of thermodynamics** states that all spontaneous chemical reactions and physical changes involve an overall increase in entropy. It is this entropy change that makes it possible for endothermic reactions to be spontaneous.

Changes in the entropy of a system, as with the enthalpy of the system, depend only on the initial and final states of the system and not on the route by which the change is achieved. When a system goes through a cycle the entropy change is zero, $\Sigma_{cycle}\Delta S_{steps} = 0$.

Entropy and chemical change

Standard entropy changes for systems undergoing chemical change are calculated in the same way as standard enthalpy changes. One significant difference, however, is that absolute entropy values can be assigned to every substance under standard conditions.

$$\Delta S = \Sigma S_{products}^{\ominus} - \Sigma S_{reactants}^{\ominus}$$

The entropy change when graphite is burnt in a good supply of air is given by:

$$C(s) + O_{2}(g) \rightarrow CO_{2}(g)$$

$$\Sigma S_{reactants}^{\ominus} = 5.7 + 205$$
$$= 210.7\,J\,mol^{-1}K^{-1}$$

$$\Sigma S_{products}^{\ominus} = 213.6\,J\,mol^{-1}K^{-1}$$

$$\Delta S = \sum S^{\ominus}_{products} - \sum S^{\ominus}_{reactants}$$
$$= 213.6 - 210.7$$
$$= 2.9\,J\,mol^{-1}K^{-1}$$

Similarly, the entropy change for the thermal decomposition of sodium nitrate can be calculated using standard entropy values. The standard entropy values at 298 K ($S^{\ominus}_{298}\,J\,mol^{-1}K^{-1}$) are $NaNO_3(s) = 116.5$, $NaNO_2(s) = 103.8$, $O_2(g) = 205$.

$$2NaNO_3(s) \rightarrow 2NaNO_2(s) + O_2(g)$$

$$\sum S^{\ominus}_{reactants} = 2 \times 116.5$$
$$= 233.0\,J\,mol^{-1}K^{-1}$$
$$\sum S^{\ominus}_{products} = 2 \times 103.8 + 205$$
$$= 412.6\,J\,mol^{-1}K^{-1}$$

$$\Delta S = \sum S^{\ominus}_{products} - \sum S^{\ominus}_{reactants}$$
$$= 412.6 - 233.0$$
$$= 179.6\,J\,mol^{-1}K^{-1}$$

This is a large change in entropy because 1 mole of gas is formed.

Entropy and physical change

It is convenient to consider the changes in entropy of the universe in terms of the particular part that we are interested in, the **system**, and the remainder, the **surroundings**.

$$\Delta S_{universe} = \Delta S_{surroundings} + \Delta S_{system}$$

When a change to a system involves the intake or release of energy the entropy of the surroundings is also changed. When heat energy is released to the surroundings, its molecules move more quickly and become more disordered; the entropy of the surroundings increases. The converse occurs when heat is absorbed from the surroundings by the system.

Entropy change in surroundings
$$= \frac{\text{heat released or absorbed from the system}}{\text{temperature of the system in K}}$$

$$\Delta S = \frac{\Delta H}{T}$$

Since the amount of heat lost or gained by the surroundings will be equal but opposite to the heat taken in or lost by the system:

$$\Delta S_{surroundings} = +\frac{\Delta H_{surroundings}}{T_{surroundings}}$$

$$= -\frac{\Delta H_{system}}{T_{system}}$$

Many physical changes involve the gain or loss of heat energy. The melting of a solid and the boiling of a liquid are two common examples.

Standard Gibbs free energy change, $\Delta G^{\ominus}$

All spontaneous processes involve an increase in the overall entropy of the universe, or at least result in no net loss. It is not possible to measure the net gain or loss in entropy of the universe during a change; however, it is possible to write an expression involving only the properties of the system.

$$\Delta S_{universe} = \Delta S_{surroundings} + \Delta S_{system}$$

but

$$\Delta S_{surroundings} = +\frac{\Delta H_{surroundings}}{T_{surroundings}}$$

$$= -\frac{\Delta H_{system}}{T_{system}}$$

$$\therefore \quad \Delta S_{universe} = -\frac{\Delta H_{system}}{T_{system}} + \Delta S_{system}$$

Multiplying by $-T$:

$$-T\Delta S_{universe} = \Delta H_{system} - T\Delta S_{system}$$

The term $-T\Delta S_{universe}$ is known as the **Gibbs free energy change** and is given the symbol ΔG. The equation then becomes:

$$\Delta G = \Delta H_{system} - T\Delta S_{system}$$

Since, for spontaneous change, $\Delta S_{universe}$ must be greater than or equal to zero it follows that ΔG must be negative or zero for a spontaneous change to occur.

Notice that being spontaneous at temperature T does not necessarily mean that a reaction will occur. It does, however, mean that the reaction can occur provided the activation energy is not too high. However, if a reaction is not spontaneous at temperature T it means that it will not go, no matter what is done to it.

Example 1
Calculate the standard free energy change for the following reaction at 298 K:

$$Mg(s) + \tfrac{1}{2}O_2(g) \rightarrow MgO(s); \qquad \Delta H^{\ominus} = -601.7\,kJ\,mol^{-1}$$

The standard entropy values at 298 K ($S^{\ominus}_{298}\,J\,mol^{-1}K^{-1}$) required are: $Mg(s) = 32.7$, $MgO(s) = 26.9$, $O_2(g) = 205$.

$$\Delta H_{system} = -601.7\,kJ\,mol^{-1}$$
$$\Delta S_{system} = 26.9 - (32.7 + 102.5)$$
$$= -108.3\,J\,mol^{-1}K^{-1}$$
$$= -0.108\,kJ\,mol^{-1}K^{-1}$$
$$T = 298\,K$$
$$\Delta G = \Delta H_{system} - T\Delta S_{system}$$
$$= -601.7 - (298 \times -0.108)$$
$$= -569.5\,kJ\,mol^{-1}$$

ΔG is less than zero, therefore the reaction is spontaneous

ΔG and metal extraction

When some metals are heated in a stream of carbon monoxide/dioxide they act as reducing agents and are themselves oxidised.

$$2Mg(s) + CO_2(g) \rightarrow 2MgO(s) + C(s)$$

However, in the case of other metals, the metal oxide is reduced by carbon and thus the metal oxide or ion acts as an oxidising agent.

$$ZnO(s) + C(s) \rightarrow Zn(s) + CO(g)$$

These reactions are chemical opposites. In extracting metals from their oxides it is necessary to know whether a particular reducing agent will reduce a metal oxide or not. This can be predicted by considering the free energy of formation of an oxide as a measure of its stability.

In order for a reaction to occur spontaneously at a given temperature, ΔG must be less than zero. Since

$$\Delta G^{\ominus} = \Sigma \Delta G_f^{\ominus}(\text{products}) - \Sigma \Delta G_f^{\ominus}(\text{reactants})$$

the reaction will occur spontaneously if:

$$\Delta G_f^{\ominus}(\text{products}) < \Delta G_f^{\ominus}(\text{reactants})$$

In both of the equations above one of the products is an element and since the standard molar free energy of formation of an element is zero the inequality becomes:

$$\Delta G_f^{\ominus}(\text{product oxide}) < \Delta G_f^{\ominus}(\text{reactant oxide})$$

The more stable oxide, the one with the more negative $\Delta G_f^{\ominus}$ value, must therefore be the product and not the reactant. $\Delta G_f^{\ominus}$ values are given in Table 25.2.

Table 25.2 Values of $\Delta G_f^{\ominus}$ of some oxides

Oxide	$\Delta G_f^{\ominus}$/kJ mol^{-1}
CO	−137.2
CO_2	−394.4
MgO	−569.4
ZnO	−318.3

In the case of magnesium, magnesium oxide is more stable than carbon dioxide so we would expect the reaction to be spontaneous in the direction

$$2Mg(s) + CO_2(g) \rightarrow 2MgO(s) + C(s)$$

However, in the case of zinc the reaction is not spontaneous in the direction

$$ZnO(s) + C(s) \rightarrow Zn(s) + CO(g)$$

since $\Delta G_f^{\ominus}(CO)$ is not less than $\Delta G_f^{\ominus}(ZnO)$. In fact the reaction will be spontaneous in the opposite direction. Does this mean that zinc oxide cannot be reduced by carbon?

It means that zinc oxide cannot be reduced by carbon at standard temperature, 298 K. However, free energy values vary with temperature. ΔG_f values for metal oxides increase with increasing temperature whereas the ΔG_f value for carbon monoxide decreases. This can be shown conveniently on an **Ellingham diagram** (Figure 25.3).

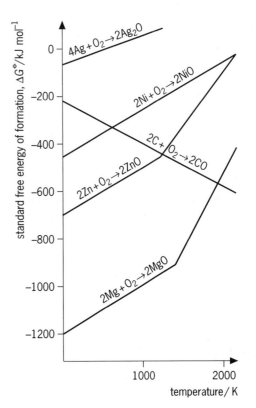

Figure 25.3 An Ellingham diagram

In these diagrams, the lower the line the more stable the oxide. For a reaction to be spontaneous the product oxide must be more stable than the reactant oxide.

Zinc oxide is more stable than carbon monoxide up to a temperature of about 1200 K so heating zinc oxide with carbon below this temperature will not produce zinc. However, beyond this temperature, zinc oxide becomes less stable than carbon monoxide, therefore it has a less negative ΔG_f value. Thus if zinc oxide is heated with carbon at a temperature above 1200 K it will be reduced to zinc.

1 Which of the following equations involves a standard enthalpy change represented by the symbol $\Delta H_{ea}^{\ominus}$?

A $Br_{(l)} + e^- \rightarrow Br^-_{(l)}$

B $Na_{(g)} \rightarrow Na^+_{(g)} + e^-$

C $I_{(g)} + e^- \rightarrow I^-_{(g)}$

D $Mg^+_{(aq)} \rightarrow Mg^{2+}_{(aq)} + e^-$

2 Which of the following correctly shows the order of increasing entropy in the solid, liquid and gas phases?

A gas > liquid > solid

B solid < liquid > gas

C solid > liquid > gas

D gas < liquid < solid

3 Graphite burns in a restricted supply of air to form carbon monoxide.

$$C_{(s)} + \tfrac{1}{2}O_{2(g)} \rightarrow CO_{(g)}$$

Using the values given in the text, the entropy change, in $J\,mol^{-1}K^{-1}$ for this reaction at 298 K is:

A -89.4

B -13.1

C 13.1

D 89.4

4 For a reaction to occur spontaneously:

A $\Delta S_{overall} \leq 0$ and $\Delta G \geq 0$

B $\Delta S_{overall} \geq 0$ and $\Delta G \geq 0$

C $\Delta S_{overall} \leq 0$ and $\Delta G \leq 0$

D $\Delta S_{overall} \geq 0$ and $\Delta G \leq 0$

5 The following equations represent two possible reactions:

$$NiO_{(s)} + C_{(s)} \rightarrow Ni_{(s)} + CO_{(g)} \qquad \text{(reaction X)}$$
$$Ni_{(s)} + CO_{(g)} \rightarrow NiO_{(s)} + C_{(s)} \qquad \text{(reaction Y)}$$

Using the Ellingham diagram in the text (Figure 25.3), state which of the following statements is correct.

A Only reaction X can be spontaneous irrespective of temperature

B Both reaction X and reaction Y could be spontaneous depending on temperature

C Only reaction Y can be spontaneous irrespective of temperature

D Neither reaction X nor reaction Y is spontaneous at any temperature

6 Explain the following:

a Lattice enthalpies are always negative.

b ΔG must be negative or zero for a spontaneous chemical reaction to occur.

c Endothermic reactions may be spontaneous.

d The entropy of water is less than the entropy of steam.

7 a Using the values given below, and those in the text, calculate the entropy changes for the following reactions.

Standard entropy at 298 K ($S_{298}^{\ominus}$ $J\,mol^{-1}K^{-1}$):
$CH_{4(g)} = 186.2$; $CaCO_{3(s)} = 92.9$; $CaO_{(s)} = 39.7$; $Ca(OH)_{2(s)} = 83.4$

i $CaCO_{3(s)} \rightarrow CaO_{(s)} + CO_{2(g)}$

ii $CaO_{(s)} + H_2O_{(l)} \rightarrow Ca(OH)_{2(s)}$

iii $Ca(OH)_{2(s)} + CO_{2(g)} \rightarrow CaCO_{3(s)} + H_2O_{(l)}$

b Comment on the values of ΔS for these reactions.

8 Determine whether the following reactions are spontaneous under standard conditions. If a reaction is not spontaneous, state what would need to be done to make it so.

a $CH_{4(g)} + 2O_{2(g)} \rightarrow 2H_2O_{(g)} + CO_{2(g)}$
($\Delta H_{298}^{\ominus} = -890.3\,kJ\,mol^{-1}$; $\Delta S_{298}^{\ominus} = -5.0\,J\,mol^{-1}K^{-1}$)

b $NaHCO_{3(s)} \rightarrow \tfrac{1}{2}Na_2CO_{3(s)} + \tfrac{1}{2}H_2O_{(g)} + \tfrac{1}{2}CO_{2(g)}$
($\Delta H_{298}^{\ominus} = +130\,kJ\,mol^{-1}$; $\Delta S_{298}^{\ominus} = +333.9\,J\,mol^{-1}K^{-1}$)

9 Use the Ellingham diagram given in the text (Figure 25.3) to answer the following questions.

a Why is it that silver oxide decomposes when heated whereas magnesium oxide does not?

b Why doesn't the following reaction occur at 1000 K?

$$2MgO_{(s)} + 2C_{(s)} \rightarrow 2Mg_{(s)} + 2CO_{(g)}$$

c At what temperature would the above reaction occur? Explain why.

10 a Define enthalpy of atomisation, $\Delta H_{at}^{\ominus}$.

b Use the following enthalpy values to construct a Born–Haber cycle and use it to find the standard lattice enthalpy for potassium bromide, $\Delta H_1^{\ominus}(K^+Br^-_{(s)})$.

$\Delta H_f^{\ominus}(K^+Br^-_{(s)}) = -393.8\,kJ\,mol^{-1}$
$\Delta H_i^{\ominus}(K_{(g)}) = +419\,kJ\,mol^{-1}$
$\Delta H_{ea}^{\ominus}(Br) = -324.6\,kJ\,mol^{-1}$
$\Delta H_{at}^{\ominus}(K_{(s)}) = +89.2\,kJ\,mol^{-1}$
$\Delta H_{at}^{\ominus}(\tfrac{1}{2}Br_{2(l)}) = +111.9\,kJ\,mol^{-1}$

1 C

2 A

3 D

$$\Delta S = \Sigma S^{\ominus}_{products} - \Sigma S^{\ominus}_{reactants}$$
$$= (197.6) - (5.7 + 102.5)$$
$$= 89.4 \, J\,mol^{-1}K^{-1}$$

4 D

5 B

6 a Energy is released when opposite charges approach each other.

b ΔG is equal to $-T\Delta S_{universe}$. The second law of thermodynamics states that all spontaneous chemical reactions involve an overall increase in entropy. If $\Delta S_{universe}$ is positive then $-T\Delta S_{universe}$ must be negative.

c The viability of a chemical reaction depends not just on the value of ΔH but also on the value of ΔS.

d Entropy is a measure of the disorder of a system. In steam the molecules are more spaced out and moving more rapidly in random motion, and thus in greater disorder than in water where they are closer together and not moving as quickly.

7 a $\Delta S = \Sigma S^{\ominus}_{products} - \Sigma S^{\ominus}_{reactants}$

i $\Delta S = (39.7 + 213.6) - (92.9)$
$= +160.4 \, J\,mol^{-1}K^{-1}$

ii $\Delta S = (83.4) - (39.7 + 69.9)$
$= -26.2 \, J\,mol^{-1}K^{-1}$

iii $\Delta S = (92.9 + 69.9) - (83.4 + 213.6)$
$= -134.2 \, J\,mol^{-1}K^{-1}$

b In reaction **i** a solid becomes a solid and a gas; there is a large increase in the entropy of the system.

In reaction **ii** a solid and a liquid become a solid; there is a small drop in the entropy of the system.

In reaction **iii** a solid and a gas become a solid and a liquid; there is a large decrease in the entropy of the system.

8 a $\Delta G^{\ominus}_{298} = -890.3 - (298 \times (-5) \times 10^{-3})$
$= -888.8 \, kJ\,mol^{-1}$

ΔG is negative indicating the reaction is spontaneous under standard conditions.

b $\Delta G^{\ominus}_{298} = +130 - (298 \times (+333.9) \times 10^{-3})$
$= +30.5 \, kJ\,mol^{-1}$

ΔG is positive indicating the reaction is not spontaneous under standard conditions. However, ΔG is not very large; if the temperature was increased the value of $T\Delta S$ would increase and the reaction would become spontaneous.

9 a The free energy change for the formation of silver oxide is negative up to around 450 K. Above this temperature the free energy value becomes greater than zero and the reverse reaction occurs. The free energy change for the formation of magnesium oxide is negative across the whole temperature range shown so the reverse reaction would never become spontaneous within that temperature range.

b For a reaction to occur spontaneously ΔG must be less than zero. The standard free energy of an oxide is a measure of its stability. At 1000 K MgO is more stable than CO since it has a lower free energy. At this temperature ΔG would be greater than zero hence it does not occur.

c At around 1900 K MgO becomes more stable than CO and the reaction becomes feasible since the change in free energy would be less than zero.

10 a The standard enthalpy change that occurs when 1 mole of gaseous atoms is formed from an element under standard conditions.

b

$$K^+(g) + e^- + Br(g)$$

$\Delta H^{\ominus}_{at} (\frac{1}{2}Br_2(l))$
$= +111.9 \, kJ\,mol^{-1}$
$K^+(g) + e^- + \frac{1}{2}Br_2(l)$

$\Delta H^{\ominus}_{ea} (Br)$
$= -324.6 \, kJ\,mol^{-1}$

$\Delta H^{\ominus}_{i} (K(g))$
$= +419 \, kJ\,mol^{-1}$

$K^+(g) + Br^-(g)$

$\Delta H^{\ominus}_{l} (K^+Br^-(s))$
$= ?$

$K(g) + \frac{1}{2}Br_2(l)$

$\Delta H^{\ominus}_{at} (K(s))$
$= +89.2 \, kJ\,mol^{-1}$
$K(s) + \frac{1}{2}Br_2(l)$

$\Delta H^{\ominus}_{f} (K^+Br^-(s))$
$= -393.8 \, kJ\,mol^{-1}$

$$K^+Br^-(s)$$

The sum of the energy changes round the cycle is zero, that is $\Sigma_{cycle}\Delta H^{\ominus}_{step} = 0$

$$\Delta H^{\ominus}_{l}(K^+Br^-(s)) = +\Delta H^{\ominus}_{f}(K^+Br^-(s)) - \Delta H^{\ominus}_{at}(K(s)) - $$
$$\Delta H^{\ominus}_{i}(K(g)) - \Delta H^{\ominus}_{at}(\frac{1}{2}Br_2(g)) - \Delta H^{\ominus}_{ea}(Br)$$

$$= (-393.8) - (+89.2) - $$
$$(+419) - (+111.9) - (-324.6)$$
$$= -689.3 \, kJ\,mol^{-1}$$

26 Acid–base equilibria

Dissociation constants, K_a and K_b

Although the Lewis theory of acids and bases is more comprehensive, the Brønsted–Lowry theory is the one which is used when considering aqueous acid equilibria.

When an acid or base is in aqueous solution it separates either wholly or partially into ions. This process is called **dissociation**.

$$HA(aq) \rightleftharpoons H^+(aq) + A^-(aq)$$

The **acid dissociation constant**, K_a, can be obtained for this reaction.

$$K_a = \frac{[H^+(aq)][A^-(aq)]}{[HA(aq)]}$$

The larger the value of K_a the more the acid dissociates and thus the stronger the acid. Strong acids are virtually fully dissociated in aqueous solution and the value of K_a is large. However, for weak acids the value of K_a may be very small. It is often more convenient to compare the strengths of acids using pK_a values, where

$$pK_a = -\log_{10} K_a$$

Similarly, for bases:

$$BOH \rightleftharpoons B^+ + OH^-$$

$$K_b = \frac{[B^+(aq)][OH^-(aq)]}{[BOH(aq)]}$$

Dissociation constants for some common acids and bases in water at 25°C are shown in Tables 26.1 and 26.2.

Table 26.1 Dissociation constants of common acids

Acid	Formula	K_a/mol dm^{-3}	pK_a
sulphuric acid	H_2SO_4	very large	—
nitric acid	HNO_3	40	−1.6
ethanoic acid	CH_3COOH	1.7×10^{-5}	4.8

Table 26.2 Dissociation constants of common bases

Base	Formula	K_b/mol dm^{-3}	pK_b
diethylamine	$(C_2H_5)_2NH$	9.6×10^{-4}	3.0
methylamine	CH_3NH_2	4.5×10^{-4}	3.3
ammonia	NH_3	1.7×10^{-5}	4.8

Ionic product of water, K_w

Pure water dissociates according to the equation

$$H_2O(l) \rightleftharpoons H^+(aq) + OH^-(aq)$$

The equilibrium constant for this reaction is given by

$$K_c = \frac{[H^+(aq)][OH^-(aq)]}{[H_2O]}$$

Very few water molecules actually dissociate, therefore the concentration of water can be assumed to be constant. Rearranging the equation,

$$K_c[H_2O] = [H^+(aq)][OH^-(aq)]$$
$$= \text{a new constant, } K_w$$

K_w is the **ionic product** of water.

$$K_w = 1 \times 10^{-14} \, mol^2 \, dm^{-6}$$
$$[H^+(aq)][OH^-(aq)] = 1 \times 10^{-14} \, mol^2 \, dm^{-6}$$

In pure water the concentration of H^+ and OH^- are equal.

$$[H^+(aq)] = [OH^-(aq)]$$
$$= 1 \times 10^{-7} \, mol \, dm^{-3}$$

The ionic product of water is linked to acid and base dissociation constants. Consider the dissociation of ethanoic acid:

$$CH_3COOH(aq) + H_2O(l) \rightleftharpoons CH_3COO^-(aq) + H_3O^+(aq)$$

$$K_a = \frac{[H_3O^+(aq)][CH_3COO^-(aq)]}{[CH_3COOH(aq)]}$$

For ethanoate, which is the conjugate base of ethanoic acid:

$$CH_3COO^-(aq) + H_2O(l) \rightleftharpoons CH_3COOH(aq) + OH^-(aq)$$

$$K_b = \frac{[CH_3COOH(aq)][OH^-(aq)]}{[CH_3COO^-(aq)]}$$

Multiplying the expressions for K_a and K_b together:

$$K_a \times K_b = [H_3O^+(aq)][OH^-(aq)]$$
$$= K_w$$

Therefore:

$$K_w = K_a \times K_b$$

Or taking logarithms:

$$pK_w = pK_a + pK_b$$
$$= 14 \quad (\text{at } 25°C)$$

pH

The **pH scale** is a scale of acidity–alkalinity. pH is a measure of the concentration of hydrogen ions, $[H^+(aq)]$.

$$pH = -\log_{10}[H^+(aq)]$$

The pH of pure water at $25\,°C$ is given by

$$
\begin{aligned}
pH &= -\log_{10}[H^+(aq)] \\
&= -\log_{10}10^{-7} \\
&= 7
\end{aligned}
$$

A strong acid such as hydrochloric acid will be completely dissociated when it is dissolved in water. Therefore $[H^+(aq)]$ will be the same as the concentration of the acid.

Example 1

Calculate the pH of hydrochloric acid of concentration $1\,mol\,dm^{-3}$.

$[HCl(aq)] = 1\,mol\,dm^{-3}$, therefore $[H^+(aq)] = 1\,mol\,dm^{-3}$.

$$
\begin{aligned}
pH &= -\log_{10}[H^+(aq)] \\
&= -\log_{10}1 \\
&= 0
\end{aligned}
$$

Notice that the pH of a strong acid of concentration greater than $1\,mol\,dm^{-3}$ will be less than zero.

A weak acid will be only partially dissociated when it dissolves in water. It is, therefore, necessary to use the value of K_a to calculate pH.

Example 2

Calculate the pH of a solution of ethanoic acid of concentration $1\,mol\,dm^{-3}$ ($K_a = 1.7 \times 10^{-5}\,mol\,dm^{-3}$) at equilibrium.

$$K_a = \frac{[H^+(aq)][CH_3COO^-(aq)]}{[CH_3COOH(aq)]}$$

Since the dissociation constant is so small the initial concentration of acid can be used as the equilibrium concentration.

$$1.7 \times 10^{-5} = \frac{[H^+(aq)][CH_3COO^-(aq)]}{1}$$

but

$$[H^+(aq)] = [CH_3COO^-(aq)]$$
$$\therefore [H^+(aq)] = \sqrt{(1.7 \times 10^{-5})} = 4.1 \times 10^{-3}$$
$$pH = -\log[H^+(aq)] = -\log_{10}(4.1 \times 10^{-3}) = 2.39$$

Alkalis have a pH value of more than 7. By analogy to pH,

$$pOH = -\log_{10}[OH^-]$$

It is therefore possible to state the strength of an alkali in terms of pOH; however, it is more meaningful and convenient to use one scale, the pH scale, for measuring the strengths of both acids and alkalis. This is possible because of the relationship between pH and pOH.

From the ionic product of water:

$$K_w = [H^+(aq)][OH^-(aq)]$$

$$\therefore \log_{10}K_w = \log_{10}[H^+(aq)] + \log_{10}[OH^-(aq)]$$
$$\therefore pK_w = pH + pOH$$

At $25\,°C$ K_w is $1 \times 10^{-14}\,mol^2\,dm^{-6}$, therefore

$$
\begin{aligned}
pK_w &= -\log_{10}(1 \times 10^{-14}) \\
&= 14
\end{aligned}
$$

$$pH + pOH = 14$$

To calculate the pH for a base, we simply calculate the pOH and subtract this from 14.

Example 3

Calculate the pH of a solution of sodium hydroxide of concentration $0.005\,mol\,dm^{-3}$.

$$
\begin{aligned}
pOH &= -\log_{10}0.005 \\
&= 2.30 \\
pH &= 14 - pOH \\
&= 14 - 2.3 \\
&= 11.7
\end{aligned}
$$

Buffer solutions

A buffer solution is one that resists changes to its pH when small volumes of acid or base are added to it. Buffer solutions are very important in living things and in experiments using enzymes, for example.

A buffer solution usually consists of a weak acid and a salt of that acid, or a weak base and a salt of that base. Within the solution two processes are taking place:

$$MA \rightleftharpoons M^+ + A^-$$

The salt MA dissociates completely, giving a large pool of A^- ions which can react with any H^+ ions which are added, thus resisting a decrease in pH.

$$HA \rightleftharpoons H^+ + A^-$$

The weak acid HA dissociates to only a small degree, leaving a large pool of HA. The HA can dissociate to provide more H^+ ions which can combine with any OH^- ions added, thus resisting an increase in pH.

When a small volume of acid is added:

$$H^+ + A^- \rightarrow HA$$

When a small volume of base is added:

$$H^+ + OH^- \rightarrow H_2O$$

pH of a buffer solution

For the acid HA:

$$K_a = \frac{[H^+(aq)][A^-(aq)]}{[HA(aq)]}$$

$$[H^+(aq)] = \frac{K_a[HA(aq)]}{[A^-(aq)]}$$

The acid dissociates very little, so the concentration of HA at equilibrium can be assumed to be the same as the concentration at the beginning. The amount of A^- ions that comes from the acid is negligible compared with the amount which comes from the salt. The concentration of the A^- ions at equilibrium can be assumed to come only from the salt.

$$[H^+] = \frac{K_a[acid]}{[salt]}$$

$[H^+(aq)]$ and consequently the pH of the buffer solution depends upon the ratio of [acid]:[salt]. It is possible to produce a buffer of any pH by choosing the correct ratio of acid to salt.

For practical purposes the equation is used in the form:

$$pH = pK_a + \log_{10}\left(\frac{[salt]}{[acid]}\right)$$

Buffer solutions can also be prepared from a weak base and one of its salts. These alkaline buffers have pH values between 7 and 10 and resist changes in pH in a similar way to weak acid/salt buffers.

A solution of ammonia in ammonium chloride is commonly used as an alkaline buffer. In the solution the ammonium chloride is fully dissociated. This ensures a large pool of NH_4^+ ions.

$$NH_4Cl(aq) \rightarrow NH_4^+(aq) + Cl^-(aq)$$

Ammonia is only partially associated.

$$NH_3(aq) + H_2O(l) \rightleftharpoons NH_4^+(aq) + OH^-(aq)$$

If acid is added, H^+ ions are removed from the solution by OH^- ions and the equilibrium in the lower equation will shift to the right. Similarly, if a base is added the equilibrium shifts to the left, removing OH^- ions from the solution.

The equation for preparing an alkaline buffer solution is

$$pH = pK_w - pK_b + \log_{10}\left(\frac{[base]}{[salt]}\right)$$

Example 4

Calculate how much sodium ethanoate must be dissolved in $1\,dm^3$ of ethanoic acid with a concentration of $0.015\,mol\,dm^{-3}$ to produce a buffer solution of pH 4.5 (ethanoic acid $pK_a = 4.75$).

$\log_{10}[\text{sodium ethanoate}] = \log_{10}[\text{ethanoic acid}] + pH - pK_a$

$\log_{10}[\text{ethanoic acid}] = \log_{10}[0.015]$

$\qquad = -1.82$

$\log_{10}[\text{sodium ethanoate}] = -1.82 + 4.5 - 4.75$

$\qquad = -2.07$

$[\text{sodium ethanoate}] = 0.0084\,mol\,dm^{-3}$

Thus 0.0084 mol of sodium ethanoate (0.69 g) must be dissolved in $1\,dm^3$ ethanoic acid to give a pH of 4.5.

1 The acid dissociation constant, K_a, for hydrochloric acid is given by

A $\dfrac{[H^+(aq)][Cl^-(aq)]}{[HCl(aq)]}$

B $\dfrac{[H^+(aq)][HCl(aq)]}{[Cl^-(aq)]}$

C $\dfrac{[HCl(aq)][Cl^-(aq)]}{[H^+(aq)]}$

D $\dfrac{[HCl(aq)]}{[H^+(aq)][Cl^-(aq)]}$

2 The ionic product of water, K_w,
A has no unit
B is measured in $mol\,dm^{-3}$
C is measured in $mol^{-1}dm^3$
D is measured in $mol^2\,dm^{-6}$

3 The pH of sodium hydroxide of concentration $0.0005\,mol\,dm^{-3}$ is
A 10.3
B 10.7
C 11.7
D 13.3

4 Water dissociates according to the equation

$$H_2O(l) \rightleftharpoons H^+(aq) + OH^-(aq); \qquad \Delta H = +56\,kJ\,mol^{-1}$$

At 25 °C water has a pH of 7. Which of the following occurs when water is heated to 30 °C?
A It remains neutral and its pH decreases
B It becomes acidic and its pH decreases
C It remains neutral and its pH increases
D It becomes acidic and its pH increases

5 Which of the following describes the composition of a buffer solution?
A a strong acid and a salt of that acid
B a strong acid and a salt of a weak acid
C a weak acid and a salt of that acid
D a weak acid and a salt of a strong acid

6 **a** Define the terms pH and pOH and derive the relationship between them at 25 °C.
b Explain why accurate pH values cannot be obtained using indicators and state how accurate values can be obtained.
c Explain why pH values for strong acids and bases can be calculated directly from their concentrations but pH values for weak acids and bases cannot.

7 **a** What is a buffer solution?
b Explain, giving appropriate equations, how a solution of ethanoic acid and sodium ethanoate acts as a buffer.

8 **a** Calculate the pH of the following:
i $0.0020\,mol\,dm^{-3}$ hydrochloric acid
ii $0.025\,mol\,dm^{-3}$ sodium hydroxide solution
iii $0.05\,mol\,dm^{-3}$ sulphuric acid.
b The pH of a strong acid, HA, is 3.7. Calculate the concentration of this acid.
c The pH of a strong base, MOH, is 12.9. Calculate the concentration of this base.

9 At 25 °C the dissociation constant (K_a) for ethanoic acid is $1.74 \times 10^{-5}\,mol\,dm^{-3}$. At the same temperature the ionic product of water (K_w) is $1 \times 10^{-14}\,mol^2\,dm^{-6}$.
a Write expressions for K_a and K_w.
b Calculate the basicity constant (K_b) for the ethanoate ion in water at 25 °C.
c Calculate the pH of a solution of ethanoic acid of concentration $0.025\,mol\,dm^{-3}$ at 25 °C.

10 An acid–phosphate buffer solution contains phosphoric acid, H_3PO_4, of concentration $0.15\,mol\,dm^{-3}$ and sodium phosphate, NaH_2PO_4, of concentration $0.25\,mol\,dm^{-3}$. The sodium phosphate is fully ionised into $Na^+(aq)$ and $H_2PO_4^-(aq)$ ions.

$$H_3PO_4(aq) \rightleftharpoons H^+(aq) + H_2PO_4^-(aq);$$
$$K_a = 7.1 \times 10^{-3}\,mol\,dm^{-3}$$

Explain how the pH of the buffer solution can be found and carry out the calculation.

1 A

2 D

3 B

$pH + pOH = 14$
$pH = 14 - pOH$
$pOH = -\log_{10}[OH^-]$
$\qquad = -\log_{10}(5 \times 10^{-4})$
$\qquad = 3.3$
$pH = 14 - 3.3$
$\qquad = 10.7$

4 A

5 C

6 a $pH = -\log_{10}[H^+]$
$pOH = -\log_{10}[OH^-]$
$K_w = [H^+_{(aq)}][OH^-_{(aq)}]$
$\log_{10}K_w = \log_{10}[H^+_{(aq)}] + \log_{10}[OH^-_{(aq)}]$
$pK_w = pH + pOH$
but at 25 °C, $\quad K_w = 1 \times 10^{-14}\,mol^2\,dm^{-6}$
$pK_w = -\log_{10}(1 \times 10^{-14})$
$\qquad = 14$
Therefore $pH + pOH = 14$

b Indicators give a colour change over a range of several pH units. For accurate measurement a pH meter must be used.

c Strong acids/bases are effectively fully ionised so the concentration of hydrogen/hydroxide ions can be calculated from the initial concentration and the formula of the acid/base.
  Weak acids/bases are only partially ionised so the concentration of hydrogen/hydroxide ions will be considerably less than the initial concentration of the acid/base.

7 a A buffer solution is one that can maintain a constant pH when small volumes of acid or base are added to it.

b Ethanoic acid partially dissociates in solution while sodium ethanoate may be considered to be fully dissociated.

$CH_3COOH_{(aq)} \rightleftharpoons CH_3COO^-_{(aq)} + H^+_{(aq)}$ (equation 1)
$CH_3COONa_{(aq)} \rightarrow CH_3COO^-_{(aq)} + Na^+_{(aq)}$ (equation 2)

Addition of hydrogen ions will force the equilibrium in equation 1 to the left. The hydrogen ions will combine with ethanoate ions to form ethanoic acid. Addition of hydroxide ions will force the equilibrium in equation 1 to the right. The hydroxide ions will combine with hydrogen ions to form water. The dissociation of sodium ethanoate ensures a large reserve of ethanoate ions.

8 a i $pH = -\log_{10}[H^+]$
$\qquad = -\log_{10}(0.0020)$
$\qquad = 2.7$
ii $pH = 14 - pOH$
$\qquad = 14 + \log_{10}[OH^-]$
$\qquad = 14 + \log_{10}(0.025)$
$\qquad = 14 - 1.6$
$\qquad = 12.4$

iii Sulphuric acid is diprotic – one mole of acid produces two moles of H^+ therefore
$pH = -\log_{10}[H^+]$
$\qquad = -\log_{10}(0.10)$
$\qquad = 1.0$
b $pH = -\log_{10}[H^+]$
$\qquad = 3.7$
$[HA] = [H^+]$
$\qquad = 2.0 \times 10^{-4}\,mol\,dm^{-3}$
c $pH = 14 - pOH$
$\therefore \ pOH = 14 - pH$
$\qquad = 14 - 12.9$
$\qquad = 1.1$
$pOH = -\log_{10}[OH^-]$
$\qquad = 1.1$
$[MOH] = [OH^-]$
$\qquad = 8.0 \times 10^{-2}\,mol\,dm^{-3}$

9 a $K_a = \dfrac{[H^+_{(aq)}][CH_3COO^-_{(aq)}]}{[CH_3COOH_{(aq)}]}$

$K_w = [H^+_{(aq)}][OH^-_{(aq)}]$

b $CH_3COO^-_{(aq)} + H_2O_{(l)} \rightleftharpoons CH_3COOH_{(aq)} + OH^-_{(aq)}$

$K_b = \dfrac{[CH_3COOH_{(aq)}][OH^-_{(aq)}]}{[CH_3COO^-_{(aq)}]}$

$K_a \times K_b = K_w$

$K_b = \dfrac{K_w}{K_a}$

$\qquad = \dfrac{1 \times 10^{-14}\,mol^2\,dm^{-6}}{1.74 \times 10^{-5}\,mol\,dm^{-3}}$

$\qquad = 5.7 \times 10^{-10}\,mol\,dm^{-3}$

c $pH = -\log_{10}[H^+]$

but this cannot be found directly since we do not know the hydrogen ion concentration. We first need to find the degree to which the ethanoic acid has dissociated.

initial concentration of ethanoic acid $= c$
degree of dissociation $= \alpha$
At equilibrium:

$\qquad\qquad CH_3COOH_{(aq)} \rightleftharpoons CH_3COO^-_{(aq)} + H^+_{(aq)}$
Concentration $\quad (1-\alpha)c \qquad\qquad \alpha c \qquad\quad \alpha c$

$K_a = \dfrac{[H^+_{(aq)}][CH_3COO^-_{(aq)}]}{[CH_3COOH_{(aq)}]}$

$\quad = \dfrac{\alpha^2 c^2}{(1-\alpha)c}$

$\quad = \dfrac{\alpha^2 c}{(1-\alpha)}$

$\quad = 1.74 \times 10^{-5}\,mol\,dm^{-3}$

From this expression, inserting $c = 0.025\,mol\,dm^{-3}$, we can obtain a quadratic equation and solve for α, giving

$\alpha = 0.026$

$\begin{aligned} pH &= -\log_{10}[H^+] \\ &= -\log_{10}(\alpha c) \\ &= -\log_{10}(0.026 \times 0.025) \\ &= 3.18 \end{aligned}$

Notice that a reasonable approximation is obtained by assuming that, since very little CH_3COO^- is formed from CH_3COOH therefore $[CH_3COOH]$ is approximately the same as the initial value.

$K_a = \dfrac{[H^+{}_{(aq)}][CH_3COO^-{}_{(aq)}]}{[CH_3COOH_{(aq)}]} = \dfrac{[H^+{}_{(aq)}]^2}{[CH_3COOH_{(aq)}]}$

$\begin{aligned} [H^+{}_{(aq)}] &= \sqrt{(K_a \times [CH_3COOH_{(aq)}])} \\ &= \sqrt{((1.74 \times 10^{-5}) \times 0.025)}\ mol^2\,dm^{-6} \\ &= 1.04 \times 10^{-4}\ mol\,dm^{-3} \\ pH &= -\log_{10}(1.04 \times 10^{-4}) \\ &= 3.98 \end{aligned}$

10 $K_a = \dfrac{[H^+{}_{(aq)}][H_2PO_4{}^-{}_{(aq)}]}{[H_3PO_4{}_{(aq)}]}$

$[H^+{}_{(aq)}] = \dfrac{K_a[H_3PO_4{}_{(aq)}]}{[H_2PO_4{}^-{}_{(aq)}]}$

Relatively little of the acid dissociates so in the above equation:

$[H_3PO_4{}_{(aq)}]$ is equal to the initial concentration of the acid

$[H_2PO_4{}^-{}_{(aq)}]$ is equal to the initial concentration of the salt

$[H^+{}_{(aq)}] = \dfrac{K_a[acid]}{[salt]}$

Taking $\log_{10}$ of each side of the equation:

$pH = pK_a + \log_{10}\left(\dfrac{[salt]}{[acid]}\right)$

$\begin{aligned} &= -\log_{10}(7.1 \times 10^{-3}) + \log_{10}\left(\dfrac{0.25}{0.15}\right) \\ &= 2.15 + 0.22 \\ &= 2.37 \end{aligned}$

Metal-aqua ions

Transition metal ions form **complexes** with water. Lone pairs of electrons from the water molecules are donated to form coordinate bonds.

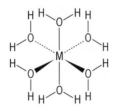

When anhydrous copper(II) sulphate is added to water its colour changes from white to blue as the hexaaquacopper(II) cation is formed.

$$\underset{\text{no colour}}{Cu^{2+}} + 6H_2O \rightarrow \underset{\text{blue}}{[Cu(H_2O)_6]^{2+}}$$

Notice that the anion plays no part in this colour change.

Similarly, if anhydrous cobalt(II) chloride is added to water, the blue anhydrous cobalt(II) ion changes colour to form the pink hexaaquacobalt(II) ion.

$$\underset{\text{blue}}{Co^{2+}} + 6H_2O \rightarrow \underset{\text{pink}}{[Co(H_2O)_6]^{2+}}$$

The characteristic colours we see in aqueous solutions of transition metal salts are due to these metal-aqua ions. The bonds between the metal ions and the water molecules are so strong that the hexaaqua ions remain when solids are obtained from their solutions.

- Solid $CuSO_4.5H_2O$ is blue because it contains $[Cu(H_2O)_6]^{2+}$.
- Solid $CoSO_4.6H_2O$ is pink because it contains $[Co(H_2O)_6]^{2+}$
- Solid $FeSO_4.7H_2O$ is green because it contains $[Fe(H_2O)_6]^{2+}$.

Acidity of metal-aqua ions

When a metal-aqua ion is placed in solution, the following equilibrium is established.

$$[M(H_2O)_6]^{2+} + H_2O \rightleftharpoons [M(H_2O)_5(OH)]^+ + H_3O^+$$

This is a **hydrolysis** reaction since a water molecule has been broken down into H^+ and OH^-.

For a metal(II) ion the equilibrium lies very much over to the left of the equation, thus a solution of a metal(II) ion will only be slightly acidic.

$$K_a = \frac{[[M(H_2O)_5(OH)]^+][H_3O^+]}{[[M(H_2O)_6]^{2+}]}$$

The acidity constant for a metal(II) ion varies between 10^{-6} and 10^{-11} (pK_a varies between 6 and 11). Metal(II) ions form very weak acids, typically around pH 6.

However, for metal(III) ions the equilibrium is more to the right, giving a significant degree of acidity.

$$[M(H_2O)_6]^{3+} + H_2O \rightleftharpoons [M(H_2O)_5(OH)]^{2+} + H_3O^+$$

In this case K_a varies between 10^{-2} and 10^{-5} (pK_a varies between 2 and 5). Metal(III) ions form weak acids, typically around pH 3–4.

In the case of metal(IV) ions the first stage of hydrolysis goes to completion. The solutions formed are strong acids, with pH < 1.

The acidity of metal ions is determined by:

- the charge on the metal ion – acidity increases with charge
- the size of the metal ion – acidity decreases with size.

The most acidic metal ions are small and highly charged. Such ions have a high polarising power to attract electron density from the oxygen atom of a coordinated water molecule. This, in turn, weakens the O—H bonds in the molecule. The weaker the O—H bond, the easier it is for H^+ to be lost.

Several stages of hydrolysis are possible. For a metal(II) ion,

$$[M(H_2O)_6]^{2+} + H_2O \rightleftharpoons [M(H_2O)_5(OH)]^+ + H_3O^+$$
$$[M(H_2O)_5OH]^+ + H_2O \rightleftharpoons [M(H_2O)_4(OH)_2] + H_3O^+$$

and for a metal(III) ion,

$$[M(H_2O)_6]^{3+} + H_2O \rightleftharpoons [M(H_2O)_5(OH)]^{2+} + H_3O^+$$
$$[M(H_2O)_5OH]^{2+} + H_2O \rightleftharpoons [M(H_2O)_4(OH)_2]^+ + H_3O^+$$
$$[M(H_2O)_4(OH)_2]^+ + H_2O \rightleftharpoons [M(H_2O)_3(OH)_3] + H_3O^+$$

When aqueous sodium hydroxide is added to a solution containing transition metal ions the equilibrium is forced to the right. The result in each case is a precipitate of the insoluble hydroxide.

$$[Cu(H_2O)_6]^{2+} + 2OH^- \rightarrow \underset{\text{blue precipitate}}{[Cu(H_2O)_4(OH)_2]} + 2H_2O$$

Table 27.1 Identifying transition metal ions

Transition metal ion	Colour of solution	Insoluble hydroxide	Colour of precipitate
$[Co(H_2O)_6]^{2+}$	pink	$[Co(H_2O)_4(OH)_2]$	blue-green
$[Cr(H_2O)_6]^{3+}$	violet*	$[Cr(H_2O)_3(OH)_3]$	green
$[Fe(H_2O)_6]^{2+}$	green	$[Fe(H_2O)_4(OH)_2]$	green
$[Fe(H_2O)_6]^{3+}$	yellow-brown*	$[Fe(H_2O)_3(OH)_3]$	brown

* The colour of the ion $[Cr(H_2O)_6]^{3+}$ is actually ruby red and the colour of $[Fe(H_2O)_6]^{3+}$ is pale violet; however, these colours are not seen in solution due to the loss of H^+, forming the ions $[Cr(H_2O)_5(OH)]^{2+}$ and $[Fe(H_2O)_5(OH)]^{2+}$, respectively.

The colours of these precipitates are often used to identify the transition metal ion (Table 27.1).

Notice that the addition of ammonia solution gives the same initial reaction since it contains OH^- ions. However, when ammonia solution is in excess a further reaction may occur.

$$\underset{\substack{\text{blue} \\ \text{solution}}}{[Cu(H_2O)_6]^{2+}} \rightarrow \underset{\substack{\text{blue} \\ \text{precipitate}}}{[Cu(H_2O)_4(OH)_2]} \rightarrow \underset{\substack{\text{deep-blue} \\ \text{solution}}}{[Cu(H_2O)_2(NH_3)_4]^{2+}}$$

Amphoteric oxides

Metal hydroxides dissolve in strong acids to give solutions of metal-aqua ions and are thus basic in nature. However, some are also able to react with OH^- ions to form anionic complexes and thus also show an acidic nature.

The ability of a metal hydroxide to dissolve both in strong acids and in strong bases is called **amphoteric character**.

When aqueous sodium hydroxide is added to a solution of an aluminium salt there is an initial white precipitate of aluminium hydroxide. However, if an excess of sodium hydroxide is added the precipitate dissolves, giving a colourless solution containing the aluminate ion, $[Al(OH)_4]^-$

$$Al^{3+}(aq) + 3OH^-(aq) \rightarrow Al(OH)_3(s)$$
$$\text{white precipitate}$$

$$Al(OH)_3(s) + OH^-(aq) \rightarrow [Al(OH)_4]^-(aq)$$
$$\text{colourless solution}$$

Solutions of zinc salts and lead salts react in a similar way, forming the zincate ion, $[Zn(OH)_4]^{2-}$, and the plumbate ion, $[Pb(OH)_4]^{2-}$, respectively.

Beryllium is atypical of the Group 2 elements in that beryllium hydroxide is amphoteric. Beryllium hydroxide dissolves in excess OH^- to form beryllates containing the ion $[Be(OH)_4]^{2-}$.

Most transition metal hydroxides are amphoteric; however, some of them only dissolve in very concentrated sodium hydroxide solution.

Reaction of metal-aqua ions with carbonate ions

Sodium carbonate reacts with acids to give carbon dioxide gas. The carbonate ion forms carbonic acid (H_2CO_3) which then breaks down into carbon dioxide and water.

$$CO_3{}^{2-}(aq) + 2H_3O^+(aq) \rightarrow H_2CO_3(aq) + 2H_2O(l)$$
$$H_2CO_3(aq) \rightleftharpoons H_2O(l) + CO_2(g)$$

Metal(II) ions are only very weakly acidic in aqueous solution. They are not sufficiently acidic to release carbon dioxide from sodium carbonate, but instead they form insoluble metal(II) carbonates.

$$M^{2+}(aq) + CO_3{}^{2-}(aq) \rightarrow MCO_3(s)$$

Metal(III) ions in aqueous solution are more acidic and will displace carbonic acid from sodium carbonate solution. The addition of aqueous sodium carbonate to a solution of a metal(III) salt results in the evolution of carbon dioxide and the formation of the metal hydroxide.

$$[M(H_2O)_6]^{3+}(aq) + H_2O(l) \rightleftharpoons [M(H_2O)_5(OH)]^{2+}(aq) + H_3O^+(aq)$$
$$CO_3{}^{2-}(aq) + 2H_3O^+(aq) \rightarrow 3H_2O(l) + CO_2(g)$$

Metal(III) carbonates cannot be prepared in aqueous solution and many are unknown.

METAL-AQUA IONS

1 Iron(III) ammonium alum, $Fe_2(SO_4)_3.(NH_4)_2SO_4.24H_2O$, is a pale violet crystalline solid. The pale violet colour is due to the presence of

A Fe^{3+}

B $[Fe(H_2O)_6]^{3+}$

C NH_4^+

D SO_4^{2-}

2 Which of the following characteristics of an ion would give the greatest acidity when the ion was in solution?

A small charge and small size

B small charge and large size

C large charge and small size

D large charge and large size

3 Which of the following ions is most acidic in aqueous solution?

A Ce^{4+}

B Fe^{3+}

C Li^+

D Mg^{2+}

4 When aqueous ammonia is added dropwise to a solution containing Cu^{2+} ions

A no reaction is observed

B a blue precipitate is formed which remains in excess ammonia

C a blue precipitate is formed which dissolves in excess ammonia to form a deep-blue solution

D a deep-blue solution is formed immediately

5 Which of the following when added to aqueous potassium carbonate would release carbon dioxide gas?

A Ba^{2+}

B Cr^{3+}

C Na^+

D Zn^{2+}

6 a Describe, including suitable equations, the reactions of aqueous magnesium sulphate and aqueous aluminium sulphate with sodium carbonate solution.

b Explain why these salts react differently with sodium carbonate solution.

7 The ion $[Fe(H_2O)_6]^{3+}$ is pale violet. By considering the first stage of hydrolysis of this ion in solution answer the following questions.

a Suggest why the pale violet crystals of $Fe(NO_3)_3.9H_2O$ dissolve in water to form a brown solution.

b Discuss the effects of adding acid and adding alkali to the above solution.

8 a Give equations for the three stages of hydrolysis of the $[Cr(H_2O)_6]^{3+}$ ion.

b Write the expression for the dissociation constant, K_a, for the first hydrolysis of the ion $[Cr(H_2O)_6]^{3+}$.

c Discuss the relative acidity of M^{2+}, M^{3+} and M^{4+} ions in respect of the values of the dissociation constants for these ions in aqueous solution.

9 The following table summarises the reactions of some metal ions with sodium hydroxide solution and aqueous ammonia.

Ion	Precipitated by $NaOH_{(aq)}$	Soluble in excess $NaOH_{(aq)}$	Precipitated by $NH_{3(aq)}$	Soluble in excess $NH_{3(aq)}$
Al^{3+}	✓	✓	✓	
Cd^{2+}	✓		✓	✓
Mg^{2+}	✓		✓	
Pb^{2+}	✓	✓	✓	
Sn^{2+}	✓	✓	✓	
Zn^{2+}	✓	✓	✓	✓

a Account for the reactions of the metal ions in the table.

b Explain how you could differentiate between aqueous solutions of magnesium sulphate and aluminium sulphate using the above information.

10 Explain the following observations, giving equations where appropriate.

a When OH^- ions are added to a pale green solution containing the hexaaquachromium(III) ion, $[Cr(H_2O)_6]^{3+}$, a pale green precipitate is formed. The precipitate dissolves both if H^+ ions are added, giving a pale green solution, and if an excess of OH^- is added, giving a dark green solution.

b Magnesium carbonate can be made by reacting solutions of magnesium sulphate and sodium carbonate but aluminium carbonate cannot be made by this method.

c Copper(II) sulphate forms a blue precipitate when sodium hydroxide solution or aqueous ammonia is added dropwise. This precipitate will dissolve in excess ammonia and in excess sodium hydroxide solution provided the latter is very concentrated.

1 B **2** C **3** A **4** C **5** B

6 a Magnesium sulphate forms a white precipitate of magnesium carbonate.

$$Mg^{2+}(aq) + CO_3^{2-}(aq) \rightarrow MgCO_3(s)$$

Aluminium sulphate releases carbon dioxide and forms a white precipitate of aluminium hydroxide.

$$2[Al(H_2O)_6]^{3+}(aq) + 3CO_3^{2-}(aq) \rightarrow$$
$$2[Al(H_2O)_3(OH)_3](s) + 3CO_2(g) + 3H_2O(l)$$

b The difference in reaction is due to the acidity of the Al^{3+} ion.

$$[Al(H_2O)_6]^{3+}(aq) + H_2O(l) \rightleftharpoons [Al(H_2O)_5(OH)]^{2+}(aq) + H_3O^+(aq)$$

The Al^{3+} is sufficiently acidic to displace carbonic acid from sodium carbonate solution.

7 a $Fe(NO_3)_3.9H_2O$ contains the ion $[Fe(H_2O)_6]^{3+}$ which is responsible for the characteristic pale violet colour. When added to water the following hydrolysis occurs:

$$[Fe(H_2O)_6]^{3+}(aq) + H_2O(l) \rightleftharpoons [Fe(H_2O)_5(OH)]^{2+}(aq) + H_3O^+(aq)$$
pale violet   brown

The brown colour is due to the presence of the $[Fe(H_2O)_5(OH)]^{2+}$ ion. Although it is in low concentration it masks the original pale violet colour.

b Addition of H^+ drives the equilibrium to the left causing the solution to become pale violet once more. Addition of OH^- ions drives the equilibrium to the right so the solution becomes deeper brown and eventually a precipitate of $[Fe(H_2O)_3(OH)_3]$ is formed.

8 a $[Cr(H_2O)_6]^{3+}(aq) + H_2O(l) \rightleftharpoons [Cr(H_2O)_5(OH)]^{2+}(aq) + H_3O^+(aq)$
$[Cr(H_2O)_5(OH)]^{2+}(aq) + H_2O(l) \rightleftharpoons [Cr(H_2O)_4(OH)_2]^+(aq) + H_3O^+(aq)$
$[Cr(H_2O)_4(OH)_2]^+(aq) + H_2O(l) \rightleftharpoons [Cr(H_2O)_3(OH)_3](aq) + H_3O^+(aq)$

b $K_a = \dfrac{[[Cr(H_2O)_5(OH)]^{2+}(aq)][H_3O^+(aq)]}{[[Cr(H_2O)_6]^{3+}(aq)]}$

c For M^{2+} the equilibrium is very much to the left. pK_a varies between 6 and 11 giving very weak acids with pH around 6.

For M^{3+} the equilibrium is more to the right. pK_a varies between 2 and 5 giving weak acids with pH around 3–4.

For M^{4+} the equilibrium is completely to the right. pK_a is less than 1 giving strong acids with pH around 1.

9 a All of the ions react with OH^- ions to form insoluble hydroxides.

Magnesium hydroxide and cadmium hydroxide are not amphoteric since they do not dissolve in excess NaOH. The remaining hydroxides are amphoteric since they dissolve in excess NaOH. This shows they have an acidic character.

Cadmium hydroxide and zinc hydroxide react with excess ammonia to form soluble complex ions in a similar way to copper(II) hydroxide.

b Both magnesium sulphate solution and aluminium sulphate form a precipitate with sodium hydroxide but only the latter will dissolve in excess sodium hydroxide.

10 a The hexaaquachromium(III) ion reacts with alkali to form insoluble $[Cr(H_2O)_3(OH)_3]$ which is the pale green precipitate. This oxide dissolves in acid to regenerate the original hexaaquachromium(III) ion. When excess alkali is added chromium(III) oxide dissolves forming an anion, showing that the oxide is amphoteric in nature.

$$[Cr(H_2O)_6]^{3+}(aq) \rightleftharpoons [Cr(H_2O)_3(OH)_3](s) \rightleftharpoons [Cr(OH)_6]^{3-}(aq)$$

b Magnesium sulphate solution contains predominantly Mg^{2+} ions which react with carbonate ions forming a precipitate of insoluble magnesium carbonate.

$$Mg^{2+}(aq) + CO_3^{2-}(aq) \rightarrow MgCO_3(s)$$

In aluminium sulphate solution the aluminium ions are hydrolysed, making the solution sufficiently acidic to react with the carbonate ions, releasing carbon dioxide and forming aluminium hydroxide.

$$[Al(H_2O)_6]^{3+}(aq) + H_2O(l) \rightleftharpoons [Al(H_2O)_5(OH)]^{2+}(aq) + H_3O^+(aq)$$
$$CO_3^{2-}(aq) + 2H_3O^+(aq) \rightarrow 3H_2O(l) + CO_2(g)$$

c Aqueous ammonia and sodium hydroxide solution both contain OH^- ions, which react with Cu^{2+} ions to form insoluble copper(II) hydroxide.

Copper(II) hydroxide dissolves in excess ammonia forming the diaquatetraamminecopper(II) complex ion $[Cu(NH_3)_4(H_2O)_2]^{2+}$.

Copper(II) hydroxide also dissolves in excess sodium hydroxide solution, provided the latter is very concentrated, to form the tetrahydroxocuprate(II) anion, $[Cu(OH)_4]^{2-}$, thus showing amphoteric character.

28 Electrochemical cells

An electrochemical cell consists of a pair of electrodes suspended in an electrolyte. The electrodes may be in the same container (Figure 28.1) or in different containers joined by a conducting **salt bridge**.

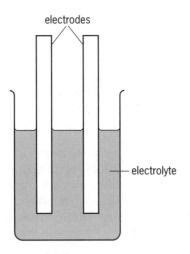

Figure 28.1 Simple electrolysis

Oxidation is defined as the loss of electrons and reduction as the gain of electrons. In an electrochemical cell these processes necessarily occur simultaneously at the electrodes, giving an overall redox reaction. The electrode at which oxidation occurs is the **anode** and the one at which reduction occurs is the **cathode**.

Electrodes

The simplest electrode is one in which a metal is in equilibrium with a solution of its ions, for example, a copper electrode suspended in an aqueous solution of copper(II) sulphate. Such an electrode is written as $Cu(s)|Cu^{2+}(aq)$. The vertical line denotes a boundary between the two phases.

A **gas electrode** consists of an inert metal, such as platinum, surrounded by a gas in equilibrium with a solution of its ions. The standard hydrogen electrode is an example that is described later in this topic.

Cell reactions

Where a cell consists of two components, it is conventional when drawing the cell diagram to show the cathode as the right-hand electrode. The cell diagram is constructed by writing the two electrodes next to each other. Two vertical lines ($||$) are drawn between them to indicate that they are joined by a salt bridge.

The following cell consists of a zinc electrode and a copper electrode.

$$Zn(s)|ZnSO_4(aq)||CuSO_4(aq)|Cu(s)$$

The cell reaction that occurs is derived as follows:

1 Write the right-hand half-reaction as a reduction.
2 Write the left-hand half-reaction as a reduction.
3 Subtract the left-hand half-reaction from the right-hand half-reaction.

- Right-hand half-reaction as a reduction

$$Cu^{2+}(aq) + 2e^- \rightarrow Cu(s)$$

- Left-hand half-reaction as a reduction

$$Zn^{2+}(aq) + 2e^- \rightarrow Zn(s)$$

- Subtract left from right

$$Cu^{2+}(aq) + 2e^- - Zn^{2+}(aq) - 2e^- \rightarrow Cu(s) - Zn(s)$$
$$Cu^{2+}(aq) - Zn^{2+}(aq) \rightarrow Cu(s) - Zn(s)$$

- Rearrange to remove negative signs

$$Cu^{2+}(aq) + Zn(s) \rightarrow Cu(s) + Zn^{2+}(aq)$$

From this equation we can see that the reaction is spontaneous; zinc displaces copper from a solution of its ions.

Remember: left-hand electrode   anode   oxidation
right-hand electrode   cathode   reduction
(Reduced on the Right)

Cell potential

In an electrochemical cell a potential difference or voltage exists between the two electrodes, with the left-hand electrode (anode) being more negative than the right-hand electrode (cathode). This potential difference is called the **cell potential**.

There are several factors that affect the cell potential.

- Solution concentration – the standard concentration is $1.0\,mol\,dm^{-3}$.
- Temperature – the standard temperature is 298 K.
- Pressure – the standard pressure is 1 bar (100 kPa).
- Current – the true cell potential must be measured in a way that does not draw current from the cell. This can be done using a high-resistance voltmeter or by applying an equal but opposite potential using a potentiometer. Under these conditions the cell potential is called the **electromotive force** (e.m.f.).

When cell potentials are measured under these standard conditions they are called **standard cell potentials** and are given the symbol $E^{\ominus}$.

Standard electrodes

It is meaningless to talk of the potential of a single electrode since it only has potential when it is connected to a second electrode. In order to provide a meaningful comparison it is necessary to have a standard electrode against which the potential of all other electrodes can be measured. The standard electrode usually adopted is the **hydrogen electrode** (Figure 28.2).

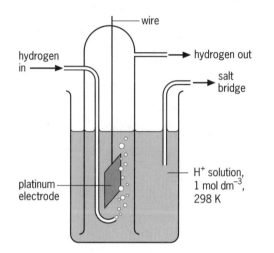

Figure 28.2 Standard hydrogen electrode

The platinum electrode is surrounded by hydrogen gas in equilibrium with a solution of hydrogen ions. It is denoted by

$$\text{Pt(s)} \,|\, \text{H}_2\text{(g, 1 bar)} \,|\, \text{H}^+\text{(aq, 1.0 mol dm}^{-3})$$

Notice that there are two phase boundaries and that they are shown by two vertical lines. When used under standard conditions, $E^{\ominus} = 0.0\,\text{V}$ for this electrode. The reduction half-reaction at the electrode is

$$2\text{H}^+\text{(aq)} + 2\text{e}^- \rightarrow \text{H}_2\text{(g)}$$

The standard electrode potential of an electrode involved in a redox system is found by measuring the potential difference between it, as the right-hand electrode, and the standard hydrogen electrode as the left-hand electrode. Both electrodes must be under standard conditions.

The hydrogen electrode is difficult to set up and use. It is often more convenient to use a secondary reference electrode whose standard electrode potential, relative to hydrogen, is known. A common choice is the **calomel electrode**, which has a standard electrode potential of $+0.27\,\text{V}$.

$$\text{Pt(s)} \,|\, \text{Hg(l)} \,|\, \text{Hg}_2\text{Cl}_2\text{(sat)}, \text{KCl}(1.0\,\text{mol dm}^{-3})$$

The reduction half-reaction for this cell is

$$\text{Hg}_2\text{Cl}_2 + 2\text{e}^- \rightarrow 2\text{Hg} + 2\text{Cl}^-$$

If the calomel cell is used as the left-hand electrode to determine the potential of another electrode, then it is necessary to add $0.27\,\text{V}$ to the value obtained in order to get the true standard electrode potential of the unknown electrode.

Table 28.1 gives the standard electrode potentials for a range of electrodes.

Table 28.1 Values of standard electrode potential

Reduction half-reaction	$E^{\ominus}/\text{V}$
$\text{MnO}_4^{2-}\text{(aq)} + 4\text{H}^+\text{(aq)} + 2\text{e}^- \rightarrow \text{MnO}_2\text{(s)} + 2\text{H}_2\text{O(l)}$	+1.55
$\text{MnO}_4^-\text{(aq)} + 8\text{H}^+\text{(aq)} + 5\text{e}^- \rightarrow \text{Mn}^{2+}\text{(s)} + 4\text{H}_2\text{O(l)}$	+1.51
$\text{Cl}_2\text{(g)} + 2\text{e}^- \rightarrow 2\text{Cl}^-\text{(aq)}$	+1.36
$\text{Br}_2\text{(g)} + 2\text{e}^- \rightarrow 2\text{Br}^-\text{(aq)}$	+1.09
$\text{Fe}^{3+}\text{(aq)} + \text{e}^- \rightarrow \text{Fe}^{2+}\text{(aq)}$	+0.77
$\text{MnO}_4^-\text{(aq)} + \text{e}^- \rightarrow \text{MnO}_4^{2-}\text{(aq)}$	+0.60
$\text{I}_2\text{(g)} + 2\text{e}^- \rightarrow 2\text{I}^-\text{(aq)}$	+0.54
$\text{Cu}^{2+}\text{(aq)} + 2\text{e}^- \rightarrow \text{Cu(s)}$	+0.34
$\text{Hg}_2\text{Cl}_2\text{(aq)} + 2\text{e}^- \rightarrow 2\text{Hg(l)} + 2\text{Cl}^-\text{(aq)}$	+0.27
$\text{AgCl(s)} + \text{e}^- \rightarrow \text{Ag(s)} + \text{Cl}^-\text{(aq)}$	+0.22
$2\text{H}^+\text{(aq)} + 2\text{e}^- \rightarrow \text{H}_2\text{(g)}$	0.00
$\text{V}^{3+}\text{(aq)} + \text{e}^- \rightarrow \text{V}^{2+}\text{(aq)}$	−0.26
$\text{Fe}^{2+}\text{(aq)} + 2\text{e}^- \rightarrow \text{Fe(s)}$	−0.44
$\text{Zn}^{2+}\text{(aq)} + 2\text{e}^- \rightarrow \text{Zn(s)}$	−0.76
$\text{Al}^{3+}\text{(aq)} + 3\text{e}^- \rightarrow \text{Al(s)}$	−1.66
$\text{Mg}^{2+}\text{(aq)} + 2\text{e}^- \rightarrow \text{Mg(s)}$	−2.36
$\text{Na}^+\text{(aq)} + \text{e}^- \rightarrow \text{Na(s)}$	−2.71
$\text{K}^+\text{(aq)} + \text{e}^- \rightarrow \text{K(s)}$	−2.93

Standard cell potential, $E^{\ominus}_{\text{cell}}$

The standard potential of a cell is found by subtracting the left-hand electrode potential from the right-hand one.

$$E^{\ominus}_{\text{cell}} = E^{\ominus}_{\text{right}} - E^{\ominus}_{\text{left}}$$

Note that the lower standard electrode potential is put to the left so that $E^{\ominus}_{\text{cell}} > 0$. This ensures that the cell reaction is spontaneous.

Consider a cell consisting of zinc ($-0.76\,\text{V}$) and iron ($-0.44\,\text{V}$) in a solution of $1.0\,\text{mol dm}^{-3}$ zinc ions and iron(II) ions. The standard e.m.f. of the cell is given by

$$\begin{aligned} E^{\ominus}_{\text{cell}} &= E^{\ominus}_{\text{right}} - E^{\ominus}_{\text{left}} \\ &= -0.44 - (-0.76) \\ &= +0.32\,\text{V} \end{aligned}$$

Spontaneous cell reactions

A cell reaction is only spontaneous if $E^{\ominus}_{cell}$ is positive.

Since $\Delta G^{\ominus}$ is negative for a spontaneous reaction it follows that positive cell potential is linked to negative $\Delta G^{\ominus}$. For the transfer of z electrons

$$\Delta G^{\ominus} = -zFE^{\ominus}_{cell}$$

where F is the Faraday constant and is equal to $96\,500\,C\,mol^{-1}$.

In respect of the electrochemical series, the half-reaction with the greater (more positive) potential oxidises the half-reaction with the lesser potential.

Example 1

A cell consists of silver and iron in a solution of $1.0\,mol\,dm^{-3}$ silver ions and iron(II) ions. Determine the direction of spontaneous reaction and calculate $E^{\ominus}_{cell}$.

$$Ag^{+}(aq) + e^{-} \rightarrow Ag(s); \qquad +0.80\,V$$
$$Fe^{2+}(aq) + 2e^{-} \rightarrow Fe(s); \qquad -0.44\,V$$

A simple diagram of the cell often helps (Figure 28.3).

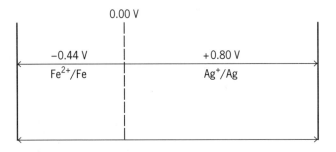

Figure 28.3

Ag^{+}/Ag is the more positive and is therefore the cathode (right-hand side) in the cell.

$$E^{\ominus}_{cell} = E^{\ominus}_{right} - E^{\ominus}_{left}$$
$$= +0.80 - (-0.44)$$
$$= +1.24\,V$$

The spontaneous reaction is:

- Right-hand (cathode) half-reaction as a reduction

$$2Ag^{+}(aq) + 2e^{-} \rightarrow 2Ag(s)$$

- Left-hand (anode) half-reaction as a reduction

$$Fe^{2+}(aq) + 2e^{-} \rightarrow Fe(s)$$

- The overall reaction is given by subtracting the left-hand half-reaction from the right-hand half-reaction:

$$2Ag^{+}(aq) + Fe(s) \rightarrow 2Ag(s) + Fe^{2+}(aq)$$

It is also possible to use this technique in order to predict whether a particular redox reaction will occur spontaneously.

Example 2

Will zinc reduce vanadium(III) ions to vanadium(II) ions?

If this reaction were to occur the half-reactions for reduction would be (V^{3+}/V^{2+}, $-0.26\,V$) and (Zn^{2+}/Zn, $-0.76\,V$).

- The following reduction would occur at the cathode

$$2V^{3+}(aq) + 2e^{-} \rightarrow 2V^{2+}(aq)$$

- The following oxidation would occur at the anode

$$Zn(s) \rightarrow Zn^{2+}(aq) + 2e^{-}$$

- The overall reaction would be

$$2V^{3+}(aq) + Zn(s) \rightarrow 2V^{2+}(aq) + Zn^{2+}(aq)$$

$$E^{\ominus}_{cell} = E^{\ominus}_{right} - E^{\ominus}_{left}$$
$$= -0.26 - (-0.76)$$
$$= +0.50\,V$$

$E^{\ominus}_{cell} > 0$ so we predict that this reaction will occur spontaneously.

Example 3

Similarly, will iodine oxidise silver to silver ions spontaneously?

If this reaction is spontaneous, reduction occurs at the cathode

$$I_2(aq) + 2e^{-} \rightarrow 2I^{-}(aq)$$

and oxidation at the anode

$$2Ag(aq) \rightarrow 2Ag^{+}(aq) + 2e^{-}$$

The standard electrode potentials are I_2/I^{-}, $+0.54\,V$ and Ag^{+}/Ag, $+0.80\,V$.

$$E^{\ominus}_{cell} = E^{\ominus}_{right} - E^{\ominus}_{left}$$
$$= +0.54 - (+0.80)$$
$$= -0.26\,V$$

$E^{\ominus}_{cell}$ is not >0 so we predict that this reaction will not occur spontaneously.

Note: the standard electrode potentials given in the text are required for these questions.

1 $E^{\ominus}$ for the standard hydrogen electrode is
  A 0.00 V
  B 0.27 V
  C 0.50 V
  D 1.00 V

2 Which of the following is not a standard condition under which cell potentials are obtained?
  A current $= 0$ A
  B concentration $= 1.0$ mol dm^{-3}
  C pressure $= 1$ bar
  D temperature $= 273$ K

3 $E^{\ominus} > 0$ corresponds to
  A $\Delta G^{\ominus} > 0$
  B $\Delta S^{\ominus} > 0$
  C $\Delta G^{\ominus} < 0$
  D $\Delta S^{\ominus} < 0$

4 A standard cell has an e.m.f. of $+0.07$ V with calomel as the left-hand electrode and the redox couple M^{2+}/M as the right-hand one. The standard potential of the couple M^{2+}/M is
  A -0.20 V
  B -0.07 V
  C $+0.07$ V
  D $+0.34$ V

5 Which of the following can oxidise Fe(II) to Fe(III) in a spontaneous reaction?
  A chlorine
  B iodine
  C vanadium(III)
  D zinc

6 a Complete the following table showing the conventions used in drawing cell diagrams.

Left-hand electrode		Oxidation
	Cathode	

  b The following cell consists of a magnesium electrode and a copper electrode.

    $Mg(s) | MgSO_4(aq) || CuSO_4(aq) | Cu(s)$

  Give the half-reactions as reductions and derive the overall cell reaction.

7 Calculate the e.m.f. value of the following cells.
  a Left-hand electrode: Al^{3+}/Al, $E^{\ominus} = -1.66$ V
    Right-hand electrode: Ag^+/Ag, $E^{\ominus} = +0.80$ V
  b Left-hand electrode: Cu^{2+}/Cu, $E^{\ominus} = +0.34$ V
    Right-hand electrode: Zn^{2+}/Zn, $E^{\ominus} = -0.76$ V
  c Left-hand electrode: Fe^{3+}/Fe^{2+}, $E^{\ominus} = +0.77$ V
    Right-hand electrode: $AgCl/Ag$, $E^{\ominus} = +0.22$ V

8 Give the reactions at the right-hand and left-hand electrodes and the overall reaction for the following and predict whether each will occur spontaneously.
  a Magnesium(II) reducing chlorine to chloride ions.
  b Silver reducing hydrogen ions to hydrogen.
  c Zinc reducing manganese(VII) to manganese(VI) ions.

9 Explain the displacement reactions that occur between chlorine, bromine and iodine and their ions in terms of e.m.f. Comment on the magnitude of the standard electrode potential for $F_2(g) + 2e^- \rightarrow 2F^-(aq)$.

10 Explain in terms of e.m.f. why manganate(VI) ions in aqueous solution undergo a disproportionation reaction, in which some of the ions are reduced to manganese(IV) while others are oxidised to manganese(VII) simultaneously.

ELECTROCHEMICAL CELLS

1 A

2 D

3 C

4 D

5 A

6 a

Left-hand electrode	Anode	Oxidation
Right-hand electrode	Cathode	Reduction

b Right-hand half-reaction as a reduction:

$Cu^{2+}_{(aq)} + 2e^- \rightarrow Cu_{(s)}$

Left-hand half-reaction as a reduction:

$Mg^{2+}_{(aq)} + 2e^- \rightarrow Mg_{(s)}$

Subtract left from right:

$Cu^{2+}_{(aq)} + 2e^- - Mg^{2+}_{(aq)} - 2e^- \rightarrow Cu_{(s)} - Mg_{(s)}$

Simplifies to:

$Cu^{2+}_{(aq)} + Mg_{(s)} \rightarrow Cu_{(s)} + Mg^{2+}_{(aq)}$

7 a $E^{\ominus}_{cell} = E^{\ominus}_{right} - E^{\ominus}_{left}$
$= +0.80 - (-1.66) = +2.46\,V$

b $E^{\ominus}_{cell} = E^{\ominus}_{right} - E^{\ominus}_{left}$
$= -0.76 - (+0.34) = -1.10\,V$

c $E^{\ominus}_{cell} = E^{\ominus}_{right} - E^{\ominus}_{left}$
$= +0.22 - (+0.77) = -0.55\,V$

8 a The half-cells would be ($Cl_2/2Cl^-$, $+1.36\,V$) and (Mg^{2+}/Mg, $-2.36\,V$).
At the cathode (right):

$Cl_{2(g)} + 2e^- \rightarrow 2Cl^-_{(aq)}$

At the anode (left):

$Mg_{(s)} \rightarrow Mg^{2+}_{(aq)} + 2e^-$

Overall:

$Mg_{(s)} + Cl_{2(g)} \rightarrow Mg^{2+}_{(aq)} + 2Cl^-_{(aq)}$

$E^{\ominus}_{cell} = E^{\ominus}_{right} - E^{\ominus}_{left}$
$= +1.36 - (-2.36) = +3.72\,V$

$E^{\ominus}_{cell} > 0$ so this reaction will occur spontaneously.

b The half-cells would be ($2H^+/H_2$, $0.00\,V$) and (Ag^+/Ag, $+0.80\,V$).
At the cathode (right):

$2H^+_{(aq)} + 2e^- \rightarrow H_{2(g)}$

At the anode (left):

$2Ag_{(s)} \rightarrow 2Ag^+_{(aq)} + 2e^-$

Overall:

$2Ag_{(s)} + 2H^+_{(aq)} \rightarrow 2Ag^+_{(aq)} + H_{2(g)}$

$E^{\ominus}_{cell} = E^{\ominus}_{right} - E^{\ominus}_{left}$
$= 0.00 - (+0.80) = -0.80\,V$

$E^{\ominus}_{cell} < 0$ so this reaction will not occur spontaneously.

c The half-cells would be (MnO_4^-/MnO_4^{2-}, $+0.60\,V$) and (Zn^{2+}/Zn, $-0.76\,V$).
At the cathode (right):

$2MnO_4^-_{(aq)} + 2e^- \rightarrow 2MnO_4^{2-}_{(aq)}$

At the anode (left):

$Zn_{(s)} \rightarrow Zn^{2+}_{(aq)} + 2e^-$

Overall:

$Zn_{(s)} + 2MnO_4^-_{(aq)} \rightarrow Zn^{2+}_{(aq)} + 2MnO_4^{2-}_{(aq)}$

$E^{\ominus}_{cell} = E^{\ominus}_{right} - E^{\ominus}_{left}$
$= +0.60 - (-0.76) = +1.36\,V$
$E^{\ominus}_{cell} > 0$ so this reaction will occur spontaneously.

9 For a displacement reaction to occur between two halogens X and Y the following must occur:

$2X^-_{(aq)} + Y_{2(g)} \rightarrow 2Y^-_{(aq)} + X_{2(g)}$

This can be separated into a pair of half-reactions:
X^- is oxidised and is $\therefore$ the left-hand electrode in the cell.

$2X^-_{(aq)} \rightarrow X_{2(g)} + 2e^-$
(written as a reduction $X_{2(g)} + 2e^- \rightarrow 2X^-_{(aq)}$)

Y_2 is reduced and is $\therefore$ the right-hand electrode in the cell.

$Y_{2(g)} + 2e^- \rightarrow 2Y^-_{(aq)}$

If a displacement reaction is to occur then $E^{\ominus}_{cell} > 0$
$\therefore E^{\ominus}_Y - E^{\ominus}_X > 0$
In other words a halogen will only displace the ions of another halogen from aqueous solution if it has a higher standard electrode potential.
Since fluorine will displace the ions of all of the other halogens the standard electrode potential for

$F_{2(g)} + 2e^- \rightarrow 2F^-_{(aq)}$

must be greater than $+1.36\,V$, the standard electrode potential for chlorine.

10 The reactions are oxidation (VI) to (VII), and reduction (VI) to (IV).
The half-reactions written as reductions are:
Right-hand electrode (reduction)

$MnO_4^{2-}_{(aq)} + 4H^+_{(aq)} + 2e^- \rightarrow MnO_{2(s)} + 2H_2O_{(l)};$
$$E^{\ominus} = +1.55\,V$$

Left-hand electrode (oxidation)

$2MnO_4^-_{(aq)} + 2e^- \rightarrow 2MnO_4^{2-}_{(aq)};\qquad E^{\ominus} = +0.60\,V$

Overall cell reaction (right − left)

$MnO_4^{2-}_{(aq)} + 4H^+_{(aq)} + 2e^- - 2MnO_4^-_{(aq)} - 2e^- \rightarrow$
$$MnO_{2(s)} + 2H_2O_{(l)} - 2MnO_4^{2-}_{(aq)}$$

which simplifies to:

$3MnO_4^{2-}_{(aq)} + 4H^+_{(aq)} \rightarrow MnO_{2(s)} + 2MnO_4^-_{(aq)} + 2H_2O_{(l)}$

$E^{\ominus}_{cell} = E^{\ominus}_{right} - E^{\ominus}_{left}$
$= +1.55 - (+0.60)$
$= +0.95\,V$
The e.m.f. > 0 so this disproportionation can take place.

29 Qualitative analysis

Inorganic compounds

The qualitative analysis of inorganic compounds is concerned with detecting the presence of cations and anions. The identity, or partial identity, of a compound may be evident simply from its colour and/or its solubility in water.

Table 29.1 shows some of the common ions responsible for coloured solutions. It should be borne in mind that the colour often varies with the concentration of the ions.

Table 29.1 Using colour to identify substances

Colour of solution	Possible ion or other substance present
pink	Co^{2+}, MnO_4^- (very dilute)
orange	$Cr_2O_7^{2-}$, Br_2
yellow	Fe^{3+}, CrO_4^{2-}, Br_2 (dilute), I_3^- (dilute)
brown	Fe^{3+} (dilute), CrO_4^{2-}, Br_2 (dilute), I_3^-
green	Ni^{2+}, Cr^{2+}, Fe^{2+}, Cu^{2+} (concentrated solutions of some salts)
blue	Co^{2+}, Cu^{2+}, $Ni(NH_3)_6^{2+}$ (deep blue), $Cu(NH_3)_4(H_2O)_2^{2+}$ (deep blue)
purple	Cr^{3+} (concentrated), MnO_4^-

Table 29.2 gives some general guidance about solubility.

Table 29.2 Solubility rules

Soluble	Insoluble
all salts of ammonium, potassium and sodium	
all nitrates	
most bromides, chlorides and iodides	lead and silver bromides, chlorides and iodides
most sulphates (calcium sulphate is slightly soluble)	barium and lead sulphates
ammonium, potassium and sodium carbonates	most other carbonates
ammonium, potassium and sodium hydroxides (calcium hydroxide is slightly soluble)	most other hydroxides

The tests described in this topic give additional information.

Flame tests

A flame test may be carried out on a solid or a solution. A small amount of the solid (or solution) is introduced into a non-luminous Bunsen flame on a clean platinum or nichrome wire. The colour observed in the flame indicates the cation present (Table 29.3).

Table 29.3 Flame colours

Colour of flame	Likely identity of the cation
crimson	Sr^{2+}
green (apple)	Ba^{2+}
green (blue)	Cu^{2+}
lilac	K^+
red	Li^+
red (brick)	Ca^{2+}
yellow	Na^+

The yellow colour of sodium is very intense and if present, even in very small amounts as an impurity, may mask other colours. If potassium is suspected, the flame should be looked at through cobalt-blue glass. This filters out wavelengths that interfere with observation, particularly the yellow of sodium.

Other tests for cations

Not all cations can be identified by a flame test. Table 29.4 gives details of other tests used to identify cations.

Table 29.4 Tests for cations

Cation	Test and result
Ba^{2+}	Addition of potassium chromate solution forms a yellow precipitate of barium chromate.
Cr^{3+}	Addition of sodium hydroxide solution forms a grey-green precipitate that dissolves in excess giving a deep-green solution containing the chromate(III) ion. Subsequent addition of hydrogen peroxide and gentle warming oxidises chromate(III) to chromate(VI), giving a yellow solution.
Cu^{2+}	Addition of ammonia solution initially forms a blue precipitate of copper(II) hydroxide. This dissolves in excess ammonia solution to form a deep-blue solution containing the complex $[Cu(H_2O)_2(NH_3)_4]^{2+}$.
Fe^{2+}	Addition of sodium hydroxide solution forms a muddy-green gelatinous precipitate of iron(II) hydroxide. This rapidly turns brown at the surface due to oxidation.
Fe^{3+}	Addition of sodium hydroxide solution forms a red-brown precipitate of iron(III) hydroxide.
NH_4^+	All ammonium compounds evolve ammonia when heated with a slight excess of sodium hydroxide solution. Ammonia has a characteristic smell, and turns moist red litmus paper blue.
Ni^{2+}	Addition of ammonia solution initially forms a pale green precipitate of nickel(II) hydroxide. This dissolves in excess ammonia solution to form a deep-blue solution containing the complex $[Ni(H_2O)_2(NH_3)_4]^{2+}$. The colour of this solution is similar to the one obtained from copper(II).

Reaction with aqueous sodium hydroxide and ammonia solution

The reactions of aluminium, magnesium and zinc salts with sodium hydroxide solution and ammonia solution provide a way of distinguishing between these cations (Table 29.5).

Table 29.5 Distinguishing between Al^{3+}, Mg^{2+} and Zn^{2+}

Hydroxide	Precipitated by $NaOH_{(aq)}$	Dissolves in excess $NaOH_{(aq)}$	Precipitated by $NH_{3(aq)}$	Dissolves in excess $NH_{3(aq)}$
$Al(OH)_3$	✓	✓	✓	
$Mg(OH)_2$	✓		slight	
$Zn(OH)_2$	✓	✓	✓	✓

In the case of magnesium this is particularly useful as there is no satisfactory test for the Mg^{2+} ion.

Reaction with aqueous silver nitrate

When silver nitrate solution is added to a solution of a metal halide, a precipitate of the silver halide is formed. The colour of the precipitate and its solubility in dilute and concentrated aqueous ammonia provide a way of identifying the halide (Table 29.6).

Table 29.6 Identifying halides

Halide	Colour of precipitate formed with $AgNO_{3(aq)}$	Precipitate soluble in concentrated $NH_{3(aq)}$	Precipitate soluble in dilute $NH_{3(aq)}$
Cl^-	white	✓	✓
Br^-	cream	✓	
I^-	yellow		

The presence of halide ions can also be detected using displacement reactions. Chlorine will displace bromide and iodide ions from solution by oxidising them to bromine and iodine, respectively. A small amount of chlorine water and hexane is shaken with a solution of the unknown substance. If the solution contains bromide ions the hexane will be coloured orange due to the production of bromine. Similarly, if the solution contains iodide ions the hexane will turn pink/purple due to the iodine produced.

Other tests for anions

Table 29.7 gives details of other tests used to identify anions.

Table 29.7 Tests for anions

Anion	Test and result
HCO_3^- / CO_3^{2-}	Addition of an acid results in effervescence due to the evolution of carbon dioxide gas. This gas turns lime water milky.
NO_3^-	Addition of freshly prepared iron(II) sulphate solution followed by concentrated sulphuric acid poured down the inside of the test-tube forms two layers. A brown ring forms at the interface.
SO_3^{2-}	Addition of dilute hydrochloric acid with warming results in the evolution of sulphur dioxide gas. This gas is pungent and turns acidified potassium dichromate solution green.
SO_4^{2-}	Addition of barium chloride solution forms a dense white precipitate of barium sulphate.

The ions HCO_3^- and CO_3^{2-} can be distinguished using a solution of magnesium sulphate.

- Addition of magnesium sulphate to a solution of a hydrogencarbonate forms no precipitate, since magnesium hydrogencarbonate is soluble.
- Addition of magnesium sulphate to a solution of a carbonate gives a white precipitate of magnesium carbonate.

Organic compounds

The qualitative analysis of organic compounds is concerned with detecting the presence of functional groups. Some indication of the nature of the functional group in a compound and the length of the carbon chain may be evident simply from its solubility in water and/or organic solvents: long-chain compounds tend to be less soluble in water. The tests described in this topic give additional information.

Triiodomethane (iodoform) test

A small amount of the test substance is added to a solution of iodine in potassium iodide. Then, sodium hydroxide solution is added dropwise until the colour of the iodine is just discharged. A yellow precipitate and an antiseptic smell indicate the formation of triiodomethane (CHI_3).

continues

Triiodomethane is produced if either of the following structural groups is present in the test substance:

$$CH_3-\underset{OH}{\overset{H}{\underset{|}{\overset{|}{C}}}}- \quad \text{or} \quad CH_3-\overset{O}{\overset{\|}{C}}-$$

Compounds that give a positive triiodomethane test include ethanol, propan-2-ol, butan-2-ol, ethanal and all methyl ketones.

Alcohols

Alcohols react with solid PCl_5 with the evolution of hydrogen chloride gas, which is acidic in aqueous solution and thus turns damp blue litmus paper red.

Primary, secondary and tertiary alcohols can be distinguished by the ease of oxidation using the colour change of potassium dichromate, from orange to green, as an indication that oxidation has taken place.

Primary and secondary alcohols are readily oxidised, while tertiary alcohols can only be oxidised destructively:

- primary alcohol → aldehyde → carboxylic acid
- secondary alcohol → ketone
- tertiary alcohol → products of destructive oxidation, for example carbon dioxide.

Alkenes

Alkenes are identified by shaking with an excess of bromine water. The bromine water is decolorised as an addition reaction takes place.

$$RHC{=}CHR + Br_2 \rightarrow RHBrC{-}CBrHR$$

Alkenes also decolorise acidified potassium manganate(VII) solution, often forming a diol.

Care must be taken with the interpretation of tests carried out with the latter reagent since there are other compounds, such as alcohols and aldehydes, which are readily oxidised and will thus also decolorise potassium manganate(VII) solution. Other observations, such as miscibility with water, should also be made when an unknown compound is being tested. Alkenes are generally immiscible with water.

Aldehydes and ketones

These carbonyl compounds are identified using 2,4-DNPH and Tollens' reagent.

2,4-dinitrophenylhydrazine (2,4-DNPH)

A small amount of the substance is added to a solution of 2,4-DNPH. A yellow or yellow-orange precipitate indicates that the substance is either an aldehyde or a ketone.

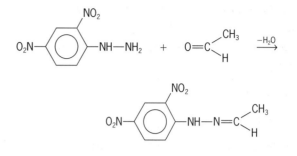

This test involves the formation of a condensation product involving the C=O group of an aldehyde or ketone. The test does not work with a carboxylic acid.

Silver mirror test – Tollens' reagent

Tollens' reagent is an ammoniacal solution of silver nitrate. It is prepared by adding aqueous sodium hydroxide dropwise to silver nitrate solution in a test-tube to form a grey–brown precipitate. The supernatant liquid is decanted off and then aqueous ammonia is added until the remaining solid just dissolves.

A few drops of the compound are added to the solution and the mixture is left to stand for a few minutes in hot water. If the compound is an aldehyde a silver mirror is deposited on the inside of the test-tube.

The ammoniacal solution of silver nitrate contains the ion $[Ag(NH_3)_2]^+$. If an aldehyde is added the following reduction takes place:

$$Ag(NH_3)_2{}^+(aq) + e^- \rightarrow Ag(s) + 2NH_3(aq)$$

A similar test to this may also be carried out using Fehling's solution. In this case the result is a brick-red precipitate of copper(I) oxide.

Carboxylic acids

Carboxylic acids show the typical reactions of an acid. Short-chain carboxylic acids are soluble in water, forming solutions that turn blue litmus paper red and liberate carbon dioxide from a carbonate or hydrogencarbonate.

Haloalkanes

Haloalkanes are dissolved in water (or ethanol if insoluble in water) and then aqueous silver nitrate is added. The halide ion liberated by the hydrolysis of the haloalkane forms a precipitate of the silver halide that can be identified by its colour and its solubility in ammonia solution.

Phenols

Phenols dissolve in water to form weakly acidic solutions. A phenolic solution is just about acidic enough to turn blue litmus paper red but will not liberate carbon dioxide from a carbonate or hydrogencarbonate.

1 Which of the following cations gives a lilac coloured flame test?
   A Ba^{2+}
   B K^+
   C Sr^{2+}
   D Na^+

2 Which of the following compounds gives a positive result for an iodoform test?
   A 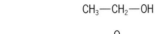 $CH_3—CH_2—OH$

   B
   $$CH_3—CH_2—\overset{\overset{\displaystyle O}{\|}}{C}—CH_2—CH_3$$

   C $CH_3—OH$

   D $CH_3—CH_2—C\overset{\displaystyle O}{\underset{\displaystyle H}{}}$

3 When heated with dilute hydrochloric acid, which of the following ions produces a gas that turns acidified potassium dichromate solution from orange to green?
   A CO_3^{2-} **B** NO_3^- **C** SO_3^{2-} **D** SO_4^{2-}

4 Which of the following results would indicate that an organic compound is an aldehyde?

	Positive test with Tollens' reagent	Positive test with 2,4-DNPH
A	no	yes
B	no	no
C	yes	no
D	yes	yes

5 When reacted with aqueous ammonia, which of the following ions initially gives a pale green precipitate that dissolves in excess reagent to form a deep-blue solution?
   A Cu^{2+} **B** Fe^{2+} **C** Fe^{3+} **D** Ni^{2+}

6 Briefly describe simple laboratory tests you could carry out to differentiate between:
   a methanol and ethanol
   b pentanal and pentan-1-ol
   c bromoethane and chloroethane
   d ethanoic acid and phenol.

7 Briefly describe simple laboratory tests you could carry out to differentiate between:
   a magnesium sulphate and zinc sulphate
   b barium chloride and strontium chloride
   c potassium bromide and potassium iodide
   d sodium carbonate and sodium hydrogencarbonate.

8 A mixture is composed of two substances. The following is a summary of the tests carried out on this mixture. Comment on the results of each test and identify the components in the mixture.

- Test 1: appearance and solubility
  The mixture was a black powder and was insoluble in water.
- Test 2: flame test
  The mixture gave a green-blue coloured flame.
- Test 3: with dilute hydrochloric acid and heating
  The mixture produced a green solution and a black residue; these were separated by filtration.
- Test 4: filtrate with aqueous ammonia
  Initially a pale blue precipitate that dissolved in excess ammonia to give a deep-blue solution.
- Test 5: heating the mixture strongly for several minutes
  The contents glowed red hot even outside the flame; a gas was given off that turned lime water milky; a dark brown solid was formed.
- Test 6: dilute sulphuric acid added to cooled residue from test 5
  Solution became tinted slightly blue but not enough to indicate that a reaction had taken place.

9 A double salt contains two cations and one anion. The following is a summary of the tests carried out on this substance. Comment on the results of each test and identify the ions present.

- Test 1: solubility in water
  Readily dissolved in water giving a colourless solution.
- Test 2: flame test
  Lilac colour observed in flame.
- Test 3: addition of sodium hydroxide solution
  Formed a white precipitate that dissolved in excess reagent.
- Test 4: addition of ammonia solution
  Formed a white precipitate that did not dissolve in excess reagent.
- Test 5: addition of silver nitrate solution
  No reaction was observed.
- Test 6: addition of barium chloride solution
  Formed a white precipitate.

10 Two organic liquids, X and Y, are structural isomers which have a relative molecular mass of 58. The following is a summary of the tests carried out on these substances. Comment on the results of each test and identify the substances.

- Test 1: addition of an equal volume of water
  X is immiscible; Y is miscible.
- Test 2: addition of 2,4-DNPH
  X forms an orange precipitate; Y forms a yellow precipitate.
- Test 3: iodoform test
  X no observed reaction; Y yellow precipitate.
- Test 4: addition of ammoniacal silver nitrate solution
  X silver mirror formed on inside of test-tube; Y no observed reaction.
- Test 5: addition of acidified potassium manganate(VII) solution
  X reagent decolorised; Y no observed reaction.

1 B

2 A

3 C

4 D

5 D

6 a A small amount of each substance is added separately to a solution of iodine in potassium iodide followed by just sufficient sodium hydroxide solution to discharge the iodine colour. Ethanol gives a yellow precipitate that has a characteristic antiseptic smell while methanol does not.

b A few drops of a dry sample of each substance are placed in separate test-tubes and a small amount of phosphorus pentachloride is added to each. Pentan-1-ol reacts with the phosphorus pentachloride releasing hydrogen chloride gas which turns damp blue litmus paper red. Pentanal does not react with phosphorus pentachloride.

c On hydrolysis, bromoethane releases bromide ions. Addition of silver nitrate solution produces a cream precipitate that is soluble in concentrated aqueous ammonia but insoluble in dilute ammonia. On hydrolysis, chloroethane releases chloride ions. Addition of silver nitrate solution produces a white precipitate that is soluble in both concentrated and dilute aqueous ammonia.

d Ethanoic acid reacts with sodium carbonate to produce carbon dioxide gas, which can be identified by turning lime water milky. An aqueous solution of phenol does not react with sodium carbonate.

7 a The addition of sodium hydroxide solution results in the precipitation of both hydroxides; however, only zinc hydroxide is soluble in excess reagent.

b Barium gives an apple-green flame test while for strontium the flame is crimson.

c Addition of aqueous silver nitrate to potassium bromide solution gives a cream precipitate that is soluble in concentrated ammonia solution. The same test carried out on potassium iodide gives a yellow precipitate that is insoluble in concentrated ammonia solution.

d If solutions of the two salts are added to a solution of magnesium sulphate the carbonate gives a white precipitate while the hydrogencarbonate gives no precipitate.

8 • Test 1: insolubility indicates Group 1 salts are not present; the black colour suggests perhaps transition metal compound(s).

• Test 2: green-blue flame indicates the presence of copper(II) ions.

• Test 3: it appears that one component of the mixture reacts with the hydrochloric acid while the other does not; the green coloured solution is perhaps copper(II) chloride; the black residue may be carbon.

• Test 4: the pale blue precipitate is consistent with copper(II) hydroxide while the deep-blue solution is the result of the tetraamminecopper(II) complex.

• Test 5: the red glow indicates a strongly exothermic reaction; the gas given off is carbon dioxide; the dark brown solid may be copper.

• Test 6: little reaction has taken place, which would be expected if the brown solid was copper.

Overall: both substances are black and one is a compound of copper(II); heating copper(II) oxide with carbon would bring about a reduction of the former to copper and the oxidation of the latter forming carbon dioxide; copper does not react with dilute sulphuric acid. The substances are copper(II) oxide and carbon.

9 • Test 1: the absence of colour indicates that there are no transition metal ions present.

• Test 2: the lilac colour observed in the flame indicates the presence of potassium ions, K^+.

• Test 3: the white precipitate formed on addition of sodium hydroxide solution indicates that one of the cations forms an insoluble hydroxide; this could be Al^{3+}, Mg^{2+} or Zn^{2+}. As the precipitate dissolves in excess sodium hydroxide solution it is amphoteric and must therefore be Al^{3+} or Zn^{2+}.

• Test 4: of the two possible cations indicated in the previous test, only aluminium hydroxide is insoluble in excess ammonia solution so the second cation is Al^{3+}.

• Test 5: no reaction with silver nitrate solution indicates that the anion is not a halide.

• Test 6: the white precipitate indicates the presence of the sulphate ion, SO_4^{2-}.

Overall, the ions present are K^+, Al^{3+} and SO_4^{2-}.

10 • Test 1: miscibility with water gives some indication of carbon chain length and functional group. Since X and Y are structural isomers there is not likely to be much difference in carbon chain length. Miscibility with water suggests that Y perhaps has an —OH or —COOH group, but there are other possibilities.

• Test 2: as both X and Y give a precipitate with 2,4-DNPH they must both contain —C=O and be either an aldehyde or a ketone.

• Test 3: since Y gives a positive result with the iodoform test it must contain $CH_3C=O$ or $CH_3CH(OH)$. These groups must be absent from X.

• Test 4: since X gives a silver mirror it must be capable of being easily oxidised so it is an aldehyde rather than a ketone. Conversely, since Y is not easily oxidised it must be a ketone rather than an aldehyde.

• Test 5: the result of this test confirms that of test 4; X is readily oxidised while Y is not.

Overall: since X is known to be an aldehyde it must contain the group —CH=O. From the relative molecular mass the remainder of the molecule must be $58 - 29 = 29$ and so is CH_3CH_2—. Therefore X is propanal, CH_3CH_2CHO. Since Y is a structural isomer of X and is a ketone, it must be propanone, CH_3COCH_3.

Synoptic questions

1 a Define the term **relative atomic mass**.
  b i State the mass of and charge carried by an electron, a neutron and a proton.
  ii Draw a diagram to show how these particles are arranged in an atom of ^7Li.
  iii Explain how the structure of ^7Li differs from that of ^6Li.
  c Describe the important features of a mass spectrometer and explain how this device is able to detect different isotopes of the same element.
  d The following graph shows detail from the mass spectrum of an element.

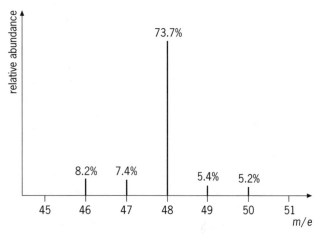

Use the information in the spectrum to calculate the relative atomic mass of the element and identify it.
  e The mass spectrum of an organic compound has a molecular ion at $m/e = 58$. The infrared spectrum of the compound shows a large absorption around $1710\,cm^{-1}$. Analysis of the compound shows it consists of 62.07% carbon, 10.34% hydrogen and the remainder is oxygen.
  i Calculate the molecular formula of this compound.
  ii Draw two possible structural formulae for this compound and name them.
  iii State what additional information you could obtain from the mass spectrum that would enable you to differentiate between the two structures you have drawn in part **ii**.
  iv State one chemical test you could use to differentiate between the two structures you have drawn in part **ii**.

2 a i Draw a diagram to show the shape of the ammonia molecule and the size of the H—N—H angle.
  ii Explain why the H—N—H angle in ammonia is different from the H—C—H angle in methane.
  b Explain the terms **ligand** and **bidentate**.
  c 1,2-diaminoethane (en) forms complexes with many metal ions. In what shape are the donor sites arranged around the metal ion in:
  i $[Fe(en)_2]^{2+}$
  ii $[Co(en)_3]^{3+}$?
  d Name each of the following ions, giving the oxidation state of the central atom in each case.
  i ClO_3^-
  ii $Ni(CN)_4^{2-}$
  iii $Pt(NH_3)_4^{2+}$
  e Iodide ions reduce copper(II) to copper(I). The liberated iodine can be titrated against standard sodium thiosulphate ($Na_2S_2O_3$) solution.
    A sample of 2.810 g of impure copper(II) sulphate ($CuSO_4.5H_2O$) was dissolved in water and made up to $250\,cm^3$ of solution. An excess of potassium iodide solution was added to $25\,cm^3$ of the impure copper(II) sulphate solution. The resulting liberated iodine reacted with $21.6\,cm^3$ of $0.05\,mol\,dm^{-3}$ sodium thiosulphate solution.
  i Give ionic equations for the two reactions involved.
  ii Estimate the purity of the copper(II) sulphate.

3 a Explain the terms **ionisation energy** and **oxidation state**.

  b The following table contains information about the ionisation energies of sodium and magnesium.

Element	First ionisation energy/kJ mol^{-1}	Second ionisation energy/kJ mol^{-1}
sodium	496	4563
magnesium	738	1451

Explain why the first ionisation energy of magnesium is greater than that of sodium but the second ionisation energy is much smaller.

continues

c The following graph shows the melting points of the Period 3 elements.

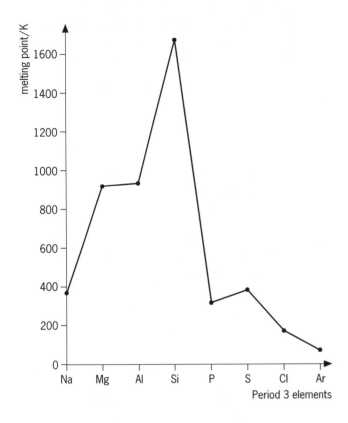

Explain the variation in melting point across Period 3 in terms of structure and bonding.

d Explain why one of the following is a redox reaction and the other is not.

$$ZnO + 2HCl \rightarrow ZnCl_2 + H_2O \qquad \text{(reaction 1)}$$
$$Zn + 2HCl \rightarrow ZnCl_2 + H_2 \qquad \text{(reaction 2)}$$

e In acidic solution the following redox reaction takes place:

$$Cr_2O_7^{2-} \rightarrow Cr^{3+} \qquad \text{and} \qquad SO_3^{2-} \rightarrow SO_4^{2-}$$

Identify which of these is an oxidation and which is a reduction. Write half-equations for the oxidation and reduction and use these to derive an equation for the overall reaction.

4 a Ethanoic acid dissolves in water according to the following equation.

$$CH_3COOH(l) + H_2O(l) \rightleftharpoons H_3O^+(aq) + CH_3COO^-(aq)$$

Use this reaction to explain the terms **conjugate acid** and **conjugate base**.

b Write the acid dissociation constant, K_a, for the above reaction.

c A solution of ethanoic acid has $pK_a = 4.75$. Calculate the value of the dissociation constant and state its units.

d Calculate the pH of a $0.15\,mol\,dm^{-3}$ solution of ethanoic acid.

e Sketch a graph to show how the pH changes when sodium hydroxide solution is added to a solution of ethanoic acid.

f Explain why methyl orange would be unsuitable for determining an accurate end-point for a titration involving sodium hydroxide and ethanoic acid.

5 a Analysis of a bromoalkane **A** shows it to contain 35.0% carbon and 6.57% hydrogen. Calculate the empirical formula of the compound **A**.

b The mass spectrum of compound **A** shows molecular ions at $m/e = 136$ and $m/e = 138$, while its n.m.r. spectrum shows only one peak. Draw the structural formula of compound **A**.

c When gently warmed, compound **A** reacts with aqueous potassium hydroxide to form compound **B**. Name the type of reaction involved and draw a mechanism.

d Discuss whether the above reaction is likely to be zero order or first order with respect to the two reactants and give the rate equation.

e When compound **A** reacts with hot ethanolic potassium hydroxide a different compound, **C**, is the main product. Name the type of reaction involved. Draw the structural formula of compound **C** and give its name.

f A structural isomer of compound **A** is optically active. Draw the two optical isomers of this compound.

6 Methane burns in air according to the following equation.

$$CH_4(g) + 2O_2(g) \rightarrow CO_2(g) + 2H_2O(g)$$

a Define the term **standard enthalpy of formation**.

b The following table contains some standard enthalpies of formation, $\Delta H_f^\ominus$.

Substance	$CH_4(g)$	$O_2(g)$	$CO_2(g)$	$H_2O(g)$
$\Delta H_f^\ominus/kJ\,mol^{-1}$	−74.8	0	−393.5	−241.8

i Why is the standard enthalpy of formation of oxygen equal to zero?

ii Use the standard enthalpies of formation in the table to calculate the standard enthalpy change for the combustion of methane.

c Define the term **mean bond enthalpy**.

d The following table contains some mean bond enthalpies.

Bond	Mean bond enthalpy/kJ mol^{-1}
C—H	435
C=O	805
O—H	464
O=O	498.3

Use the mean bond enthalpy values in the table to calculate the standard enthalpy change for the combustion of methane.

e Suggest why the values you have calculated in parts **b ii** and **d** are not the same.

7 The following sequence shows a two-step synthesis of propylbenzene.

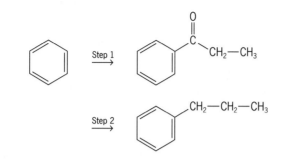

a i Step 1 involves an acylation reaction using an acyl chloride. State the reagents needed to bring about Step 1 and write an equation for the formation of the reactive intermediate involved.
ii Name and draw the mechanism for the reaction of this reactive intermediate with benzene.
b State the reagent used to bring about Step 2.
c Give details of two chemical tests on the carbonyl group that you could use to show that the product of Step 1 is a ketone and not an aldehyde.
d What major difference would you expect to see in the infrared spectra of this ketone and propylbenzene?
e Explain why propylbenzene is not made directly from benzene using a similar method to Step 1 using 1-chloropropane.

8 Hydrogen reacts with the halogens to form hydrogen halides.

$$H_2(g) + X_2(g) \rightleftharpoons 2HX(g)$$

a i Write an expression for the equilibrium constant, K_c, for this general reaction.
ii The following table gives the order of value of equilibrium constants for three halogens at a particular temperature.

X	Order of equilibrium constant
Cl	10^{16}
Br	10^8
I	10^1

Comment on the position of the equilibrium in the three reactions.

b i Write an expression for the equilibrium constant K_p for the reaction

$$H_2(g) + I_2(g) \rightleftharpoons 2HI(g)$$

ii Hydrogen and iodine vapour, both at a pressure of 100 kPa, were mixed together at 500 K. At equilibrium, the partial pressure of each of these gases had dropped to 13.7 kPa. Calculate the equilibrium constant K_p for this reaction at 500 K.
c i Explain why the values of K_c and K_p for the formation of HI are the same but this is not always true for reactions that take place in the gaseous phase.
ii Show that $K_c \neq K_p$ for the reaction:

$$N_2(g) + 3H_2(g) \rightleftharpoons 2NH_3(g)$$

9 **a** Read the following passage carefully.

Compound **A** is a black solid that dissolves in water to form a blue solution.
When concentrated hydrochloric acid is added to the aqueous solution of **A** its colour changes to yellow-green due to the formation of species **B**. If water is added to this solution the original blue colour returns.
When aqueous ammonia is added dropwise to the aqueous solution of **A**, a blue precipitate is formed. When more ammonia is added the precipitate dissolves forming a deep-blue solution due to the formation of species **C**.
When aqueous silver nitrate is added to the aqueous solution of **A**, a cream precipitate is formed. This precipitate is insoluble in dilute aqueous ammonia but dissolves in concentrated aqueous ammonia forming species **D**.

i Identify compound **A**.
ii Give the names and formulae of species **B**, **C** and **D**.
b The following table gives some standard electrode potentials.

Reaction at electrode	$E^\ominus$/V
$Cu^{2+}(aq) + 2e^- \rightarrow Cu(s)$	+0.34
$2H^+(aq) + 2e^- \rightarrow H_2(g)$	0.00
$Mg^{2+}(aq) + 2e^- \rightarrow Mg(s)$	−2.37
$NO_3^-(aq) + 4H^+(aq) + 3e^- \rightarrow NO(g) + 2H_2O(l)$	+0.96

i Write a cell diagram to represent the overall reaction in an electrochemical cell consisting of copper and magnesium electrodes.
ii Calculate the e.m.f. of the cell.
iii Calculate the standard free energy change, $\Delta G^\ominus$, for the reaction that occurs in the cell. (Faraday constant = 96 500 C mol^{-1})
c i Give equations for the reactions of copper and magnesium with dilute nitric acid.
ii Use the standard electrode potentials given in part **b** to explain why these metals react differently with dilute nitric acid.

10 Methanol can be synthesised from carbon dioxide and hydrogen according to the following equation.

$$CO_2(g) + 3H_2(g) \rightleftharpoons CH_3OH(g) + H_2O(g);$$
$$\Delta H = -49\,kJ\,mol^{-1}$$

a i From thermodynamic considerations, explain the conditions that favour the formation of methanol.
ii What practical problems may be experienced in operating under these conditions and how may the conditions be modified to overcome these problems?
b Here are the standard molar entropies for the substances in the above reaction.

Substance	$S^{\ominus}/JK^{-1}mol^{-1}$
$CH_3OH(g)$	239.7
$CO_2(g)$	213.6
$H_2(g)$	130.6
$H_2O(g)$	188.7

i Calculate the standard entropy change for this reaction.
ii Explain how the feasibility of a reaction can be predicted and find the lowest temperature at which this reaction will occur.

11 The following diagram shows the n.m.r. spectrum of an amine.

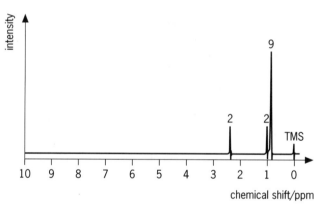
chemical shift/ppm

a The mass spectrum of the amine has a molecular ion at $m/e = 87$. Deduce the molecular formula of the amine.
b Use the n.m.r. spectrum to deduce the structural formula of the amine.
c Explain how shaking with D_2O can be used in conjunction with n.m.r. spectroscopy to determine the structure of an organic compound. Comment on what effect it would have on the n.m.r. spectrum of this amine.
d i Give an equation for the reaction of ammonia with water and write an expression for the base dissociation constant (K_b) of ammonia.

ii Explain the differences in the pK_b values of the following substances.

Substance	pK_b
methylamine	3.4
ammonia	4.8
phenylamine	9.4

e i Give the name and give details of the mechanism of the reaction between methylamine and ethanoyl chloride. What type of substance is formed?
ii Why is ethanoyl chloride preferred to ethanoic anhydride for the preparation of this compound?

ethanoic anhydride

12 a Explain why the boiling points of the halogens increase down the group.
b The following table gives some information about the hydrogen halides.

Hydrogen halide	Boiling point/°C	Bond dissociation energy/ kJ mol⁻¹	Acid dissociation constant, K_a/mol dm⁻³
HF	20	568	5.6×10^{-4}
HCl	−85	432	1×10^7
HBr	−67	366	1×10^9
HI	−35	298	1×10^{11}

Explain:
i the trend in bond dissociation energies down the group
ii why hydrogen fluoride has a higher boiling point than might be expected looking at the pattern of the remaining hydrogen halides
iii why the value of K_a for hydrogen fluoride is much lower than for the remaining hydrogen halides.
c Describe and explain the different properties of hydrogen chloride gas dissolved in water, and hydrogen chloride gas dissolved in methylbenzene.
d Using your knowledge of the reaction of solid sodium halides with concentrated sulphuric acid predict the products formed when sodium astatide reacts with concentrated sulphuric acid, apart from sodium hydrogensulphate and water, and account for their formation.

Answers to synoptic questions

1 a The ratio of the average mass per atom of a naturally occurring form of an element to one-twelfth the mass of an atom of carbon-12.

b i Electron: mass $\frac{1}{1836}$ (negligible), charge -1

neutron: mass 1, no charge
proton: mass 1, charge $+1$

ii

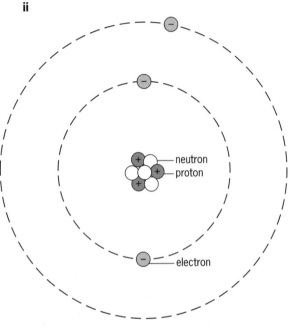

neutron
proton

electron

iii The nucleus of an atom of ^7Li contains four neutrons whereas the nucleus of an atom of ^6Li contains only three.

c In a mass spectrometer a sample is bombarded with a stream of electrons causing the formation of positive ions. The ions are accelerated by a negatively charged plate and deflected by a magnetic field into a detector. The amount of deflection depends on the mass (and charge) of the ion. By varying the strength of the magnetic field it is possible to separate ions of different mass.

d Relative atomic mass
$$= (0.082 \times 46) + (0.074 \times 47) + (0.737 \times 48) +$$
$$(0.054 \times 49) + (0.052 \times 50)$$
$$= 47.87$$
The element is titanium.

e i

	Carbon	Hydrogen	Oxygen
by mass (%)	62.07	10.34	$100 - (62.07 + 10.34) = 27.59$
by mol mass	$\frac{62.07}{12}$	$\frac{10.34}{1}$	$\frac{27.59}{16}$
	$= 5.17$	$= 10.34$	$= 1.72$
$\div$ by smallest	$\frac{5.17}{1.72}$	$\frac{10.34}{1.72}$	$\frac{1.72}{1.72}$
	$= 3$	$= 6$	$= 1$

Molecular formula is C_3H_6O.

ii

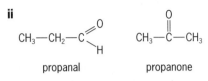

propanal   propanone

iii If the compound is propanal you might expect large peaks at $m/e = 28$ and $m/e = 29$ corresponding to the ions $[CO]^+$ and $[CH_3CH_2]^+$, respectively. However, if the compound is propanone you might expect large peaks at $m/e = 15$ and $m/e = 43$ corresponding to the ions $[CH_3]^+$ and $[CH_3CO]^+$, respectively.

iv Silver mirror test will differentiate between an aldehyde, which reduces an ammoniacal solution of silver nitrate to silver, and a ketone, which does not.

2 a i

$$\underset{107°}{\overset{\cdot\cdot}{H-N}}\begin{smallmatrix}H\\H\end{smallmatrix}$$

ii The repulsion between a lone pair of electrons and a bonding pair of electrons is greater than the repulsion between two bonding pairs of electrons. In the ammonia molecule the hydrogen atoms are forced closer together, hence the bond angle is less.

b Ligand: an ion or molecule that donates a pair of electrons to a metal atom or ion to form a coordination complex. Bidentate: describes a ligand that donates two pairs of electrons.

c i tetrahedral   **ii** octahedral

d i chlorate(v)
ii tetracyanonickelate(II)
iii tetraammineplatinum(II)

e i $2Cu^{2+}(aq) + 4I^-(aq) \rightarrow 2CuI(s) + I_2(aq)$
$I_2(aq) + 2S_2O_3^{2-}(aq) \rightarrow 2I^-(aq) + S_4O_6^{2-}(aq)$
ii $2Cu^{2+} \equiv I_2 \equiv 2S_2O_3^{2-}$ therefore
$CuSO_4.5H_2O \equiv Na_2S_2O_3$
$21.6 cm^3$ of $0.05 mol dm^{-3}$ sodium thiosulphate solution contains

$$0.05 mol dm^{-3} \times 21.6 \times 10^{-3} dm^{-3} = 0.00108 mol$$

If the copper(II) sulphate was pure the mass of the sample would be

$$0.00108 \times 10 \times 249.5 = 2.695 g$$

The purity of the sample of copper(II) sulphate is

$$\frac{2.695 \times 100}{2.810} = 95.9\%$$

3 a Ionisation energy: the energy needed to remove an electron from 1 mole of free atoms or ions in the gaseous state. Oxidation state of an element in a compound is the number of electrons lost (or partially lost in covalent bonds) per atom, compared to an atom in the element..

b In both elements the first electron removed comes from the 3s orbital. The nucleus of the magnesium atom carries a greater charge than that of sodium so the force of attraction between the positively charged nucleus and the 3s electron will be greater.

In magnesium, the second electron removed also comes from the 3s orbital while in sodium the second electron comes from the full 2p orbital. The electrons in the 2p orbital are nearer to the nucleus and less shielded therefore the force of attraction between the nucleus and the 2p electrons is greater.

c Sodium, magnesium and aluminium are metals. As the charge on the nucleus increases so does the strength of the metal–metal bonding hence there is an increase in melting point from sodium to aluminium.

Silicon has a giant structure in which each atom is bonded covalently to four other atoms. A large amount of energy is needed to break these bonds thus silicon has a very high melting point.

Phosphorus, sulphur and chlorine each have a simple covalent structure. They form molecules containing only a small number of atoms, S_8, P_4 and Cl_2. Relatively little energy is needed to overcome the weak van der Waals forces between the molecules.

Argon exists as single atoms attracted together only by weak van der Waals forces, thus the melting point of argon is low.

d In reaction 1 the oxidation state of zinc does not change; it is +2 in both ZnO and $ZnCl_2$.

In reaction 2 the oxidation state of zinc changes from 0 to +2, and the oxidation state of hydrogen changes from +1 to 0 thus reaction 2 is a redox reaction.

e Oxidation: $SO_3^{2-} \rightarrow SO_4^{2-}$
Half-equation: $SO_3^{2-} + H_2O \rightarrow SO_4^{2-} + 2H^+ + 2e^-$
Reduction: $Cr_2O_7^{2-} \rightarrow Cr^{3+}$
Half-equation $Cr_2O_7^{2-} + 14H^+ + 6e^- \rightarrow 2Cr^{3+} + 7H_2O$
Overall equation:
$Cr_2O_7^{2-} + 3SO_3^{2-} + 3H_2O + 14H^+ + 6e^- \rightarrow$
$\qquad\qquad 2Cr^{3+} + 7H_2O + 3SO_4^{2-} + 6H^+ + 6e^-$
$Cr_2O_7^{2-}(aq) + 3SO_3^{2-}(aq) + 8H^+(aq) \rightarrow$
$\qquad\qquad 2Cr^{3+}(aq) + 4H_2O(l) + 3SO_4^{2-}(aq)$

4 a According to the Brønsted–Lowry theory an acid is a proton donor and a base is a proton acceptor. When an acid loses a proton it forms a base which will accept a proton in the reverse reaction, thus CH_3COOH is the conjugate acid of the base CH_3COO^- and CH_3COO^- is the conjugate base of the acid CH_3COOH.

b $K_a = \dfrac{[H^+(aq)][CH_3COO^-(aq)]}{[CH_3COOH(aq)]}$

c $pK_a = -\log_{10}K_a$
$K_a = \text{antilog}_{10}(-pK_a)$
$\quad = \text{antilog}_{10}(-4.75)$
$\quad = 1.78 \times 10^{-5}\,mol\,dm^{-3}$

d Since ethanoic acid is a weak acid the degree of ionisation is small and $[CH_3COOH(aq)]$ at equilibrium is close in value to the initial concentration.

$K_a = \dfrac{[H^+(aq)]^2}{[CH_3COOH(aq)]}$

$[H^+] = \sqrt{(K_a[CH_3COOH(aq)])}$
$\quad = \sqrt{(1.78 \times 10^{-5} \times 0.15)} = 1.63 \times 10^{-3}\,mol\,dm^{-3}$
$pH = -\log_{10}[H^+]$
$\quad = -\log_{10}(1.63 \times 10^{-3}) = 2.79$

e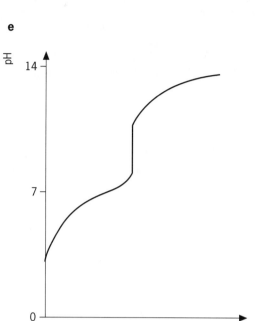

f Methyl orange changes colour in the range pH 3.2–4.4. The end-point for a titration involving a weak acid and a strong base, such as ethanoic acid and sodium hydroxide solution, at pH greater than 7 so the methyl orange would change colour before the actual end-point.

5 a

	C	H	Br
by mass (%)	35.0	6.57	100 − (35.0 + 6.57) = 58.43
by mol mass	$\dfrac{35.0}{12}$	$\dfrac{6.57}{1}$	$\dfrac{58.43}{80}$
	= 2.92	= 6.57	= 0.73
÷ by smallest	$\dfrac{2.92}{0.73}$	$\dfrac{6.57}{0.73}$	$\dfrac{0.73}{0.73}$
	= 4	= 9	= 1

The empirical formula is C_4H_9Br

b The relative molecular mass is 137, therefore the molecular formula is C_4H_9Br. The n.m.r. spectrum indicates that the compound contains only one type of proton therefore it must be 2-bromo-2-methylpropane.

c Substitution reaction – S_N1 mechanism.

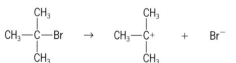

d The mechanism involves the formation of a stable carbocation. The rate-determining step will therefore depend only on the concentration of 2-bromo-2-methylpropane and not on the concentration of the potassium hydroxide solution. The reaction is first order with respect to 2-bromo-2-methylpropane and zero order with respect to potassium hydroxide.

rate = k[2-bromo-2-methylpropane]

e Under these conditions an elimination reaction takes place and the main product is 2-methylpropene.

f

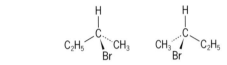

6 a The enthalpy change when 1 mole of a compound is produced from its elements under standard conditions, all reactants and products being in their standard states.

b **i** Oxygen is an element.

ii $\Delta H^{\ominus} = \Delta H_f^{\ominus}(\text{products}) - \Delta H_f^{\ominus}(\text{reactants})$
$= (-393.5 + 2 \times -241.8) - (-74.8)\,\text{kJ mol}^{-1}$
$= -802.3\,\text{kJ mol}^{-1}$

c The amount of energy needed to break 1 mole of a bond in the gaseous state, taken as an average value for breaking the bond in many different compounds.

d $\Delta H^{\ominus}$ = energy needed to break bonds −
energy released when bonds are formed
$= (4 \times C\text{—}H + 2 \times O\text{=}O) - (2 \times C\text{=}O + 4 \times O\text{—}H)$
$= (4 \times 435 + 2 \times 498.3) -$
$(2 \times 805 + 4 \times 464)\,\text{kJ mol}^{-1}$
$= -729.4\,\text{kJ mol}^{-1}$

e Mean bond enthalpies are only approximate values taken for the same bond in a range of compounds in which it occurs. They can be used to give only an approximate value for the enthalpy change in a reaction.

7 a **i** Propanoyl chloride and aluminium chloride (Friedel–Crafts acylation).

$CH_3\text{—}CH_2\text{—}C\overset{O}{\underset{Cl}{\diagup}} + AlCl_3 \rightarrow CH_3\text{—}CH_2\text{—}\overset{O}{\underset{+}{C}}\diagup + AlCl_4^-$

ii Electrophilic substitution.

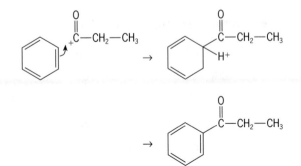

b Zinc amalgam and concentrated hydrochloric acid.

c Reaction with 2,4-dinitrophenylhydrazine solution: a yellow/orange precipitate shows the presence of a carbonyl group in a ketone or aldehyde.

Reaction with an ammoniacal solution of silver nitrate: a silver mirror deposited on the inside of the test-tube shows the presence of an aldehyde.

With the ketone formed in Step 1 expect the first test to be positive and the second test to be negative.

d The ketone would show a C=O stretch around 1680–1750 cm^{-1} that would be absent from the infrared spectrum of propylbenzene.

e There are two problems associated with the use of haloalkanes.

• A monoalkylated benzene is more electron rich than benzene because of the electron-releasing nature of the alkyl group. The result is a tendency for polyalkylation to occur.

• If a Friedel–Crafts alkylation is carried out using 1-chloropropane, propylbenzene is only the minor product; the major product is 1-methylethylbenzene.

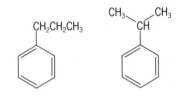

propylbenzene   1-methylethylbenzene

The problem arises because the initially formed primary carbocation rearranges into the thermodynamically more stable secondary species.

$CH_3\text{—}\overset{\overset{H}{|}}{CH}\text{—}\overset{+}{CH_2} \quad \rightarrow \quad CH_3\text{—}\overset{+}{CH}\text{—}CH_3$

8 a **i** $K_c = \dfrac{[HX_{(g)}]^2}{[H_{2(g)}][X_{2(g)}]}$

ii The high value for HCl indicates that the equilibrium is well over to the right and that the reaction has all but gone to completion.

The low value for HI indicates that the reaction is far from completion with around half of the hydrogen and iodine combined to form hydrogen iodide at equilibrium.

The situation for hydrogen bromide is intermediate between these.

b **i** $K_p = \dfrac{p(HI)^2}{p(H_2)p(I_2)}$

ii Partial pressure of HI at equilibrium
$= 2 \times 100\,\text{kPa} - 2 \times 13.7\,\text{kPa}$
$= 172.6\,\text{kPa}$

$K_p = \dfrac{(172.6)^2}{(13.7)(13.7)} = \dfrac{31\,046.44}{187.69} = 165.4$

units: $K_p = \dfrac{(\text{kPa})^2}{(\text{kPa})(\text{kPa})} = \text{no unit}$

c **i** From the ideal gas equation $pV = nRT$ and so $p = (n/V)RT$. The term (n/V) is the concentration of a gas in $mol\,dm^{-3}$ and may be written as [gas]. Therefore the partial pressure of a gas, $p = [gas]RT$. To convert from K_c to K_p each [gas] must be replaced by [gas]RT. Provided that there are the same number of moles on each side of the equation the factors RT will cancel out so K_c is equal to K_p.

ii $$K_c = \frac{[NH_{3(g)}]^2}{[N_{2(g)}][H_{2(g)}]^3}$$

$$K_p = \frac{p(NH_3)^2}{p(N_2)p(H_2)^3} = \frac{[NH_{3(g)}]^2(RT)^2}{[N_{2(g)}](RT)[H_{2(g)}]^3(RT)^3}$$

$$= \frac{[NH_{3(g)}]^2(RT)^{-2}}{[N_{2(g)}][H_{2(g)}]^3}$$

9 a **A** = copper(II) bromide, $CuBr_2$
B = tetrachlorocuprate(II) ion, $[CuCl_4]^{2-}$
C = tetraamminecopper(II) ion, $[Cu(H_2O)_2(NH_3)_4]^{2+}$
D = diamminesilver(I), $[Ag(NH_3)_2]^+$

b **i** $Mg_{(s)}|Mg^{2+}_{(aq)}||Cu^{2+}_{(aq)}|Cu_{(s)}$
ii $E^{\ominus}_{cell} = +0.34 - (-2.37)$
$= +2.71\,V$
iii $\Delta G^{\ominus} = -zFE^{\ominus}_{cell}$ where z is the number of electrons transferred
$= -2 \times 96\,500\,C\,mol^{-1} \times 2.71\,V$
$= -5.23 \times 10^5\,CV\,mol^{-1}$
$= -5.23 \times 10^2\,kJ\,mol^{-1}$

c **i** $Mg_{(s)} + 2HNO_{3(aq)} \rightarrow Mg(NO_3)_{2(aq)} + H_{2(g)}$
$3Cu_{(s)} + 8HNO_{3(aq)} \rightarrow 3Cu(NO_3)_{2(aq)} + 4H_2O_{(l)} + 2NO_{(g)}$
ii When magnesium reacts with dilute nitric acid to produce hydrogen:

$Mg_{(s)}|Mg^{2+}_{(aq)}||H^+_{(aq)}|H_{2(g)}$

$E^{\ominus}_{cell} = 0.00 - (-2.37) = +2.37\,V$
$E^{\ominus}_{cell} > 0$ therefore the reaction is spontaneous.
In the case of copper:

$Cu_{(s)}|Cu^{2+}_{(aq)}||H^+_{(aq)}|H_{2(g)}$

$E^{\ominus}_{cell} = 0.00 - (+0.34) = -0.34\,V$
$E^{\ominus}_{cell} < 0$ therefore the reaction is not spontaneous.
However for:

$Cu_{(s)}|Cu^{2+}_{(aq)}||NO_3^-_{(aq)}|NO_{(g)}$

$E^{\ominus}_{cell} = +0.96 - (+0.34)$
$= +0.62\,V$
$E^{\ominus}_{cell} > 0$ therefore the reaction is spontaneous.

10 a **i** Low temperature: since the reaction is exothermic, removal of heat will drive the equilibrium to the right.
High pressure: since there is a drop from 4 moles of gas to 2 moles of gas, an increase in pressure will drive the equilibrium to the right.

ii At low temperature the reaction would take a long time to reach equilibrium. Operating at high pressure requires expensive plant equipment. The process could be run at moderate temperature and pressure using a catalyst.

b **i** $\Delta S = \Sigma S^{\ominus}_{products} - \Sigma S^{\ominus}_{reactants}$
$= (239.7 + 188.7)\,JK^{-1}\,mol^{-1} -$
$(231.6 + (3 \times 130.6))\,JK^{-1}\,mol^{-1}$
$= -195.0\,JK^{-1}\,mol^{-1}$
ii In order for a reaction to be feasible $\Delta G^{\ominus} < 0$.

$\Delta G^{\ominus} = \Delta H^{\ominus} - T\Delta S^{\ominus}$ therefore $T = \dfrac{\Delta H^{\ominus} - \Delta G^{\ominus}}{\Delta S^{\ominus}}$

When $\Delta G^{\ominus} = 0$, $T = \dfrac{-49\,000\,J\,mol^{-1}}{-195.0\,JK^{-1}\,mol^{-1}}$

$= 251\,K$
This reaction is only feasible above 251 K.

11 a An amine contains C, H and N atoms. If we assume that there is only one N atom in this molecule, the n.m.r. spectrum shows 13 H atoms in the molecule, therefore;

$87 - 14 = 73$; $73 = 60 + 13$; $\dfrac{60}{12} = 5$; $C_5H_{13}N$

b There are three types of proton in the ratio $9:2:2$ and no coupling. This is consistent with the structural formula $(CH_3)_3CCH_2NH_2$.

c When a sample is shaken with D_2O, labile protons will exchange with D^+. Deuterium has no nuclear spin so any signal due to labile protons in the n.m.r. spectrum will be lost after shaking with D_2O.
In the case of this amine, the two protons attached to the N atom would be exchanged and one of the peaks would be lost.

d **i** $NH_3 + H_2O \rightleftharpoons NH_4^+ + OH^-$

$$K_b = \frac{[NH_4^+][OH^-]}{[NH_3]}$$

ii pK_b gives a measure of the strength of a base. The smaller the value the stronger the base.
In methylamine the methyl group is electron-donating. It releases electron charge towards the nitrogen atom in the C—N bond, increasing the electron density on the nitrogen atom and making it easier for the lone pair of electrons to form a bond with H^+.
In phenylamine the phenyl group is electron-withdrawing. The lone pair of electrons on the nitrogen atom become involved in delocalisation around the ring, decreasing the electron density on the nitrogen atom and making it more difficult for the lone pair of electrons to form a bond with H^+.

continues

e i Nucleophilic substitution.

The product is an amide.
ii If ethanoic anhydride was used, ethanoic acid would also be formed as a co-product and the two products would have to be separated.

12 a The trend of increasing boiling point down the group is due to the weak attractive van der Waals forces between halogen molecules. Temporary fluctuations in electron density within molecules result in these temporary induced dipole–dipole attractions between molecules. The size of the attractive force increases with the total number of electrons and the surface area of contact between molecules. Since both atomic and molecular radii increase with increasing atomic number, the boiling point increases.

b i Passing down the group the electronegativity of the halogens decreases and the electrons in the H—X bond are further from the nucleus of the halogen atom.
ii In all of the hydrogen halides the molecules are attracted to each other by relatively weak van der Waals and permanent dipole–dipole interactions. In hydrogen fluoride the molecules are also attracted to each other by stronger hydrogen bonding that must be overcome before boiling occurs.
iii Hydrogen fluoride only forms a weak acid in aqueous solution because the H—F bond is stronger than the other H—X bonds and doesn't dissociate as readily. Also hydrogen bonding between H—F molecules inhibits dissociation.

c Hydrogen chloride dissociates in water forming H^+ and Cl^- ions. The solution has the properties of a typical ionic solution (for example it conducts an electric current) and an acid (for example its pH is less than 7).
   Hydrogen chloride does not dissociate in methylbenzene but remains as a covalent compound.

d $H_2SO_4 + NaAs \longrightarrow HAs$ – displacement reaction
   $\longrightarrow SO_2$, S, H_2S – sulphuric acid is reduced, oxidation state of sulphur in sulphuric acid is reduced from +6 to +4, 0 and −2
   $\longrightarrow As_2$ – astatide oxidised to astatine

Examination questions

In this book the A level Chemistry specification is divided into a series of short topics in order to allow you to study the contents more effectively. However, in the examination papers set by awarding bodies it is not likely that you will find many questions directed at a single area of the specification. It is the job of the examiner to test what you know from across the whole specification. As a result, it is far more likely that you will encounter questions requiring expertise from several different areas.

This section contains a series of questions taken from recent examination papers of all the awarding bodies. After each question you are advised which topics relate to the question. You should use this information to help you answer the question.

This Periodic Table may be useful in answering some of the questions in this section.

Key

relative atomic mass — AA
atomic number — 22 **E** name — element

Period	Group																	0
	1	2										3	4	5	6	7		4 **He** 2 helium
1								1 **H** 1 hydrogen										
2	7 **Li** 3 lithium	9 **Be** 4 beryllium										11 **B** 5 boron	12 **C** 6 carbon	14 **N** 7 nitrogen	16 **O** 8 oxygen	19 **F** 9 fluorine	20 **Ne** 10 neon
3	23 **Na** 11 sodium	24 **Mg** 12 magnesium										27 **Al** 13 aluminium	28 **Si** 14 silicon	31 **P** 15 phosphorus	32 **S** 16 sulphur	35.5 **Cl** 17 chlorine	40 **Ar** 18 argon
4	39 **K** 19 potassium	40 **Ca** 20 calcium	48 **Sc** 21 scandium	48 **Ti** 22 titanium	51 **V** 23 vanadium	52 **Cr** 24 chromium	55 **Mn** 25 manganese	56 **Fe** 26 iron	59 **Co** 27 cobalt	59 **Ni** 28 nickel	63.5 **Cu** 29 copper	65 **Zn** 30 zinc	70 **Ga** 31 gallium	73 **Ge** 32 germanium	75 **As** 33 arsenic	79 **Se** 34 selenium	80 **Br** 35 bromine	84 **Kr** 36 krypton
5	35 **Rb** 37 rubidium	88 **Sr** 38 strontium	89 **Y** 39 yttrium	91 **Zr** 40 zirconium	93 **Nb** 41 niobium	96 **Mo** 42 molybdenum	99 **Tc** 43 technetium	101 **Ru** 44 ruthenium	103 **Rh** 45 rhodium	106 **Pd** 46 palladium	108 **Ag** 47 silver	112 **Cd** 48 cadmium	115 **In** 49 indium	119 **Sn** 50 tin	122 **Sb** 51 antimony	128 **Te** 52 tellurium	127 **I** 53 iodine	131 **Xe** 54 xenon
6	133 **Cs** 55 caesium	137 **Ba** 56 barium		178.5 **Hf** 72 hafnium	181 **Ta** 73 tantalum	184 **W** 74 tungsten	186 **Re** 75 rhenium	190 **Os** 76 osmium	192 **Ir** 77 iridium	195 **Pt** 78 platinum	197 **Au** 79 gold	201 **Hg** 80 mercury	204 **Tl** 81 thallium	207 **Pb** 82 lead	209 **Bi** 83 bismuth	209 **Po** 84 polonium	210 **At** 85 astatine	222 **Rn** 86 radon
7	223 **Fr** 87 francium	226 **Ra** 88 radium		(261) **Rf** 104 rutherfordium	(262) **Db** 105 dubnium	(266) **Sg** 106 seaborgium	(264) **Bh** 107 bohrium	(269) **Hs** 108 hassium	(268) **Mt** 109 meitnerium	(269) **Uun** 110 ununnillium	(272) **Uuu** 111 unununium	(277) **Uub** 112 ununbium		(285) **Uuq** 114 ununquadium		(289) **Uuh** 116 ununhexium		(293) **Uno** 118 ununoctium

139 **La** 57 lanthanum	140 **Ce** 58 cerium	141 **Pr** 59 praseodymium	144 **Nd** 60 neodymium	(145) **Pm** 61 promethium	150 **Sm** 62 samarium	152 **Eu** 63 europium	157 **Gd** 64 gadolinium	159 **Tb** 65 terbium	163 **Dy** 66 dysprosium	165 **Ho** 67 holmium	167 **Er** 68 erbium	169 **Tm** 69 thulium	173 **Yb** 70 ytterbium	175 **Lu** 71 lutetium
(227) **Ac** 89 actinium	232 **Th** 90 thorium	(231) **Pa** 91 protactinium	238 **U** 92 uranium	(237) **Np** 93 neptunium	(244) **Pu** 94 plutonium	(243) **Am** 95 americium	(247) **Cm** 96 curium	(247) **Bk** 97 berkelium	(251) **Cf** 98 californium	(252) **Es** 99 einsteinium	(257) **Fm** 100 fermium	(258) **Md** 101 mendelevium	(259) **No** 102 nobelium	(262) **Lr** 103 lawrencium

1 a Write equations to show the chemical processes which occur when the first and the second ionisation energies of lithium are measured. (3)

b i Explain why helium has a much higher first ionisation energy than lithium.

ii Explain why beryllium has a higher first ionisation energy than boron.

iii Explain why the second ionisation energy of beryllium is greater than the first ionisation energy. (6)

AQA, AS/A level, CH01, June 2000
Topic 2

2 a Sketch the shapes of each of the following molecules, showing any lone pairs of electrons. In each case, state the bond angle(s) present in the molecule and name the shape.

Molecule	Sketch of shape	Bond angle(s)	Name of shape
BF_3			
NF_3			
ClF_3			

(9)

b State the types of intermolecular force which exist, in the liquid state, between pairs of BF_3 molecules and between pairs of NF_3 molecules. (3)

c Name the type of bond which you would expect to be formed between a molecule of BF_3 and a molecule of NF_3. Explain how this bond is able to form. (3)

AQA, AS/A level, CH01, June 2000
Topics 3 and 5

3 a Sketch a graph to show how the melting points of the elements vary across Period 3 from sodium to argon. Account for the shape of the graph in terms of the structure of, and the bonding in, the elements. (21)

b Give the electronic configuration of the lithium ion. Explain fully, with the aid of two examples, how and why the lithium ion is responsible for some lithium compounds being atypical of Group 1 compounds. (9)

AQA, AS/A level, CH01, June 2000
Topics 2 and 6

4 Carbon dioxide is used as a coolant gas in some nuclear reactors. Unlike hydrogen or helium, carbon dioxide shows marked deviations from the ideal behaviour that is predicted by the kinetic theory of gases.

a State **two** assumptions of the kinetic theory, and use these to explain why you might expect the behaviour of carbon dioxide to be less ideal than that of hydrogen. (4)

b Under what conditions of temperature and pressure would you expect the behaviour of carbon dioxide to be most like that of an ideal gas? (2)

c The volume of 1 mol of carbon dioxide was measured at various pressures but at a constant temperature of 285 K. The following results were obtained.

Pressure, p /Pa	Volume, V /m³	Pressure × volume, pV/m³ Pa
4.0×10^5	5.80×10^{-3}	2320
8.0×10^5	2.85×10^{-3}	
15.0×10^5	1.46×10^{-3}	
20.0×10^5	1.07×10^{-3}	

i Complete the calculations for the third column and use these data to plot a graph of the product pV against pressure p.

ii State how the value of the product pV **should** change with pressure for an ideal gas.

iii Use the graph to calculate the volume of 1 mol of carbon dioxide at a pressure of 10×10^5 Pa. Calculate the volume at 285 K that the ideal gas equation predicts for this pressure and comment on the difference between the two values. (7)

OCR, A level, 9254/1, June 2000
Topics 4 and 8

5 Poly(2-hydroxypropanoic acid) is used to make absorbable stitches for surgical procedures. The formula of the repeating unit is shown below.

a The monomer contains two functional groups which react to link the monomers together to form the polymer. Complete the diagram below to show the full structural formula of the monomer. (1)

b What type of polymerisation has taken place in order to produce poly(2-hydroxypropanoic acid)? (1)

c The average relative molecular mass of a polymer chain of poly(2-hydroxypropanoic acid) was found to be 144 000.

How many repeating units does a polymer chain of this relative molecular mass contain? (A_r: C, 12.0; H, 1.0; O, 16.0) (2)

Nylon is another material that is used for stitches. However, the stitches are non-absorbable. Two examples of nylon are shown below.

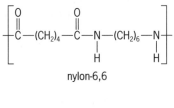

nylon-6,6

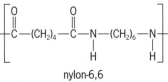

nylon-6,6

d Show, by means of a diagram, the **strongest** intermolecular force between the polymer chains of nylon-6,6. A section of one chain is drawn for you. (2)

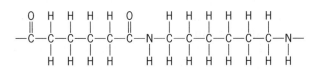

e Nylon fibres have high melting points and high tensile strengths. They are good for stitches because they are flexible.
i Suggest a reason why nylon fibres are flexible. (1)
ii Nylon-6,10 melts at a lower temperature than nylon-6,6. Suggest a reason for this. (2)
f Around 10% by mass of solid household waste in EU countries consists of plastics. One possible method of disposal is by incineration. Steam produced by using the heat given off could be used to generate electricity. State **one** advantage and **one** disadvantage of this method of disposing of waste plastic. (2)

OCR, AS/A level, 5682, January 2000
Topic 10

6 Crude oil is largely a mixture of alkanes.
  a Describe how crude oil is used as a source of aliphatic and aromatic hydrocarbons. (2)
  b Discuss how 'cracking' can be used to obtain more useful alkanes and alkenes of lower M_r from larger hydrocarbon molecules. (2)
Illustrate your answers in **a** and **b** with suitable examples.

Quality of language (4)
OCR, A level, 9254/2, June 2000
Topic 11

7 a Copy and complete the boxes below to show the electronic configuration of the magnesium atom and the magnesium ion, using the convention shown. (2)

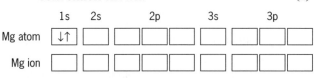

b Define the term **second ionisation energy**. Illustrate your answer with an equation. (3)
c The first three ionisation energies of magnesium are 736, 1450 and 7740 kJ mol^{-1} respectively.
i Explain why the second ionisation energy of magnesium is bigger than the first one. (2)
ii Account for the considerable difference between the values of the second and third ionisation energies of magnesium. (2)

d

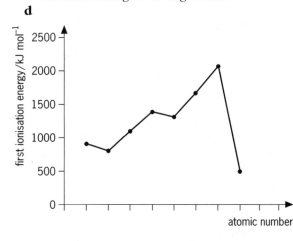

A plot of **first** ionisation energy against atomic number for **eight consecutive elements** in the periodic table is shown above.
  Give the names of the first and last elements to which the plot might apply. (2)
e i Which of the following hydroxides is the most soluble in water?

$Mg(OH)_2$ or $Ca(OH)_2$ or $Sr(OH)_2$ or $Ba(OH)_2$ (1)

ii Which of the following carbonates will decompose at the lowest temperature?

Na_2CO_3 or K_2CO_3 or $MgCO_3$ or $CaCO_3$ (1)

iii Suggest which of the following metals is likely to react **most** vigorously with water.

Na or K or Mg or Ca (1)

iv Which of the following chlorides gives a lilac flame colour?

NaCl or KCl or $CaCl_2$ or $SrCl_2$ (1)

f Explain why magnesium chloride, $MgCl_2$, has more covalent character than magnesium fluoride, MgF_2. (2)

London, AS/A level, Module Test 1, January 2000
Topics 2, 6 and 13

EXAMINATION QUESTIONS

8 a The amount of carbon monoxide present in a gas mixture can be determined by utilising its reaction with iodine(v) oxide, I_2O_5.

$$I_2O_5 + 5CO \rightarrow I_2 + 5CO_2$$

A **1.00 dm³** sample of a gas mixture produced iodine which completely reacted with 32.00 cm³ of sodium thiosulphate solution of concentration 0.500 mol dm⁻³.

Calculate the number of moles of iodine produced and hence the percentage of carbon monoxide in the gas mixture.
(1 mol of any gas has a volume of 24.0 dm³ under the reaction conditions used.) (3)

b When chlorine is passed over heated carbon and silicon(IV) oxide, carbon monoxide and silicon(IV) chloride are formed as the only products.
i Construct the balanced equation for the reaction. (1)
ii State the type of bonding present in silicon(IV) chloride **and** how the compound reacts with water, giving a balanced equation for the reaction. (2)
iii Explain why silicon(IV) chloride can be made by heating silicon in a stream of dry chlorine but any attempt to prepare lead(IV) chloride by this method results only in lead(II) chloride being produced. (2)

c i Briefly explain the following:
I   The elements of Group 7 (the halogens) show a marked tendency towards the formation of ions of type **X⁻**, where **X** represents the symbol of a Group 7 element. (1)

II   In anions of type XO⁻, the X—O bond is predominantly covalent. (1)
III A ClO_3^- ion is a stronger oxidising agent than an IO_3^- ion. (1)
ii State how the oxidation state of the halogen **X** changes in the reaction

$$3XO^- \rightarrow 2X^- + XO_3^-$$

and explain why this reaction can be described as a redox process. (2)
iii The table below shows the standard electrode potentials for the Group 7 (halogen) elements.

Half-equation	$E^\ominus$/V
$\frac{1}{2}F_2(g) + e^- \rightarrow F^-_{(aq)}$	+2.87
$\frac{1}{2}Cl_2(aq) + e^- \rightarrow Cl^-_{(aq)}$	+1.36
$\frac{1}{2}Br_2(l) + e^- \rightarrow Br^-_{(aq)}$	+1.07
$\frac{1}{2}I_2(aq) + e^- \rightarrow I^-_{(aq)}$	+0.54
$\frac{1}{2}At_2(aq) + e^- \rightarrow At^-_{(aq)}$	+0.30

State whether or not the halogen astatine (At, atomic number 85) will be a product of the reaction between sodium astatide (NaAt) and concentrated sulphuric acid. Give a reason in support of your answer. (2)

WJEC, A level, 933/01, June 2000
Topics 7, 8 and 14

9 a Naphtha is one of the components making up crude oil.
i State the name of the process by which naphtha is obtained from crude oil. (1)
ii Pentane is one of the hydrocarbons found in naphtha. Write an equation to show how pentane is **cracked** to give methane as one of the products. (1)
b Methane reacts with chlorine to produce chloromethane.
i State the conditions necessary for this reaction to take place. (1)
ii Write equations to show the initiation step and a propagation step for this reaction. (2)
c i Explain what is meant by the term **dynamic equilibrium**. (1)
ii The reaction of methane with iodine is very much slower than the reaction with chlorine and the equilibrium position lies well to the left under normal conditions.

$$CH_4(g) + I_2(s) \rightleftharpoons CH_3I(l) + HI(g)$$

Explain why, by reference to Le Chatelier's principle, the yield of iodomethane, CH_3I, can be increased in this reaction by removing hydrogen iodide as it is formed. (2)
iii The table shows the standard molar enthalpy changes of formation of methane, iodomethane and hydrogen iodide.

Compound	$\Delta H^\ominus_f$/kJ mol⁻¹
$CH_4(g)$	−74.8
$CH_3I(l)$	−15.5
$HI(g)$	+26.5

By means of an energy diagram or otherwise, calculate the enthalpy change for the reaction

$$CH_4(g) + I_2(s) \rightarrow CH_3I(l) + HI(g) \quad (2)$$

iv State, giving a reason in support of your answer, and using the enthalpy values given in part **iii**, the enthalpy change for the process

$$2HI(g) \rightarrow H_2(g) + I_2(s) \quad (2)$$

WJEC, AS/A level, 931/01, June 2000
Topics 10, 11 and 15

10 a Dinitrogen tetraoxide, N_2O_4, dissociates at 298 K into nitrogen dioxide, NO_2.

$$N_2O_4(g) \rightleftharpoons 2NO_2(g)$$

A sample of dinitrogen tetraoxide, in a sealed container, is allowed to reach equilibrium at this temperature. The partial pressure of each gas, plotted against time, is shown below.

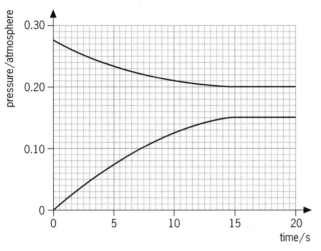

i State the partial pressures of the two gases at equilibrium. (1)

ii Write the expression for the equilibrium constant, K_p, and use the partial pressures obtained in part **i** to calculate the value of K_p, at 298 K, giving the units. (2)

iii State and explain the effect, if any, on the partial pressures of the two gases at equilibrium if the temperature remains at 298 K for a further period of time. (1)

iv At 350 K the value of K_p is found to have increased from its value at 298 K. State what effect this would have on the partial pressure of **each** gas. (1)

b Explain why anhydrous solid aluminium chloride exists as dimeric molecules, Al_2Cl_6, and not as the monomer $AlCl_3$. Your answer should include a description of the bonding in the dimeric molecule. (2)

c The formula of basic magnesium carbonate, used in some antacid tablets, is $MgCO_3.Mg(OH)_2.3H_2O$.

Give the balanced equation for the thermal decomposition of this compound and use it to show that the mass of solid remaining after heating is 41% of the original mass. (3)

WJEC, A level, 933/01, June 2000
Topics 3 and 16

11 a i The ground state electronic configuration of scandium may be written as

$$1s^2 2s^2 2p^6 3s^2 3p^6 3d^1 4s^2$$

Write down the ground state electronic configuration of nickel in the same way. (1)

ii State **three** general **chemical** properties of nickel characteristic of the d-block elements. (3)

b A **particular** sample of the element nickel contains three isotopes with abundances as shown below.

Relative isotopic mass	Percentage abundance
58	60.0
60	30.0
61	10.0

Calculate the relative atomic mass of nickel for **this** sample. (2)

c In the manufacture of nickel, the impure element is converted into the volatile compound nickel tetracarbonyl

$$4CO(g) + Ni(s) \xrightarrow{60\,°C} Ni(CO)_4(g)$$

On heating to around 200 °C the nickel tetracarbonyl decomposes to leave pure nickel

$$Ni(CO)_4(g) \rightarrow 4CO(g) + Ni(s)$$

Calculate the minimum volume, in cm^3, of carbon monoxide, measured at 0 °C and 101 kPa, which is required to produce 500 kg of pure nickel. ($A_r(Ni) = 58.71$)

(The molar gas volume at 0 °C and 101 kPa is $2.241 \times 10^4\,cm^3$) (3)

WJEC, AS/A level, 930/01, June 2000
Topics 1 and 17

12 The element bromine is very dangerous. It has been suggested that it should be turned into calcium bromide in order to transport it safely.

a i Describe the appearance of the element bromine at room temperature. (1)

ii State **two** reasons why bromine is dangerous if it is spilt in an accident. (2)

b i Draw a dot–cross diagram for calcium bromide, $CaBr_2$, showing the outer electron shells only. (2)

ii Calculate the mass of calcium bromide which would contain 1.0 kg of bromine. (A_r: Ca, 40; Br, 80) (2)

c Calcium bromide can be dissolved in water.

i In solution, calcium ions are said to be **hydrated**. Draw a labelled diagram of a hydrated calcium ion to show what this means. (2)

ii The ionic radii of Ca^{2+} and Na^+ ions are about the same, yet calcium ions are more hydrated than sodium ions. Suggest an explanation for this. (1)

iii Calculate the enthalpy change of solution of calcium bromide given the following data.

lattice enthalpy of $CaBr_2(s)$	$-2180\,kJ\,mol^{-1}$
solvation enthalpy of $Ca^{2+}(g)$	$-1562\,kJ\,mol^{-1}$
solvation enthalpy of $Br^-(g)$	$-338\,kJ\,mol^{-1}$ (2)

d Some reactions of solid and aqueous calcium bromide are given below.

$$CaBr_{2(aq)} + 2AgNO_{3(aq)} \rightarrow Ca(NO_3)_{2(aq)} + 2AgBr_{(s)}$$
(equation 1)

$$CaBr_{2(s)} + H_2SO_{4(l)} \rightarrow CaSO_{4(s)} + 2HBr_{(g)}$$
(equation 2a)

followed by:

$$2HBr_{(g)} + H_2SO_{4(l)} \rightarrow Br_{2(l)} + SO_{2(g)} + 2H_2O_{(l)}$$
(equation 2b)

$$Cl_{2(g)} + CaBr_{2(aq)} \rightarrow CaCl_{2(aq)} + Br_{2(aq)}$$
(equation 3)

i From equations 1 to 3, identify a precipitation reaction and write an **ionic** equation to represent it (including state symbols). (2)
ii From equations 1 to 3, identify a redox reaction and state what is being oxidised. (2)
iii Aqueous calcium bromide can be **electrolysed**. Write the half-equation for the reaction in which bromide ions are turned into the element bromine. (1)
iv Using equations 2a, 2b and 3 and your answer to part **d iii**, suggest (with a reason) which process would be the best for the **industrial** production of pure bromine from calcium bromide.
   State, for each of the other two processes, why you rejected them. (3)
OCR, AS/A level, 5681, January 2000
Topics 3, 17 and 25

13 a The ethanedioate ion, $C_2O_4^{2-}$, acts as a bidentate ligand.
i What is meant by the term **bidentate ligand**?
ii This ligand forms an octahedral complex with iron(III) ions. Deduce the formula of this complex and draw its structure, showing all the atoms present. (5)
b i Give the name of a naturally occurring complex compound which contains iron.
ii What is an important function of this complex compound? (2)
c You are provided with a mixture of iron(III) hydroxide and aluminium hydroxide. Describe, giving essential experimental details, how you could obtain a pure sample of aluminium hydroxide from this mixture. (4)
AQA, A level, CH05, June 2000
Topic 18

14 a Discuss how **three** characteristics of compounds of **transition elements** are displayed by copper(II) ions, by
i comparing the aqueous solutions of copper(II) nitrate and magnesium nitrate, and
ii their reactions with
I   aqueous ammonia
II aqueous potassium iodide. (5)

b An iron(II) sulphate tablet of mass 0.200 g was dissolved in excess dilute sulphuric acid. The Fe^{2+} ions were completely oxidised by 28.4 cm³ of 0.010 mol dm⁻³ potassium manganate(VII) solution.
i Write the **ionic** equation for the reaction of the manganate(VII) ions with the Fe^{2+} ions. (1)
ii Determine the **number of moles** of Fe^{2+} ions and hence the **mass** of Fe^{2+} ions in the tablet. (3)
iii Give **one** reason why iron is important in our diet. (1)
iv State what is observed when sodium hydroxide solution is added gradually to separate solutions of Fe^{2+} and Zn^{2+} ions. (3)
c i Place the elements Cl, Br and I in order of
I   increasing electronegativity
II increasing stability of the ions of the elements in the higher oxidation states. (2)
ii Explain why hydrogen chloride can be prepared by reacting sodium chloride with concentrated sulphuric acid but hydrogen iodide cannot be prepared by a similar method. (2)
iii Write the balanced equation for the reaction of chlorine with dilute aqueous sodium hydroxide. State the oxidation number of chlorine in chlorine gas and in each of the chlorine containing products. Comment on any change(s) in the oxidation state of the chlorine in the reaction. (3)
WJEC, A level, 009/02, June 2000
Topics 14, 17 and 18

15 a During exercise, lactic acid, $CH_3CH(OH)COOH$, is produced in the muscles.
i Copy the following and draw spatial representations to show the relationship between the two optical isomers of lactic acid.

isomer 1   isomer 2

(2)
ii Lactic acid dissociates in aqueous solution to form the lactate and hydrogen ions:

$$\text{lactic acid} \rightleftharpoons \text{lactate}^- + H^+$$

Write down the K_a expression for this reaction. (1)

iii Deduce whether the dissociated or undissociated form of this weak acid ($K_a = 1.4 \times 10^{-4}$ mol dm⁻³) is mainly present at the pH of the human body (pH = 7.4). (3)
iv Write the balanced equation for the reaction between lactic acid and sodium hydroxide. (1)

v Which of the following curves, showing pH changes, would occur during the titration of $0.1\,mol\,dm^{-3}$ sodium hydroxide solution into $25\,cm^3$ of $0.1\,mol\,dm^{-3}$ lactic acid? Explain your choice of diagram.

A

B

C

(2)

vi State and explain which of the indicators listed below would be most suitable for use in the above titration.

Indicator	pH range
methyl orange	3.5–4.5
methyl red	4.2–6.3
phenolphthalein	8.0–10.0

(1)

b i When solid KCl is dissolved in water the change may be represented by the equation below:

$$KCl(s) \rightarrow K^+(aq) + Cl^-(aq)$$

An enthalpy change accompanies this process. State the names of the two enthalpies which contribute to this change. (1)

ii The enthalpy changes of solution, in $kJ\,mol^{-1}$, at 25°C when the following chlorides are dissolved in water are given below:

Compound	Enthalpy change of solution
AgCl	+61
NaCl	+3.9
CaCl$_2$	−83
KCl	+17

State which compound is the most soluble at 25°C and give a reason for your choice. (2)

iii Explain **briefly** the insolubility of pentane in water. (1)

WJEC, A level, 009/02, June 2000
Topics 12, 19 and 25

16 Bromate(v) ions react with bromide ions in acid solution according to the following equation.

$$BrO_3^-(aq) + 5Br^-(aq) + 6H^+(aq) \rightarrow 3Br_2(aq) + 3H_2O(l)$$

A series of experiments was carried out to find the order of reaction by measuring the initial rates of formation of bromine for different concentrations of reactants. The results of the experiments are given in the table below.

pH	[Br$^-$]/mol dm^{-3}	[BrO$_3^-$]/mol dm^{-3}	Initial rate/mol dm^{-3}s^{-1}
1.00	2.50×10^{-3}	1.25×10^{-3}	2.00×10^{-6}
1.00	1.25×10^{-3}	1.25×10^{-3}	1.00×10^{-6}
1.00	2.50×10^{-3}	6.25×10^{-4}	1.00×10^{-6}
2.00	2.50×10^{-3}	1.25×10^{-3}	2.00×10^{-8}

a Define the term **order of reaction**. (1)
b Deduce the order of reaction with respect to Br$^-$, BrO$_3^-$ and H$^+$, explaining your reasoning clearly. (4)
c Write a rate equation for the reaction, and use it to calculate the value of the rate constant, k, including the units. (3)
d i Copy and complete the graph below to show how the value of the rate constant, k, varies with temperature, T. (1)

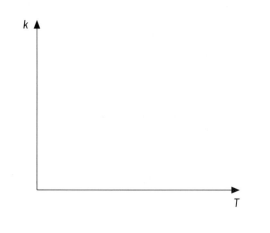

ii Explain why the value of the rate constant varies with temperature. (1)

e The reaction of bromate(v) ions with bromide ions above is thought to proceed via the following reaction mechanism.

Step 1 $H^+ + Br^- \rightarrow HBr$ fast
Step 2 $H^+ + BrO_3^- \rightarrow HBrO_3$ fast
Step 3 $HBr + HBrO_3 \rightarrow HBrO + HBrO_2$ slow
Step 4 $HBrO_2 + HBr \rightarrow 2HBrO$ fast
Step 5 $HBrO + HBr \rightarrow H_2O + Br_2$ fast

Show how this reaction mechanism is consistent with the experimentally determined rate equation.   (2)
OCR, A level, 6824/6288 (Further Physical),
March 2000
Topic 20

17 a i State the ideal gas equation.
Calculate the value of the gas constant, R, from the molar gas volume. State the units. (The molar volume of an ideal gas at $0\,°C$ and $101.3\,kPa$ is $22.4 \times 10^{-3}\,m^3\,mol^{-1}$)   (3)
ii State under which conditions of temperature and pressure, real gases deviate most from ideal gas behaviour. Explain your answer.   (2)
iii Discuss the bonding in the gases hydrogen and hydrogen chloride and explain which of these you would expect to deviate more from ideal gas behaviour.   (4)
b State **one** reaction where the rate can be monitored by the evolution of a gas. **Briefly** outline how you would do this in practice.   (2)
c Sulphur trioxide is formed in the Contact process.

$$2SO_2(g) + O_2(g) \rightleftharpoons 2SO_3(g)$$
$$\Delta H^{\ominus}(298\,K) = -197\,kJ\,mol^{-1}$$

i Draw the reaction profile of the uncatalysed forward reaction and label on your sketch the activation energy and enthalpy change of reaction.   (2)
ii Explain, by using your diagram, how a catalyst affects the reaction profile.   (1)
iii State the conditions of temperature and pressure which favour the **forward** reaction. (1)
iv Draw a diagram to show the distribution of the speeds of gas molecules at two different temperatures T_1 and T_2, where T_2 is greater than T_1. Explain why a temperature of $420\,°C$ is actually used in the industrial processes.   (3)
v The K_p of the above reaction is $3 \times 10^4\,atm^{-1}$ at $420\,°C$ and 1 atmosphere pressure. Write the expression for K_p and explain whether or not the value of K_p would be different at room temperature and pressure.   (2)
WJEC, A level, 009/02, June 2000
Topics 8 and 20

18 a Iodine reacts with pent-1-ene in a similar way to the reaction of bromine with ethene.
i State the type of reaction which has occurred.   (1)

ii The initial rate of the reaction is measured in a number of experiments, varying the concentration of one reactant whilst keeping the other constant.
The results are shown in the graphs below.

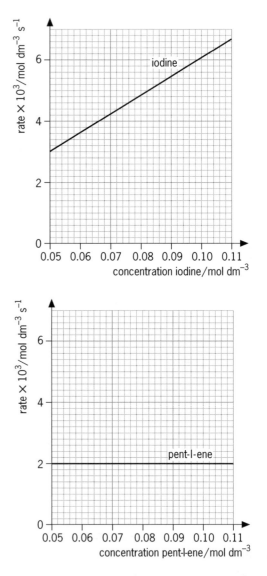

State, giving a reason, the order of reaction with respect to each reactant.   (3)

continues

b But-2-ene can be produced by heating 2-iodobutane at high temperatures.

$$CH_3CH_2CHICH_3 \rightarrow CH_3CH{=}CHCH_3 + HI$$

i 2-iodobutane is said to have a **chiral centre**. Explain what is meant by the term **chiral centre**. Copy the structure of 2-iodobutane drawn below and identify this feature, using an asterisk (*).

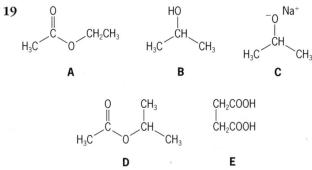

State the particular type of isomerism associated with the term **chiral centre**. (2)

ii But-2-ene exists in *cis* and *trans* forms. Draw the structures of these two forms. (2)

c Discuss the positive contribution that chemistry has made to society, illustrating your answer with **two** particular examples. (3)

WJEC, AS/A level, 931/01, June 2000
Topics 12 and 20

19

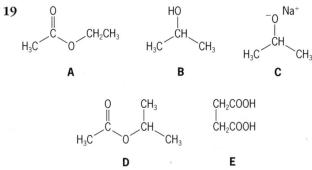

a Draw a mechanism for the hydrolysis of ethyl ethanoate, **A**, by aqueous sodium hydroxide. Explain why 1 mole of sodium hydroxide is needed for the hydrolysis of 1 mole of **A**. (4)

b The hydrolysis of **A** to the corresponding acid and alcohol can be carried out in $H_2^{18}O$ (water with isotopic enrichment in ^{18}O). On the basis of the mechanism you have drawn in part **a**, predict, giving your reasons, which product will contain the ^{18}O. (3)

c When **A** is treated with a large excess of alcohol **B**, in the presence of **C**, ester **D** is formed. On the basis of your mechanism for the hydrolysis of **A**, suggest a mechanism for the formation of **D**. Explain why only a catalytic amount of **C** is needed. (3)

d When the acid **E** is heated with a large excess of alcohol **B** in the presence of a catalytic amount of concentrated sulphuric acid, a compound **F**, $C_{10}H_{18}O_4$, is formed. Draw the structural formula of **F**. (2)

OCR, A level, 6825 (Further Organic), June 2000
Topic 22

20 a Explain the term **geometrical isomerism**. Illustrate your explanation by drawing the **two** geometrical isomers of the complex ion of formula $[Cr(NH_3)_4Cl_2]^+$. (4)

b Explain the term **structural isomerism**. Illustrate your answer by drawing the graphical formulae of **three** alkenes which have the formula C_4H_8. (5)

c State which of the isomers you have drawn in part **b** can show geometrical isomerism. (1)

d The low resolution ^1H n.m.r. spectrum of compound **X**, with the molecular formula C_4H_9Br, is shown in **Figure 1**.

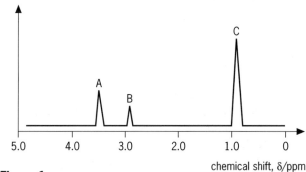

Figure 1

Type of proton	Chemical shift, δ/ppm
$R{-}CH_3$	0.9
$R{-}CH_2{-}R$	1.4
$R_2{-}CH{-}$	2.9
$R{-}CH_2{-}Br$	3.5

Figure 2

i Use the data given in **Figure 2** to deduce the type of proton responsible for each peak in **Figure 1**. (3)

ii Draw the graphical formula of compound **X**. (1)

e Compound **Y** is a structural isomer of C_4H_9Br that can exist as optical isomers. Draw formulae to represent the **two** optical isomers of **Y**. (2)

AQA, A level, Module Paper 9, June 2000
Topics 12 and 23

continues

21 a i An organic liquid **X** has the following elemental percentage composition by mass.

carbon 71.94;   hydrogen 12.08;   oxygen 15.98.

The molar mass of **X** is 100.16 g mol^{-1}.
Determine the molecular formula of **X**.   (3)

ii The infrared spectrum of **X** is shown below.

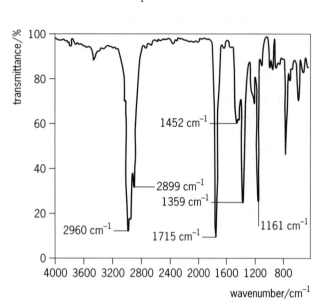

Some characteristic infrared absorptions are given in the table below.

Bond	Wavenumber/cm^{-1}
H—O stretch (in alcohol group)	3400 strong and broad
H—O stretch (in carboxyl group)	3300–3200 variable and broad
C—H alkane stretch	2965–2850 strong
—CH$_2$— bend	1465 weak/medium
—CH$_3$ bend	1450 medium
—CH$_3$ oscillation	1360 strong
C—C alkane	700–1200 weak (not usually useful)
C=O aldehyde	1740–1720 strong
C=O ketone	1725–1705 strong

Use the information given to determine a possible structure for **X** and state **one** additional experiment that could be carried out to establish the unique identity of **X**.   (6)

b i Secondary alcohols may be prepared by reduction of ketones.
Name the ketone which on reduction forms butan-2-ol and **name** a suitable reducing agent.   (2)

ii Draw **two** possible structures for butan-2-ol, clearly showing how they are related to each other.   (2)

c i Give **one** important industrial use of ethanol.   (1)

ii Give **one** reason why many consider the social use of ethanol as a drug to be harmful to society as a whole.   (1)

WJEC, A level, 932/01, June 2000
Topics 21, 22 and 23

22

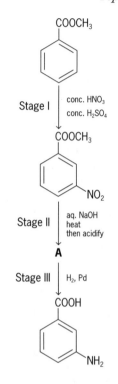

a i State the electrophile involved in Stage I.   (1)

ii Draw the mechanism for the reaction of the electrophile with C$_6$H$_5$COOCH$_3$ in Stage I.   (3)

iii Explain the function of the concentrated H$_2$SO$_4$ in this reaction, giving an equation in your answer.   (2)

b i Draw a full structure for compound **A**.   (2)

ii Calculate the volume of hydrogen in cm^3 required for reaction of 1.67 g of **A** in Stage III. (Under the conditions of the experiment one mole of hydrogen occupies 22.4 dm^3.)   (3)

OCR, A level, 6825 (Further Organic), March 2000
Topic 24

23 a Design a **two stage** synthesis of benzocaine, a modern local anaesthetic, from 4-nitrobenzenecarboxylic acid. State the names of the functional groups formed and the reagent(s) and condition(s) required at each stage.   (4)

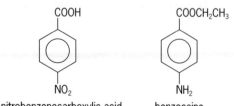

4-nitrobenzenecarboxylic acid   benzocaine

b The following diagram shows the mass spectrum of an organic compound.

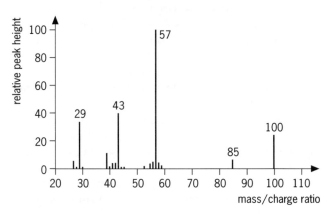

Information
The molecular ion is present.
The compound:

- reacts with a warm solution of sodium hydroxide and iodine to give a yellow crystalline precipitate
- forms a secondary alcohol on reduction by sodium tetrahydridoborate(III) (sodium borohydride), $NaBH_4$
- contains only one functional group.

i Draw a possible structure for the fragment ion of mass/charge ratio 57, which contains 4 carbon atoms in a branched chain. (1)
ii Draw the full structural (graphic) formula of the organic compound which contains the group in part **b i**. Give a brief reasoning for your choice. (3)

c i Experimentally determined values of the standard molar enthalpy changes of combustion, $\Delta H_c^{\ominus}$, for benzene, carbon and hydrogen were found to be -3280, -393 and $-286\,kJ\,mol^{-1}$, respectively. Using these values, calculate the standard molar enthalpy change of formation, $\Delta H_f^{\ominus}$, for benzene. **Show your working.** (3)
ii Using average bond energy data, the theoretical value for the enthalpy change of combustion for benzene is $-3429\,kJ\,mol^{-1}$. The experimentally determined value is $-3280\,kJ\,mol^{-1}$. Explain the difference between these two values. (2)

WJEC, A level, 009/02, June 2000
Topics 15, 23 and 24

24 a Construct a Born–Haber cycle for the formation of sodium oxide, Na_2O, and use the data given below to calculate the second electron affinity of oxygen. (9)

$$Na(s) \rightarrow Na(g) \qquad \Delta H^{\ominus} = +107\,kJ\,mol^{-1}$$
$$Na(g) \rightarrow Na^+(g) + e^- \qquad \Delta H^{\ominus} = +496\,kJ\,mol^{-1}$$
$$O_2(g) \rightarrow 2O(g) \qquad \Delta H^{\ominus} = +249\,kJ\,mol^{-1}$$
$$O(g) + e^- \rightarrow O^-(g) \qquad \Delta H^{\ominus} = -141\,kJ\,mol^{-1}$$
$$O^-(g) + e^- \rightarrow O^{2-}(g) \qquad \Delta H^{\ominus}\text{ to be calculated}$$
$$2Na^+(g) + O^{2-}(g) \rightarrow Na_2O(s)$$
$$\Delta H^{\ominus} = -2478\,kJ\,mol^{-1}$$
$$2Na(s) + \tfrac{1}{2}O_2(g) \rightarrow Na_2O(s) \quad \Delta H^{\ominus} = -414\,kJ\,mol^{-1}$$

b Explain why the second affinity of oxygen has a large positive value. (2)

c Explain, by reference to steps from relevant Born–Haber cycles, why sodium forms a stable oxide consisting of Na^+ and O^{2-} ions but not oxides consisting of Na^+ and O^- or Na^{2+} and O^{2-} ions. (4)

AQA, A level, CH04, June 2000
Topic 25

25 a The reaction given below does not occur at room temperature.

$$CO_2(g) + C(s) \rightarrow 2CO(g)$$

Use the data given below to calculate the lowest temperature at which this reaction becomes feasible. (8)

	$C_{(s)}$	$CO_{(g)}$	$CO_{2(g)}$
$\Delta H_f^{\ominus}/kJ\,mol^{-1}$	0	-110.5	-393.5
$S^{\ominus}/J\,K^{-1}\,mol^{-1}$	5.7	197.6	213.6

b The reaction shown below is very endothermic.

$$N_2(g) + O_2(g) \rightarrow 2NO(g) \quad \Delta H^{\ominus} = +180.4\,kJ\,mol^{-1}$$

i Use the expression given below to explain why the reaction does not occur to a significant extent at room temperature.

$$\ln K_p = \frac{-\Delta H^{\ominus}}{RT} + \frac{\Delta S^{\ominus}}{R}$$

ii Give one example of where this reaction occurs and explain why it causes environmental problems. (4)
c When an electrical heating coil was used to supply $3675\,J$ of energy to a sample of water which was boiling at $373\,K$, $1.50\,g$ water were vaporised. Use this information to calculate the entropy change for the process

$$H_2O(l) \rightarrow H_2O(g)$$ (3)

AQA, A level, CH04, June 2000
Topic 25

26 a Define the term **lattice enthalpy**. (2)

b Using the following data, construct a Born–Haber cycle for sodium fluoride and from it determine the lattice enthalpy of sodium fluoride. (5)

Process	The value of the energy change/kJmol^{-1}
$Na_{(g)} \rightarrow Na^+_{(g)} + e^-$	+494
$F_{2(g)} \rightarrow 2F_{(g)}$	+158
$F_{(g)} + e^- \rightarrow F^-_{(g)}$	−348
$Na_{(s)} + \frac{1}{2}F_{2(g)} \rightarrow NaF_{(s)}$	−569
$Na_{(s)} \rightarrow Na_{(g)}$	+109

c The table below gives some information about the hydroxides of the Group 2 elements.

Salt	Lattice enthalpy/ kJmol^{-1}	Hydration enthalpy/ kJmol^{-1}	Solubility in water/ g per 100g of water
magnesium hydroxide	−2383	−2380	0.9×10^{-4}
calcium hydroxide	−2094	−2110	156×10^{-4}
strontium hydroxide	−1894	−1940	800×10^{-4}
barium hydroxide	−1768	−1820	3900×10^{-4}

i Explain why energy is required to break up an ionic lattice. (1)

ii Suggest why the lattice enthalpies of the hydroxides of Group 2 metals become more exothermic from $Ba(OH)_2$ to $Mg(OH)_2$. (2)

iii Suggest why the lattice enthalpy of beryllium hydroxide, $Be(OH)_2$, cannot be predicted from the data in the table. (1)

iv Explain why energy is released when ions are hydrated. (2)

v Hence, account for the trend in solubilities from $Ba(OH)_2$ to $Mg(OH)_2$. (3)

London, A level, Module Test 3, January 2000
Topic 25

27 Sulphuric acid is manufactured by the Contact process. One stage of the process is the oxidation of sulphur dioxide according to the equation:

$$2SO_{2(g)} + O_{2(g)} \rightleftharpoons 2SO_{3(g)} \qquad \Delta H = -196 \, kJ \, mol^{-1}$$

Typical operating conditions are a temperature of about 450 °C, close to atmospheric pressure and the presence of a catalyst. The use of a converter containing four catalyst beds results in a 99.5% conversion.

a Write an expression for the equilibrium constant, K_p, for the oxidation of sulphur dioxide by oxygen. (2)

b The values for the equilibrium constant, K_p, at 25 °C and 427 °C are $4.0 \times 10^{24} \, atm^{-1}$ and $3.0 \times 10^4 \, atm^{-1}$, respectively.
i Explain why K_p varies with temperature. (2)
ii Suggest why the conversion is carried out at about 450 °C. (2)

c i State why the conversion of sulphur dioxide into sulphur trioxide would be favoured by high pressure. (1)
ii Suggest **two** reasons why the conversion is carried out at close to atmospheric pressure. (2)

d Give the name of a catalyst used in this process. (1)

e The value of the entropy change, ΔS, for the reaction between sulphur dioxide and oxygen is $-188 \, J \, mol^{-1} \, K^{-1}$.
i Explain why there is a decrease in entropy of the system. (2)
ii Use the equation

$$\Delta G = \Delta H - T\Delta S$$

to calculate the value of ΔG for the reaction between sulphur dioxide and oxygen at 25 °C. Explain the significance of the sign of the value obtained. (4)

AQA, A level, Module Paper 9, Summer 2000
Topics 16 and 25

28 a At 25 °C, the constant K_w has the value $1.00 \times 10^{-14} \, mol^2 \, dm^{-6}$. Define the term K_w. (1)
b Define the term **pH**. (1)
c Calculate the pH at 25 °C of $2.00 \, mol \, dm^{-3}$ HCl. (1)
d Calculate the pH at 25 °C of $2.50 \, mol \, dm^{-3}$ NaOH. (2)
e Calculate the pH at 25 °C of the solution that results from mixing $19.0 \, cm^3$ of $2.00 \, mol \, dm^{-3}$ HCl with $16.0 \, cm^3$ of $2.50 \, mol \, dm^{-3}$ NaOH. (6)

AQA, AS/A level, CH02, June 2000
Topic 26

29 a Phenol is a **weak acid**. The dissociation of phenol in aqueous solution is represented by the following equation:

$$C_6H_5OH_{(aq)} + H_2O_{(l)} \rightleftharpoons H_3O^+_{(aq)} + C_6H_5O^-_{(aq)}$$

i What is meant by the term **weak acid**? (1)
ii Give the formula of the conjugate base of phenol. (1)
iii Phenol is only very slightly soluble in water at 25 °C but it readily dissolves in aqueous sodium hydroxide. Consider the equilibrium shown above and suggest an explanation for the much greater solubility of phenol in sodium hydroxide. (3)

b i Write an expression for the acid dissociation constant, K_a, for phenol. (1)
ii Write an expression linking K_a with pK_a. (1)
iii The value of the acid dissociation constant, K_a, for phenol is $1 \times 10^{-10} \, mol \, dm^{-3}$. Calculate the pK_a value of phenol. (1)

iv Ethanoic acid is a stronger acid than phenol. State whether the pK_a value for ethanoic acid will be greater or smaller than that of phenol. (1)
v Account for the greater acidity of ethanoic acid compared with phenol. (3)
vi Describe a test using sodium hydrogencarbonate solution that would distinguish between phenol and ethanoic acid. (2)

c The indicator phenolphthalein is a weak acid which can be represented by the formula HIn. It dissociates in solution and has a pK_a value of 9.3.

$$HIn_{(aq)} \rightleftharpoons H^+_{(aq)} + In^-_{(aq)}$$
colourless   red

i Suggest and explain, with reference to the pK_a value, the pH range of phenolphthalein. (2)
ii State the colour change that would be observed at the end-point in an acid–base titration using phenolphthalein if sodium hydroxide solution were being added from the burette. Explain, in terms of the species present, why this colour is formed. (2)
iii State why phenolphthalein is unsuitable for a titration between a strong acid and a weak base. (1)

AQA, A level, Module Paper 8, Summer 2000
Topics 19 and 26

30 a Define the term **Brønsted–Lowry acid**. (1)
b Write an equation for the reaction between gaseous hydrogen chloride and water. State the role of water in this reaction, using the Brønsted–Lowry definition. (2)
c Write an equation for the reaction between gaseous ammonia and water. State the role of water in this reaction, using the Brønsted–Lowry definition. (2)
d The ion $H_2NO_3^+$ is formed in the first stage of a reaction between concentrated nitric acid and an excess of concentrated sulphuric acid. In this first stage the two acids react in a 1:1 molar ratio. In the second stage, the $H_2NO_3^+$ ion decomposes to form the nitronium ion, NO_2^+. Write equations for these two reactions and state the role of nitric acid in the first reaction. (3)
e i Explain the term **weak acid**.
ii Write an expression for the acid dissociation constant, K_a, of HA, a weak monoprotic acid.
iii The value of the acid dissociation constant for the monoprotic acid HX is $144\,mol\,dm^{-3}$. What does this suggest about the concentration of undissociated HX in dilute aqueous solution?
iv State whether HX should be classified as a strong acid or a weak acid. Justify your answer. (5)

AQA, AS/A level, CH02, June 2000
Topics 19 and 26

31 The table below shows some values for standard electrode potentials. These data should be used, where appropriate, to answer the questions that follow concerning the chemistry of copper and iron.

Electrode reaction	$E^\ominus/V$
$Fe^{2+}_{(aq)} + 2e^- \rightleftharpoons Fe_{(s)}$	−0.44
$2H^+_{(aq)} + 2e^- \rightleftharpoons H_{2(g)}$	0.00
$Cu^{2+}_{(aq)} + 2e^- \rightleftharpoons Cu_{(s)}$	+0.34
$O_{2(g)} + 2H_2O_{(l)} + 4e^- \rightleftharpoons 4OH^-_{(aq)}$	+0.40
$NO_3^-_{(aq)} + 4H^+_{(aq)} + 3e^- \rightleftharpoons NO_{(g)} + 2H_2O_{(l)}$	+0.96

a Write an equation to show the reaction that occurs when iron is added to a solution of a copper(II) salt. (1)
b A similar overall reaction to that shown in part **a** would occur if an electrochemical cell was set up between copper and iron electrodes.
i Write down the cell diagram to represent the overall reaction in the cell. (2)
ii Calculate the e.m.f. of the cell. (1)
iii Calculate the standard free energy change, $\Delta G^\ominus$, for the reaction occurring in the cell. (Faraday constant $= 96\,500\,C\,mol^{-1}$.) (3)
c i Use the standard electrode potential data given to explain why copper reacts with dilute nitric acid but has no reaction with dilute hydrochloric acid. (3)
ii Write an equation for the reaction between copper and dilute nitric acid. (2)
d Although iron is a widely used metal, it has a major disadvantage in that it readily corrodes in the presence of oxygen and water. The corrosion is an electrochemical process which occurs on the surface of the iron.
i Use the standard electrode potential data given to write an equation for the overall reaction that occurs in the electrochemical cell set up between iron, oxygen and water. (1)
ii State, with a reason, whether the iron acts as the anode or cathode of the cell. (2)
iii Predict and explain whether or not you would expect a similar corrosion reaction to occur with copper in the presence of oxygen and water. (2)

AQA, A level, Module Paper 9, Summer 2000
Topic 28

32 a Define the term **standard electrode potential** for a metal/metal ion system. (2)

b The apparatus drawn below was used to measure the standard electrode potential of the $Cu^{2+}(aq)/Cu(s)$ electrode.

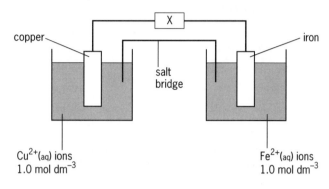

copper — iron
salt bridge
X
$Cu^{2+}(aq)$ ions 1.0 mol dm^{-3}
$Fe^{2+}(aq)$ ions 1.0 mol dm^{-3}

The conventional way of representing this cell is

$$Fe(s)\,|\,Fe^{2+}(aq)\,||\,Cu^{2+}(aq)\,|\,Cu(s)$$

i What instrument would you use at position X to measure the e.m.f. of the cell? (1)

ii The standard electrode potential of the $Fe^{2+}(aq)/Fe(s)$ electrode is $-0.44\,V$. The e.m.f. of the above cell is $+0.78\,V$. Calculate the standard electrode potential of the $Cu^{2+}(aq)/Cu(s)$ electrode. (2)

iii Indicate by an arrow on a copy of the diagram the direction in which electrons would flow between the two metals if the external circuit was completed. (1)

iv What would you expect to happen if a small piece of copper was placed in an aqueous solution of iron(II) sulphate? Give your reasoning. (2)

c The standard electrode potentials of the reactions involved in the first stage in the rusting of iron are

$Fe^{2+}(aq) + 2e^- \rightleftharpoons Fe(s)$ $-0.44\,V$
$O_2(g) + 2H_2O(l) + 4e^- \rightleftharpoons 4OH^-(aq)$ $+0.40\,V$

i Write an overall equation for the first stage in the rusting of iron. (2)

ii Explain how magnesium metal attached to a sheet of iron prevents it from rusting. (2)

London, A level, Module Test 3, January 2000
Topic 28

33 A student was carrying out an investigation into manganese in the +3 oxidation state, which is normally unstable.

a The student made manganese(III) ions by reduction of aqueous manganate(VII) ions with aqueous thallium(I) ions in the presence of a large excess of fluoride ions.

By titration, the student found that $25.0\,cm^3$ of $0.02\,mol\,dm^{-3}$ $Tl^+(aq)$ reacted with $25.0\,cm^3$ of $0.01\,mol\,dm^{-3}$ $MnO_4^-(aq)$.

i What is the change in oxidation state of manganese in this reaction? (1)

ii Calculate the value of the final oxidation state of thallium in this reaction. (3)

iii Suggest why there is a need for a large excess of fluoride ions in the reaction mixture. (2)

b Two oxidation states of manganese are commonly available: Mn(II) in compounds containing Mn^{2+} ions, and Mn(VII) in compounds containing MnO_4^- ions. The student looked at a table of standard electrode potentials to see if Mn(III) could be made by reaction between compounds containing Mn^{2+} ions and MnO_4^- ions. The relevant standard electrode potentials are given below.

$$MnO_4^-(aq) + 8H^+(aq) + 4e^- \rightarrow Mn^{3+}(aq) + 4H_2O(l)$$
$$E^{\ominus} = +0.85\,V$$
$$Mn^{3+}(aq) + e^- \rightarrow Mn^{2+}(aq) \qquad E^{\ominus} = +1.49\,V$$

i Explain why $Mn^{3+}(aq)$ would not be expected to form from reaction between $Mn^{2+}(aq)$ and $MnO_4^-(aq)$ under standard conditions. (2)

ii Suggest why $Mn^{3+}(aq)$ can be formed from $Mn^{2+}(aq)$ and $MnO_4^-(aq)$ in the presence of a high concentration of acid. (3)

c In aqueous solution, Mn^{2+} is present as the $[Mn(H_2O)_6]^{2+}$ complex, and MnO_4^- as the $[MnO_4]^-$ complex. Draw likely structures for these two complexes. (4)

d Absorption spectra in the visible region for $Mn^{3+}(aq)$ and $MnO_4^-(aq)$ are shown below. At the concentrations used in the student's investigation, $Mn^{2+}(aq)$ is colourless.

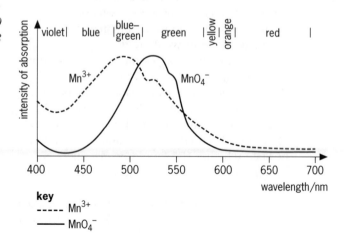

key
----- Mn^{3+}
——— MnO_4^-

i What colour is a solution of Mn^{3+}(aq)? Explain your answer. (2)

ii The student decided to use a colorimeter to investigate the effect of MnO_4^-(aq) concentration on the rate of the reaction between MnO_4^-(aq) and Mn^{2+}(aq). Describe the key steps the student should take in carrying out this investigation. (4)

OCR, A level, 5683, January 2000
Topics 5, 18 and 28

34 a State the reagents and the conditions required for:

i the elimination of HBr from 2-bromopropane (3)

ii the reduction of butanoic acid to butan-1-ol (2)

iii the conversion of an animal fat into soap. (2)

b Give the functional group identified by a positive test with each of the following reagents:

i 2,4-dinitrophenylhydrazine reagent

ii ammoniacal silver nitrate

iii phosphorus pentachloride

iv bromine water. (4)

c Three compounds **P**, **Q** and **R** all have the same molecular formula C_4H_8O.

When 2,4-dinitrophenylhydrazine reagent was added to each, **P** gave a precipitate.

When ammoniacal silver nitrate was warmed with each, **P** gave a silver mirror.

When phosphorus pentachloride, PCl_5, was added to dry samples of each, **Q** and **R** gave off steamy fumes.

When bromine water was shaken with each, **Q** and **R** turned it colourless.

i Draw **two** possible structural formulae for **P**. (2)

ii **Q** is chiral and can be oxidised to a ketone. Draw the structures to show the **two** optical isomers of **Q**. (2)

iii **R** can be oxidised to an acid **S** which has geometric isomers. Draw the formula of the **two** geometric isomers of the acid **S**. (2)

Edexcel, AS/A level, Module Test 2, June 2000
Topics 21 and 29

35 Citral is a colourless natural product, which gives lemons their characteristic flavour and smell. Its structural formula is:

$$CH_3-\underset{\underset{CH_3}{|}}{C}=CH-CH_2-CH_2-\underset{\underset{CH_3}{|}}{C}=CH-CHO$$

a i How would you show that citral has a carbonyl group, C=O? Give the reagent you would use and the observation you would make. (2)

ii How would you show that citral is an aldehyde? Give the reagent you would use and the observation you would make. (2)

b Citral has geometric isomers. Draw them and explain why they are not easily interconvertible. (3)

c Bromine, dissolved in tetrachloromethane, was added slowly to a solution of citral.

i Describe what you would observe when this was done. (1)

ii Draw the structural formula of the product of this reaction when excess bromine is added. (1)

iii This product is chiral. Mark all the atoms which cause chirality with a * on the structural formula that you have given in part **ii** above. (2)

d Citral can be reduced to an alcohol.

i State a reagent and the conditions that could be used for this reaction. (2)

ii How would you test the product to show that it has an —OH group? (2)

e Citral can be oxidised to an acid $C_9H_{15}COOH$ ($M_r = 168$) which ionises in water

$$C_9H_{15}COOH + H_2O \rightleftharpoons H_3O^+ + C_9H_{15}COO^-$$

4.62 g of this acid was dissolved in water to give a solution of volume 250 cm^3. This solution had a pH of 2.91.

i Write the expression for K_a for this acid. (1)

ii Calculate the concentration of the acid in mol dm^{-3}. (1)

iii Calculate the value of K_a of the acid. (3)

London, AS/A level, Module Test 2, January 2000
Topics 22, 26 and 29

Answers to examination questions

1 a $Li_{(g)} \rightarrow Li^+_{(g)} + e^-$ (2; 1 mark for equation and 1 mark for standard states)
$Li^+_{(g)} \rightarrow Li^{2+}_{(g)} + e^-$ (1)

b i Electron is in 1s orbital (or innermost orbital) (1)
Electron is much closer to the nucleus (or has no shielding) (1)
ii Beryllium electron is in 2s (1)
This is lower in energy than 2p (1)
iii Electron is removed from a positive ion (1)
There is stronger attraction (1)

2 a

Molecule	Sketch of shape	Bond angle(s)		Name of shape	
BF_3	F–B with F, F (1)	120°	(1)	trigonal planar	(1)
NF_3	N with lone pair, F, F, F (1)	107°	(1)	tetrahedral or pyramidal	(1)
ClF_3	F–Cl with lone pairs, F, F (1)	87–89°	(1)	trigonal bipyramidal	(1)
	or			or	
	F–Cl with lone pairs, F, F (1)	120°	(1)	trigonal bipyramidal or trigonal planar	(1)

b BF_3 van der Waals (1)
NF_3 van der Waals and dipole–dipole (2)
c Coordinate bond (1)
The lone pair on NF_3 (1) donates to empty orbital/shell on B (1)

3 a

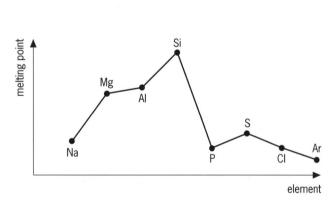

Shape
- Mg above Na (1)
- Al similar to Mg (1)
- Si significantly higher (1)
- P lower than Na to Si (1)
- S above P, but lower than Na to Si (1)
- Ar lowest (1)

Bonding
- Na and Mg metallic (1)
- Mg has stronger metallic bonding (1) because more protons (1) mean more attraction for delocalised electrons (1)
- Al metallic, like Mg (1)
- Si covalent (1), a macromolecule (1)
Strong bonds (or lots of energy needed to break down structure) (1)
- S, P, Cl molecular (2 out of 3) (1)
S_8 (1)
P_4 (1)
Cl_2 (1)
Van der Waals forces between molecules (1), with stronger forces between larger molecules (1)
- Ar free atoms (1)
Very weak van der Waals forces (1)

b $1s^2$ (1)
Li^+ ion has high charge to size ratio (1)
This polarises negative ions (1)

Examples
- Li_2CO_3 is anomalous (1)
Decomposes on heating (1) to Li_2O and CO_2 (1)

$Li_2CO_3 \rightarrow Li_2O + CO_2$ (1)

- LiI is anomalous (1)
Covalent character (1), dissolves in organic solvents (1), for example methanol (1)

4 a Assumptions
- molecules/particles/atoms have zero/negligible size/volume or can be considered as points (1)
- no intermolecular attractions or attractions between particles not between atoms (1)
- elastic collisions/no loss of kinetic energy on collision (1)

Any 2 for (2)

- CO_2 /molecules/particles have a large size (or larger than H_2) (1)
- CO_2 molecules have large intermolecular forces/attractions between particles (1)
- CO_2 molecules experience inelastic collisions (1)

Any 2 for (2)

b Low pressure (1)
High temperature (1)

c i

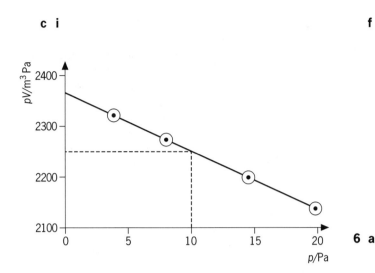

On graph, labelled axes (1), plotting of 4 points
$(2\,cm \leq 50\ pV$ units) (1)
Values of pV/m^3Pa: (2320), 2280, 2190, 2140 (1)
ii pV should be constant or a horizontal line (1)
iii From graph: $pV = 2250$
$V = 2.25 \times 10^{-3}\,m^3$ (or $2.25\,dm^3$) (1)
$V_{ideal} = nRT/P$
$\quad = 1.00\ mol \times 8.31\,Pa\,m^3\,mol^{-1}K^{-1}$
$\quad\quad \times 285\ mol^{-6} \times 1.0 \times 10^{-6}\,Pa$
$\quad = 2.37 \times 10^{-3}\,m^3$ (2.36, 2.4 but not 2.3)   (1)
Intermolecular attractions occur (1)

5 a

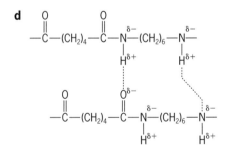

(1)

b Condensation (polymerisation) (1)
c 1 unit has a mass of $(3 \times 12) + 4 + (2 \times 16) = 72$ (1)

$$\frac{144\,000}{72} = 2000\ units$$ (1)

d

O   O
‖   ‖
—C—(CH₂)₄—C—N—(CH₂)₆—N—
                   |ᵟ⁻ |ᵟ⁻
                   Hᵟ⁺   Hᵟ⁺
```
 O          O
 ‖          ‖ᵟ⁻
—C—(CH₂)₄—C—N—(CH₂)₆—N—
            |ᵟ⁻        |ᵟ⁻
            Hᵟ⁺        Hᵟ⁺
```

Attraction shown clearly between H attached to N on
one chain to either O attached to C on another chain or
N on another chain (1)
Charges correctly labelled, $\delta+$ on H attached to N, $\delta-$
on N or O (1)
e i No bulky side groups so chains can move past each
other/hydrogen bonds can stretch without breaking (1)
ii H bonds (1) in nylon-6,10 are further apart on
chain/Intermolecular forces are more easily overcome
(1)

f Advantages
- Saves on crude oil/fossil fuels
- Energy in the bonds is utilised/energy produced is
  not wasted
- Saves on landfill
- Cheaper than recycling of plastic/using fossil fuels
                                                1 for (1)

Disadvantages
- Potentially toxic gases may be produced
- Temperatures need to be high to incinerate the
  plastic
- Need to redesign the plant to cope with waste
                                                1 for (1)

6 a • Fractional distillation used to separate crude oil into
        boiling point/M_r fractions (1)
      • Catalysts used to dehydrogenate for aromatic (or
        cyclic, reformation or branch chain) compounds (1)
      • Give an example (1)
                                                Any 2 for (2)

b A suitable example:

$$C_{16}H_{34} \rightarrow C_8H_{18} + C_6H_{12} + C_2H_4$$ (1)

Discussion of usefulness of a product, for example
C_8H_{18} as gasoline or alkenes for polymers, chemicals,
and so on (1)
Separate marks are awarded for quality of language:
Communication (2)
Spelling and punctuation (1)
Grammar (1)

7 a

	1s	2s	2p	2p	2p	3s	3p	3p	3p
Mg	↓↑	↓↑	↓↑	↓↑	↓↑	↓↑			
Mg²⁺	↓↑	↓↑	↓↑	↓↑	↓↑				

b Energy required to remove an electron from each of
1 mol (1) of gaseous (1) singly positive ions

$$M^+_{(g)} \rightarrow M^{2+}_{(g)} + e^-$$ (1)

c i Same nuclear charge but one less electron/greater
effective nuclear charge/electron closer to the
nucleus/ion is smaller than the atom (1)
Therefore it needs more energy/there is a greater
electrostatic attraction/more difficult to remove
electron from a positive species (1)
ii Inner shell/closer to the nucleus/new shell of lower
energy/second shell rather than third (1)
Less shielding for second shell/more shielding for third
shell (1)
d Beryllium (1) and sodium (1)
or magnesium (1) and potassium (1)
or zinc (1) and rubidium (1)
(any Group 1 element as last element)

e i Ba(OH)$_2$/barium (hydroxide) (1)
  ii MgCO$_3$/magnesium (carbonate) (1)
  iii K/potassium (1)
  iv KCl/potassium chloride/potassium (1)
f Cl$^-$ is a larger (1) and more polarisable anion (1) (or the reverse argument for fluoride)
  or
  Greater electronegativity difference between Mg and F (1) than between Mg and Cl (1)

8 a number of moles of S$_2$O$_3^{2-}$ used $= \dfrac{32.00}{1000} \times 0.500$ ($\frac{1}{2}$)

$= 0.0160$ ($\frac{1}{2}$)

number of moles of iodine present $= \dfrac{0.0160}{2}$

$= 0.008$ ($\frac{1}{2}$)

From equation, 5 moles of CO give 1 mole of I$_2$ ($\frac{1}{2}$)
volume of CO $= 0.008 \times 5 \times 24$
$= 0.960\,dm^3$ ($\frac{1}{2}$) (marked sequentially)

percentage of CO in mixture $= 0.960 \times \dfrac{100}{1.00}$

$= 96\%$ ($\frac{1}{2}$)

b i $2Cl_2 + 2C + SiO_2 \rightarrow SiCl_4 + 2CO$ (1)
($\frac{1}{2}$ for unbalanced equation)
  ii Covalent ($\frac{1}{2}$)
  Hydrolysed/decomposed by water ($\frac{1}{2}$)
  $SiCl_4 + 2H_2O \rightarrow SiO_2 + 4HCl$ (1)
  (SiO$_2$.2H$_2$O or Si(OH)$_4$ are correct)
  ($\frac{1}{2}$ for unbalanced equation)
  iii • SiCl$_4$ is (thermally) stable but PbCl$_4$ is (thermally) unstable (1) (comparison is needed)
    • Pb(II) compounds are more 'stable' than Pb(IV) compounds (1)
    • Inert pair effect (1)
    • For Si the +4 state is more stable (1)
    Any 2 of these for (2)
c i I Halogens are highly electronegative elements/gain of one electron gives a 'stable' noble gas configuration (1)
  II The halogen and oxygen atoms have similar electronegativities (1)
  III The IO$_3^-$ ion is more stable than the ClO$_3^-$ ion (1)
  ii +1 in XO$^-$ ($\frac{1}{2}$), to -1 in X$^-$ ($\frac{1}{2}$) and +5 in XO$_3^-$ ($\frac{1}{2}$)
  The halogen atoms (X) have undergone both oxidation and reduction/oxidation numbers for the halogen atoms have increased and decreased/disproportionation has occurred ($\frac{1}{2}$)
  iii Astatine will be a product (1)
  A less positive potential permits oxidation to the halogen/iodide is oxidised to iodine and so At$^-$ will be oxidised similarly (1)

9 a i Fractional ($\frac{1}{2}$) distillation ($\frac{1}{2}$) (not cracking)
  ii $C_5H_{12} \rightarrow C_4H_8 + CH_4$ (1)
  or $C_5H_{12} \rightarrow 2C_2H_4 + CH_4$ (1)
b i UV radiation/sunlight/high temperatures (1)
  ii Initiation: $Cl_2 \rightarrow 2Cl\cdot$ (1)
  Propagation: $Cl\cdot + CH_4 \rightarrow \cdot CH_3 + HCl$ (1)
  or $\cdot CH_3 + Cl_2 \rightarrow CH_3Cl + Cl\cdot$ (1)

c i Forward and backward reactions proceeding at the same rate/both sides are constantly reacting to form each other, but the concentrations remain the same (1)
  ii • The equilibrium is moving to the right (1)
    • Removal of a product encourages equilibrium position to be restored (1)
    • Producing more HI/CH$_3$I/products (1)
    Any 2 for (2)
  iii $\Delta H^{\ominus}_{reaction} = \Sigma\Delta H^{\ominus}_f(products) - \Sigma\Delta H^{\ominus}_f(reactants)$ (1)
  $= -15.5 + (+26.5) - (-74.8)$
  $= +85.8\,kJ\,mol^{-1}$ (1)
  iv $-53.0\,kJ\,mol^{-1}$ ($\frac{1}{2}$)
  Statement of conservation of energy ($\frac{1}{2}$)
  $\Delta H^{\ominus}_f$ is per mole ($\frac{1}{2}$), so 2 moles needed here ($\frac{1}{2}$)

10 a i N$_2$O$_4$: 0.20 atmosphere ($\frac{1}{2}$)
  NO$_2$: 0.15 atmosphere ($\frac{1}{2}$)

  ii $K_p = \dfrac{p(NO_2)^2}{p(N_2O_4)}$ ($\frac{1}{2}$)

  $= \dfrac{0.15^2}{0.20}$ ($\frac{1}{2}$)

  $= 0.11$ ($\frac{1}{2}$) atmosphere ($\frac{1}{2}$)
  (consequential to the answers from part **i**)
  iii No change in the values of the partial pressure ($\frac{1}{2}$)
  K_p is a constant (for this system) at a constant temperature ($\frac{1}{2}$)
  iv $p(NO_2)$ will be greater ($\frac{1}{2}$)
  $p(N_2O_4)$ will be smaller ($\frac{1}{2}$)
  (consequential to the answers from part **ii**)
b 'AlCl$_3$' is an 'electron deficient' molecule/only 6 electrons in the 'outer shell' ($\frac{1}{2}$), but needs 8 electrons to form stable structure ($\frac{1}{2}$)
  Chlorine lone pair ($\frac{1}{2}$) forms dative/coordinate bond to (other) aluminium atom ($\frac{1}{2}$)
c $MgCO_3.Mg(OH)_2.3H_2O \rightarrow 2MgO + CO_2 + 4H_2O$ (1)
  M_r for MgCO$_3$.Mg(OH)$_2$.3H$_2$O $= 196.7$ (1)
  M_r for $2 \times MgO = 80.6$
  Percentage of solid remaining (MgO)

  $= \dfrac{80.6 \times 100}{196.7}$ (1)

  $= 41\%$

11 a i $1s^2 2s^2 2p^6 3s^2 3p^6 3d^8 4s^2$ (1)
  ii • Variable oxidation state (1)
    • Forms complex ions/complexes/coordination compounds (1)
    • Forms coloured compounds (1)
    • Has catalytic properties (1)
    Any 3 for (3)

b Relative atomic mass

$= \dfrac{(58 \times 60.0) + (60 \times 30.0) + (61 \times 10.0)}{100}$

$= 58.9$ (2)

c $500\,kg = \dfrac{500\,000}{58.71}\,mol$ $(\tfrac{1}{2})$

$\qquad = 8516.4\,mol$ $(\tfrac{1}{2})$

volume of CO needed to give this amount of Ni

$\qquad = 4 \times 8516.4\,(\tfrac{1}{2}) \times 22\,410\,cm^3$ $(\tfrac{1}{2})$

$\qquad = 7.63 \times 10^8\,cm^3$ (1)

12 a i Dark brown/red/red-brown/orange liquid (1)

ii Corrosive (to skin)/blisters/sores/harmful to skin/burns (1)

Toxic vapour/gas (when breathed in)/respiratory problems (1)

b i

Calcium shown as 2+, accompanied by 2 bromides (1)

Bromide with correct ring of '7 + 1' electrons (1)

ii $\dfrac{200}{160} \times 1.0 = 1.25\,kg$ (2)

c

Positive calcium (not necessarily 2+) surrounded by at least 3 water molecules (1)

Structure of water molecule shown clearly with δ− in at least one oxygen (majority pointing to Ca) (1)

ii Charge on calcium ion is greater/stronger than that on sodium ion (1)

iii $2180 - 1562 - (2 \times 338) = -58\,kJ\,mol^{-1}$ (2)

d i Equation 1 (1)

$Ag^+_{(aq)} + Br^-_{(aq)} \rightarrow AgBr_{(s)}$ (1)

ii Equation 2b/3 (1)

Br/bromine/Br^-/(HBr if equation 2b chosen) (1)

iii $2Br^- \rightarrow Br_2 + 2e^-$ (1)

iv Equation 2

For: liquid bromine formed, sulphuric acid is cheap

Against: two steps, mixed products/more by-products/waste products/sulphuric acid dangerous/corrosive/SO_2/HBr toxic/acid rain/pollutant/harmful/$CaSO_4$ not useful

Any 3 for (3)

Equation 3

For: chlorine/reagent cheap/readily available/easy to obtain

Against: aqueous bromine formed/chlorine dangerous/toxic/expensive

Any 3 for (3)

Electrolysis:

For: bromine formed in a specific place/anode, no other reagents, cheap, continuous

Against: expensive, aqueous bromine formed

Any 3 for (3)

13 a i Lone pairs donated (1) from two atoms (1)

ii $[Fe(C_2O_4)_3]^{3-}$ (1)

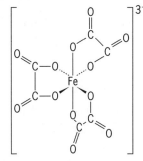

6 O octahedrally around Fe (1)

One $C_2O_4^{2-}$ shown correctly (1)

b i Haem/haemoglobin (1)

ii O_2 transport (1)

c Extract with NaOH (1), filter off $Fe(OH)_3$ (1), acidify filtrate (1) until precipitate occurs (1) (or extract with NH_3, Na_2CO_3, etc)

14 a i Aqueous solutions of Cu(II) ions are blue whilst solutions of Mg(II) ions are colourless (**transition elements form coloured compounds**) (1)

ii I Cu(II) ions form complex ions with ammonia; Mg(II) ions do not (**transition elements form complex ions**) (1)

II Cu(II) ions can react/oxidise iodide ions $(\tfrac{1}{2})$ to form iodine $(\tfrac{1}{2})$ and Cu(I) ions (1); Mg ions do not exist in oxidation states other than the (II) state/no reaction (**transition elements can exist in variable oxidation states**) (1)

b i $5Fe^{2+} + 8H^+ + MnO_4^- \rightarrow 5Fe^{3+} + Mn^{2+} + 4H_2O$ (1)

ii Number of moles $MnO_4^- = 28.4 \times 10^{-3} \times 0.010$

$\qquad = 2.84 \times 10^{-4}\,mol$ (1)

Number of moles $Fe^{2+} = 5 \times 2.84 \times 10^{-4}\,mol$

$\qquad = 1.42 \times 10^{-3}\,mol$ (1)

Mass of Fe^{2+} ions $= 1.42 \times 10^{-3} \times 55.85$

$\qquad = 0.00793\,g$ (1)

iii Iron occurs in many compounds in living cells, for example in haemoglobin (1)

iv With iron(II) ions, green precipitate seen (1), which changes to brown on standing in air $(\tfrac{1}{2})$

With zinc ions, white precipitate (1), which dissolves in excess sodium hydroxide solution $(\tfrac{1}{2})$

c i I $I < Br < Cl$ or I, Br, Cl (1)

II $Cl < Br < I$ or Cl, Br, I (1)

ii Cl is more electronegative than I/HCl is a weak reducing agent (1)

Sulphuric acid oxidises iodide to iodine/HI is a strong reducing agent (1)

iii $Cl_2 + 2NaOH \rightarrow NaCl + NaClO + H_2O$ (1)

Cl in Cl_2 is 0 $(\tfrac{1}{2})$

Cl in NaCl is -1 $(\tfrac{1}{2})$

Cl in NaClO is $+1$ $(\tfrac{1}{2})$

Chlorine is both reduced and oxidised/disproportionation has occurred $(\tfrac{1}{2})$

15 a i

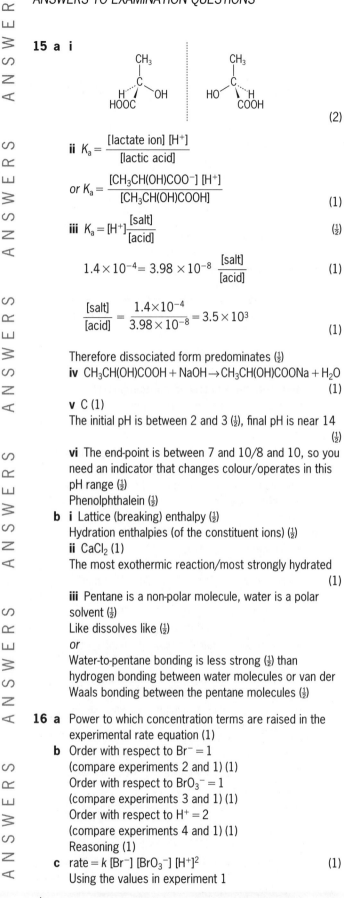

(2)

ii $K_a = \dfrac{[\text{lactate ion}]\,[H^+]}{[\text{lactic acid}]}$

or $K_a = \dfrac{[CH_3CH(OH)COO^-]\,[H^+]}{[CH_3CH(OH)COOH]}$ (1)

iii $K_a = [H^+]\dfrac{[\text{salt}]}{[\text{acid}]}$ ($\frac{1}{2}$)

$1.4 \times 10^{-4} = 3.98 \times 10^{-8}\,\dfrac{[\text{salt}]}{[\text{acid}]}$ (1)

$\dfrac{[\text{salt}]}{[\text{acid}]} = \dfrac{1.4 \times 10^{-4}}{3.98 \times 10^{-8}} = 3.5 \times 10^3$ (1)

Therefore dissociated form predominates ($\frac{1}{2}$)

iv $CH_3CH(OH)COOH + NaOH \rightarrow CH_3CH(OH)COONa + H_2O$ (1)

v C (1)

The initial pH is between 2 and 3 ($\frac{1}{2}$), final pH is near 14 ($\frac{1}{2}$)

vi The end-point is between 7 and 10/8 and 10, so you need an indicator that changes colour/operates in this pH range ($\frac{1}{2}$)

Phenolphthalein ($\frac{1}{2}$)

b i Lattice (breaking) enthalpy ($\frac{1}{2}$)

Hydration enthalpies (of the constituent ions) ($\frac{1}{2}$)

ii $CaCl_2$ (1)

The most exothermic reaction/most strongly hydrated (1)

iii Pentane is a non-polar molecule, water is a polar solvent ($\frac{1}{2}$)

Like dissolves like ($\frac{1}{2}$)

or

Water-to-pentane bonding is less strong ($\frac{1}{2}$) than hydrogen bonding between water molecules or van der Waals bonding between the pentane molecules ($\frac{1}{2}$)

16 a Power to which concentration terms are raised in the experimental rate equation (1)

b Order with respect to $Br^- = 1$
(compare experiments 2 and 1) (1)
Order with respect to $BrO_3^- = 1$
(compare experiments 3 and 1) (1)
Order with respect to $H^+ = 2$
(compare experiments 4 and 1) (1)
Reasoning (1)

c rate $= k\,[Br^-]\,[BrO_3^-]\,[H^+]^2$ (1)
Using the values in experiment 1

$k =$

$$\dfrac{2.00 \times 10^{-6}\,\text{mol}\,\text{dm}^{-3}\,\text{s}^{-1}}{(2.50 \times 10^{-3}\,\text{mol}\,\text{dm}^{-3}) \times (1.25 \times 10^{-3}\,\text{mol}\,\text{dm}^{-3}) \times (0.100\,\text{mol}\,\text{dm}^{-3})^2} = 64.0$$

calculation (1)
units: $\text{mol}^{-3}\,\text{dm}^9\,\text{s}^{-1}$ (1)

d i

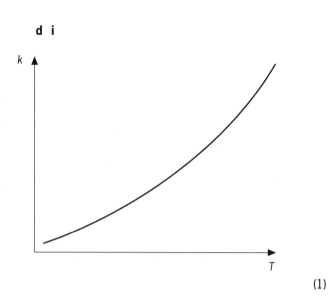

(1)

ii More particles have energy greater than or equal to E_a (1)

e Rate dependent on slow step 3 (1)
Reagents involved in rate-determining step or preceding steps will appear in the rate equation (1)

17 a i $PV = nRT$ ($\frac{1}{2}$)
Molar gas volume is $22.4 \times 10^{-3}\,\text{m}^3\,\text{mol}^{-1}$ at $273\,\text{K}$ and $101.3 \times 10^3\,\text{Pa}$ ($\frac{1}{2}$)

$R = \dfrac{PV}{nT}$ ($\frac{1}{2}$)

$R = \dfrac{101.3 \times 10^3 \times 22.4 \times 10^{-3}}{1.0 \times 273}$

$= 8.3$ (or 8.31) ($\frac{1}{2}$)
Units are $\text{J}\,\text{mol}^{-1}\,\text{K}^{-1}$ or $\text{Pa}\,\text{m}^3\,\text{mol}^{-1}\,\text{K}^{-1}$ (1)

ii Low temperature ($\frac{1}{2}$) because the particles are closer together ($\frac{1}{2}$) and attractions are more likely than at higher temperature; particle volumes more significant ($\frac{1}{2}$)
High pressures ($\frac{1}{2}$) for the same reasons as low temperatures

iii Hydrogen
- Strong covalent bonds within molecules ($\frac{1}{2}$)
- Weak van der Waals/induced dipole–induced dipole bonding between the molecules ($\frac{1}{2}$)

Hydrogen chloride
- Strong covalent bonds within molecules ($\frac{1}{2}$)
- Stronger dipole–dipole interactions between molecules ($\frac{1}{2}$) due to small size of hydrogen atom ($\frac{1}{2}$) and higher electronegativity of chlorine creating a permanent dipole within the molecule ($\frac{1}{2}$)

Hydrogen chloride more likely to deviate due to stronger intermolecular interactions/the molecules occupy a larger volume (1)

b Choose from:
- Thermal decomposition of calcium carbonate
- Hydrogen production in metal/acid reaction
- $H_2O_2 \rightarrow H_2O + \frac{1}{2}O_2$
- Decomposition of diazonium salt
- Calcium carbonate and acid

Any 1 for (1)

- Gas collection in syringe
- Gas collection in manometer
- Measurement of loss of mass

Any 1 for (1)

c i

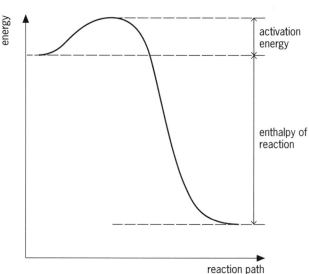

Correct sketch (1)
E_a ($\frac{1}{2}$)
ΔH_r ($\frac{1}{2}$)
ii Reduces the activation energy (1)
iii Low temperatures ($\frac{1}{2}$) and high pressure ($\frac{1}{2}$)
iv

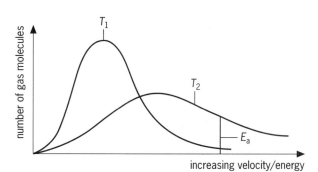

Axes labelled correctly ($\frac{1}{2}$)
Correct profile for T_1 ($\frac{1}{2}$)
Correct comparison with T_2 ($\frac{1}{2}$)
The more molecules that have the required activation energy, the faster the reaction ($\frac{1}{2}$)
Although low temperatures favour the forward reaction, the time taken to reach equilibrium would be too long and hence higher temperatures are used (1)

v $K_p = \dfrac{p(SO_3)^2}{p(SO_2)^2 \times p(O_2)}$ 　　(1)

The K_p would be higher at room temperature because the forward reaction is an exothermic reaction (1)

18 a i Electrophilic ($\frac{1}{2}$) addition ($\frac{1}{2}$)
(nucleophilic addition would get $\frac{1}{2}$, electrophilic substitution would get 0)
ii Iodine
- First order ($\frac{1}{2}$)
- When concentration of iodine doubles, rate doubles (1) or rate is proportional to concentration (1)

Pent-1-ene
- Zero order ($\frac{1}{2}$)
- Rate is independent of the alkene concentration (1)

b i Asymmetric carbon atom/carbon atom bonded to four different groups or atoms/species (1)

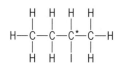 ($\frac{1}{2}$)

Optical isomerism ($\frac{1}{2}$)

ii *cis* form 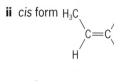 (1)

trans form (1)

c In general, for each compound
- Name compound ($\frac{1}{2}$)
- Statement of use ($\frac{1}{2}$)
- Positive contribution ($\frac{1}{2}$)

Total for 2 compounds = 3

Examples
- Silver bromide ($\frac{1}{2}$) is a light-sensitive compound which forms the basis of many types of photography ($\frac{1}{2}$). The development of photographic film has enabled events to be recorded for historical purposes and for distribution to a wider audience ($\frac{1}{2}$).
- Nitroglycerine ($\frac{1}{2}$) is a shock-sensitive compound which is used for the development of commercial explosives ($\frac{1}{2}$). It is used in blasting gelignite for removal of rocks in quarrying, which can then be used for road building ($\frac{1}{2}$).
- 'Teflon'/Poly(tetrafluoroethene) ($\frac{1}{2}$) is a polymeric material used for non-stick saucepans ($\frac{1}{2}$). Its low coefficient of friction ($\frac{1}{2}$) also enables it to be used as a bearing substitute.
- Painkillers ($\frac{1}{2}$) such as aspirin ($\frac{1}{2}$) have important uses for treating pain. Recently, aspirin has been shown to have value in treating circulatory problems ($\frac{1}{2}$).

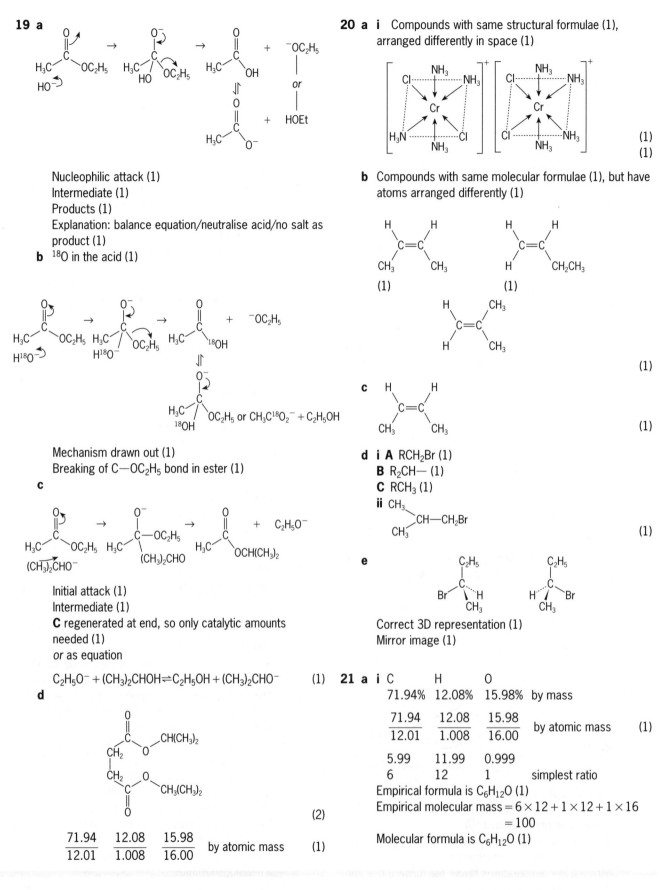

19 a

Nucleophilic attack (1)
Intermediate (1)
Products (1)
Explanation: balance equation/neutralise acid/no salt as product (1)

b ^{18}O in the acid (1)

Mechanism drawn out (1)
Breaking of $C—OC_2H_5$ bond in ester (1)

c

Initial attack (1)
Intermediate (1)
C regenerated at end, so only catalytic amounts needed (1)
or as equation

$C_2H_5O^- + (CH_3)_2CHOH \rightleftharpoons C_2H_5OH + (CH_3)_2CHO^-$ (1)

d

(2)

$$\frac{71.94}{12.01} \quad \frac{12.08}{1.008} \quad \frac{15.98}{16.00} \quad \text{by atomic mass} \quad (1)$$

20 a i Compounds with same structural formulae (1), arranged differently in space (1)

(1)
(1)

b Compounds with same molecular formulae (1), but have atoms arranged differently (1)

(1)   (1)

(1)

c

(1)

d i A RCH_2Br (1)
B $R_2CH—$ (1)
C RCH_3 (1)
ii

(1)

e

Correct 3D representation (1)
Mirror image (1)

21 a i

C	H	O	
71.94%	12.08%	15.98%	by mass

$$\frac{71.94}{12.01} \quad \frac{12.08}{1.008} \quad \frac{15.98}{16.00} \quad \text{by atomic mass} \quad (1)$$

5.99	11.99	0.999	
6	12	1	simplest ratio

Empirical formula is $C_6H_{12}O$ (1)
Empirical molecular mass $= 6 \times 12 + 1 \times 12 + 1 \times 16$
$$= 100$$
Molecular formula is $C_6H_{12}O$ (1)

ii Quality of written communication
Candidates must satisfy the following criteria:
- Selection of a form and style of writing which is appropriate to this question ($\frac{1}{2}$)
- Organisation of the relevant information clearly and coherently ($\frac{1}{2}$)
- Use of legible text with adequate spelling, grammar and punctuation ($\frac{1}{2}$).

The IR data gives the alkane stretch at around $2960\,cm^{-1}$ ($\frac{1}{2}$), the CH_3 bend at around $1452\,cm^{-1}$ (or CH_3 oscillation at $1360\,cm^{-1}$) ($\frac{1}{2}$), the $C=O$ for a ketone at $1715\,cm^{-1}$ ($\frac{1}{2}$) and the rest of the spectrum is consistent with an alkane chain ($\frac{1}{2}$)
No O—H group present ($\frac{1}{2}$)
X could be $CH_3COCH_2CH_2CH_2CH_3$ (allow any six-carbon ketone) (1)
Any sensible suggestion to confirm the identity, for example boiling point or melting point determination of DNPH derivative checked against literature values (1)
b i Butanone (1)
Sodium tetrahydridoborate(III) (or sodium borohydride) (1)
ii

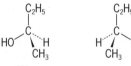

Mirror images (1)
Correct formula in 3D (1)
c i One of:
Solvent, fuel, starting material for organic synthesis (1)
ii One of:
Harmful to health, destroys brain cells, impairs reactions and so causes road traffic accidents, time lost from work, emotional problems and family breakdown (1)

22 a i NO_2^+/nitronium ion (or nitryl cation) (1)
ii

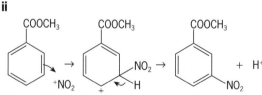

Electrophilic addition by NO_2^+ (1)
Intermediate (1)
Loss of H^+ (1)
Will also accept

as the intermediate

iii To protonate the HNO_3 (1) to make NO_2^+
Equation (1)
$HNO_3 + H^+ \rightarrow NO_2^+ + H_2O$
or
$HNO_3 + H_2SO_4 \rightarrow NO_2^+ + H_2O + HSO_4^-$
or
$HNO_3 + 2H_2SO_4 \rightarrow NO_2^+ + H_3O^+ + 2HSO_4^-$

b i

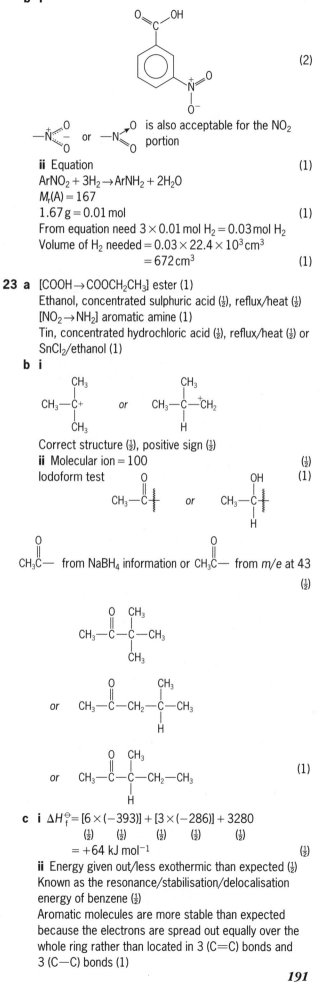

(2)

is also acceptable for the NO_2 portion

ii Equation (1)
$ArNO_2 + 3H_2 \rightarrow ArNH_2 + 2H_2O$
$M_r(A) = 167$
$1.67\,g = 0.01\,mol$ (1)
From equation need $3 \times 0.01\,mol\ H_2 = 0.03\,mol\ H_2$
Volume of H_2 needed $= 0.03 \times 22.4 \times 10^3\,cm^3$
$\qquad\qquad = 672\,cm^3$ (1)

23 a [COOH→COOCH$_2$CH$_3$] ester (1)
Ethanol, concentrated sulphuric acid ($\frac{1}{2}$), reflux/heat ($\frac{1}{2}$)
[NO_2→NH_2] aromatic amine (1)
Tin, concentrated hydrochloric acid ($\frac{1}{2}$), reflux/heat ($\frac{1}{2}$) or $SnCl_2$/ethanol (1)

b i

Correct structure ($\frac{1}{2}$), positive sign ($\frac{1}{2}$)
ii Molecular ion = 100 ($\frac{1}{2}$)
Iodoform test (1)

$CH_3\overset{O}{\overset{\|}{C}}$— from $NaBH_4$ information or $CH_3\overset{O}{\overset{\|}{C}}$— from m/e at 43 ($\frac{1}{2}$)

(1)

c i $\Delta H_f^{\ominus} = [6 \times (-393)] + [3 \times (-286)] + 3280$
$\qquad\quad$ ($\frac{1}{2}$) $\quad$ ($\frac{1}{2}$) $\quad$ ($\frac{1}{2}$) $\quad$ ($\frac{1}{2}$) $\qquad$ ($\frac{1}{2}$)
$\qquad = +64\ kJ\ mol^{-1}$ ($\frac{1}{2}$)
ii Energy given out/less exothermic than expected ($\frac{1}{2}$)
Known as the resonance/stabilisation/delocalisation energy of benzene ($\frac{1}{2}$)
Aromatic molecules are more stable than expected because the electrons are spread out equally over the whole ring rather than located in 3 ($C=C$) bonds and 3 ($C—C$) bonds (1)

ANSWERS TO EXAMINATION QUESTIONS

24 a

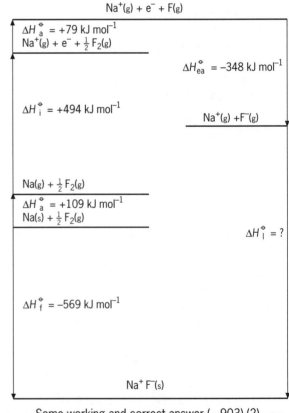

1 mark for each correct step, to a maximum of 6
Calculation (1), cycling clockwise about *

$2 \times \Delta H_a(Na) + 2 \times 1st\ IE(Na) + \frac{1}{2} \times BE(O{=}O) +$
$1st\ EA(O) + 2nd\ EA(O) + LE(formation) - \Delta H_f(Na_2O) = 0$ (1)

Correct answer (2), for example

$(2 \times 107) + (2 \times 496) + (\frac{1}{2} \times 249) + (-141) +$
$2nd\ EA(O) + (-2478) - (-414) = 0$

$2nd\ EA(O) = +874.5\,kJ\,mol^{-1}$

b The negatively charged O^- ion (1) strongly repels the incoming electron (1)

c **Na^+ and O^-**
Lattice energy less than Na_2O (1), ΔH_f less than Na_2O (1), hence less likely to form (1)
Na^{2+} and O^{2-}
Second ionisation energy of sodium very large (1), increased lattice energy insufficient to compensate (1)

Max (4)

25 a Spontaneous when $\Delta G < 0$ (1)
$\Delta G = \Delta H - T\Delta S$ (1)
Calculation of ΔH
$\Delta H_{reaction} = \Sigma \Delta H_{products} - \Sigma \Delta H_{reactants}$ (1)
$= (2 \times -110.5) - (-393.5)$
$= +172.5\,kJ\,mol^{-1}$ (1)
Calculation of ΔS
$\Delta S_{reaction} = \Sigma \Delta S_{products} - \Sigma \Delta S_{reactants}$ (1)
$= (2 \times 197.6) - (5.7 + 213.6)$
$= 175.9\,J\,mol^{-1}K^{-1}$ (1)
Calculation of T
$\Delta G = \Delta H - T\Delta S$
$\Delta G = +172.5 - \dfrac{T \times 175.9}{1000}$
$= 0$ (1)
$\dfrac{T \times 175.9}{1000} = +172.5$
$T = \dfrac{172.5 \times 1000}{175.9} = 980.7\,K$ (1)

b i No change in moles of gas, hence ΔS is very small (1)
$\dfrac{\Delta H}{RT}$ large and positive (1), hence K_p is very small (1)

ii Car engine/furnace (1)
Acid rain/respiratory problems (1)

Max (4)

c **Calculation of $\Delta H_{vaporisation}$**
3675 J vaporises 1.50 g of water
$\dfrac{3675 \times 18}{1.50}$ J vaporises 1.00 mol water
$= 44.1\,kJ\,mol^{-1}$ (1)
Calculation of ΔS
$\Delta G = 0$
$= 44.1 - \dfrac{373 \times \Delta S}{1000}$ (1)
$\Delta S = \dfrac{44.1 \times 1000}{373}$
$= 118.2\,J\,mol^{-1}K^{-1}$ (1)

26 a Enthalpy change when 1 mole of solid is formed (1) by coming together of separate ions in the gaseous state (1)
or correct symbol equation (1) for 1 mole (1)
b Born–Haber cycle correctly drawn (3)

Some working and correct answer (-903) (2)
c i Energy needed to overcome electrostatic forces/electrovalent forces/attraction between oppositely charged ions (1)
ii As cation decreases in size/the ions get closer/allows the charge density to increase (1)
More attraction so lattice enthalpy increases (1)

iii Beryllium hydroxide (Be(OH$_2$)) has considerable covalent character/Be^{2+} is highly polarising (1)
iv Bonds are formed/attractive interaction between the ions and water (1)
Bond formation releases energy (1)
v Solubility depends upon the balance between lattice and hydration enthalpies (1)
Up the group both increase/become more exothermic (1) but lattice enthalpy increases faster than hydration enthalpy (1)

27 a $K_p = \dfrac{p(SO_3)^2}{p(O_2) \times p(SO_2)^2}$

Correct pressures (1)
Correct powers (1)
b i Reaction is exothermic (1)
As temperature increases, equilibrium moves from right to left (1) (*or* as temperature decreases, equilibrium moves from left to right)
ii Conversion faster at 450 °C (1)
Catalyst less effective at lower temperatures (1)
Compromise between rate and conversion (1)

Max (2)

c i 3 mol → 2 mol of gas/smaller number of gas molecules on right-hand side (1)
ii Good yield at normal pressures (1), high cost of high pressures not justified (1)
d Vanadium(V) oxide/pentoxide (1)
e i Fewer moles of gas (1)
More 'order' in system (1)
ii 25 °C = 298 K (1)

$$\Delta S = \dfrac{-188}{1000} \text{ kJ mol}^{-1}\text{K}^{-1} \qquad (1)$$

$$\Delta G = -196 - (298 \times -0.188)$$
$$= -140 \text{ kJ mol}^{-1} \qquad (1)$$
Reaction is feasible (1)

28 a $K_w = [H^+][OH^-]$ (1)
b $pH = -\log_{10}[H^+]$ (1)
c $pH = -\log_{10}(2.00)$
$= -0.30$ (1)

d $[H^+] = \dfrac{10^{-14}}{2.50}$

$= 4.0 \times 10^{-15} \text{ mol dm}^{-3}$ (1)
Therefore pH = 14.40 (1)

e moles H$^+$ = $\dfrac{19.0 \times 2.00}{1000}$

$= 0.0380$ mol (1)

moles OH$^-$ = $\dfrac{16.0 \times 2.50}{1000}$

$= 0.0400$ mol (1)
Total volume = (19.0 + 16.0) cm^3
$= 35.0$ cm^3 (1)

There is an excess of OH$^-$ = (0.0400 − 0.0380) mol
$= 2 \times 10^{-3}$ mol

Concentration of OH$^-$ = $\dfrac{2 \times 10^{-3} \times 1000}{35.0}$

$= 0.057 \text{ mol dm}^{-3}$ (1)

$[H^+] = \dfrac{K_w}{[OH^-]}$

$= \dfrac{10^{-14}}{0.0571}$

$= 1.75 \times 10^{-13} \text{ mol dm}^{-3}$ (1)
pH = 12.76 (1)

29 a i Only partially ionised (1)
ii C$_6$H$_5$O$^-$ (1)
iii Added OH$^-$ reacts with H$_3$O$^+$ (1)
So equilibrium will move to right-hand side (1)
Equilibrium concentration of C$_6$H$_5$OH will decrease and so more will dissolve to restore the equilibrium (1)

b i $K_a = \dfrac{[C_6H_5O^-][H_3O^+]}{[C_6H_5OH]}$ (1)

ii $pK_a = -\log K_a$ (1)
iii $pK_a = 10$ (1)
iv Lower/smaller number (1)
v Acidity depends on ability to donate protons/weakness of —O—H bond (1)
O in C=O draws electrons away from C—O, so H$^+$ more easily removed (1)
Effect is greater than that of the benzene ring on the lone pair of O in phenol (1)
vi Phenol – no reaction (1)
Ethanoic acid – effervescence/bubbles (1)
c i At end-point pH = pK_a
$= 9.3$ (1)
Colour change detectable over range of 2 pH units, so range is 8.3 → 10.3 (1)
ii Colourless to pink (1)
$[In^-] \geq [HIn]$ (1)
iii Equivalence point of titration below pH 7, which is not in the range for phenolphthalein (1)

30 a A proton donor (1)
b HCl$_{(g)}$ + H$_2$O$_{(l)}$ ⇌ H$_3$O$^+_{(aq)}$ + Cl$^-_{(aq)}$ (1)
Base/proton acceptor (1)
c NH$_{3(g)}$ + H$_2$O$_{(l)}$ ⇌ NH$_4^+_{(aq)}$ + OH$^-_{(aq)}$ (1)
Acid/proton donor (1)
d H$_2$SO$_4$ + HNO$_3$ ⇌ HSO$_4^-$ + H$_2$NO$_3^+$ (1)
H$_2$NO$_3^+$ → NO$_2^+$ + H$_2$O (1)
Base/proton acceptor (1)
e i Not fully dissociated in aqueous solution (1)

ii $K_a = \dfrac{[H^+][A^-]}{[HA]}$ (1)

iii Not very big (1)
iv Strong acid (1)
Although not fully dissociated, the concentration of undissociated HX in solution is very small (1)

31 a $Fe + Cu^{2+} \rightarrow Cu + Fe^{2+}$ (1)

b i $Fe_{(s)}|Fe^{2+}_{(aq)} \vdots \vdots Cu^{2+}_{(aq)}|Cu_{(s)}$

Junctions correct (1)

Order of species correct (1)

ii e.m.f. $= +0.34 - (-0.44)$

$= +0.78V$ (1)

iii $\Delta G^{\ominus} = -zFE^{\ominus}$ (1)

$= -2 \times 96\,500 \times 0.78$ (1)

$= -151\,kJ\,mol^{-1}$ (1)

c i E.m.f. for cell must be positive for reaction to occur/be feasible (1)

Cu(s) + 2H⁺ → products

e.m.f. $= -0.34V$, so reaction won't happen (1)

Cu(s) + NO₃⁻ + 4H⁺ → products

e.m.f. $= +0.96 - 0.34$

$= +0.62V$, so reaction can occur (1)

ii $3Cu + 2NO_3^- + 8H^+ \rightarrow 3Cu^{2+} + 2NO + 4H_2O$

Species (1)

Balanced (1)

d i $2Fe + O_2 + H_2O \rightarrow 2Fe^{2+} + 4OH^-$ (1)

ii Anode (1)

Fe loses electrons, so forms negative pole (1)

iii e.m.f. $= +0.06V$ (1)

Example of reasoned argument

E.m.f. positive so reaction should occur, but the difference is so small that reaction is unlikely (1)

32 a The potential difference between a metal and a $1\,mol\,dm^{-3}$ solution of its ions (under standard conditions)/the half-cell with $1\,mol\,dm^{-3}$ solution (1) and the standard hydrogen electrode (1)

b i High resistance voltmeter/potentiometer/valve voltmeter (1)

ii $+0.34V$ (2)

iii Arrow from Fe to Cu (1) (consequential on part **ii**)

iv Nothing (1)

Explanation relates to e.m.f. of cell or electrodes/reference to relative strength of reducing agent (1)

c i $2Fe + O_2 + 2H_2O \rightarrow 2Fe(OH)_2$

or $2Fe_{(s)} + O_{2(g)} + 2H_2O_{(l)} \rightarrow 2Fe^{2+}_{(aq)} + 4OH^-_{(aq)}$

Correct species (1)

Balanced (1)

ii Magnesium reacts preferentially/reacts as sacrificial anode (1)

Magnesium has a more negative redox potential/stronger reducing agent (1)

33 a i The change in oxidation state of Mn is -4, therefore 2Tl change by $+2$ each. (1)

ii Amount of MnO_4^-(aq) $= 2.5 \times 10^{-4}\,mol$ (1)

Amount of Tl^+(aq) $= 5 \times 10^{-4}\,mol$ (1)

Therefore Tl^{3+} is formed (1)

iii It makes Mn(III) stable (1), as a result of complexing (1)

b i The MnO_4^-/Mn^{3+} electrode potential is more negative than the Mn^{3+}/Mn^{2+} electrode potential (1)

Therefore Mn^{3+} would be expected to react with itself to form MnO_4^- and Mn^{2+} under standard conditions (1)

ii There is then a high concentration of H^+ ions (1), which makes the electrode potential for MnO_4^-/Mn^{3+} more positive than that for Mn^{3+}/Mn^{2+} (1), so reaction occurs in the reverse direction (1)

c

Octahedral shape (1)

Tetrahedral shape (1)

Oxygens of the water ligands clearly bonded to the Mn (1)

Charges shown (1)

d i Red (1)

It absorbs yellow to violet light (1)

ii Use a violet filter in the colorimeter (1)

Make up several mixtures with known concentrations of MnO_4^- (1)

For each, measure the change in absorption with time (1)

Plot the initial rates of these reactions against MnO_4^- concentration (1)

34 a i KOH or NaOH (1)

Ethanol/alcoholic (1)

Heat/under reflux (1)

ii $LiAlH_4$ (1)

Dry ether (1)

iii Aqueous NaOH or KOH/NaOH or KOH solution (1)

Boil/heat (1)

b i Carbonyl/C=O/aldehyde and ketone (1)

ii Aldehyde/CHO group (1)

iii OH group/acid and alcohol/acid or alcohol (1)

iv C=C/alkene (1)

c i $CH_3CH_2CH_2CHO$ (1)

$$\begin{array}{c} CHO \\ | \\ CH_3-C-CH_3 \\ | \\ H \end{array}$$ (1)

ii

Correct identification of **Q** (1)

Isomers correctly drawn (1)

iii

Correct identification of **S** (1)

Isomers correctly drawn (1)

(Two correctly drawn isomers of **R** scores 1)

35 a i 2,4-dinitrophenylhydrazine/Brady's reagent (1)
Orange/yellow precipitate (1)
ii Ammoniacal silver nitrate/Tollens' reagent/Fehling's
solution/Benedict's solution (1)
Silver mirror/red precipitate (1)

b

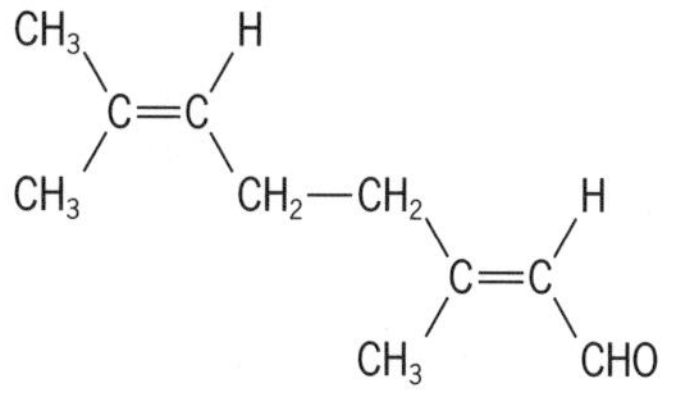

(1)

and

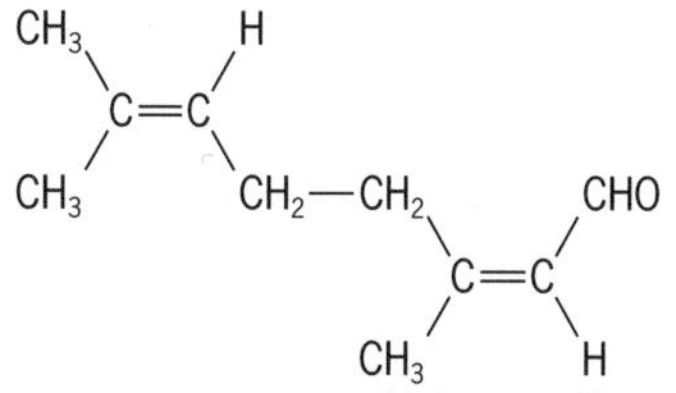

(1)

No/restricted rotation about double bond (1)

c i Bromine colour goes from brown to colourless (1)

ii

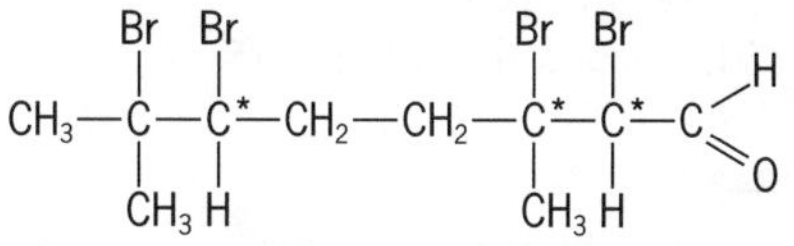

(1)

iii Correct identification of 3 chiral centres (2)
(Correct identification of 1 or 2 chiral centres 1)

d i $NaBH_4$ (1) in aqueous or ethanoic solution (1)
or sodium (1) in ethanol (1)
or $LiAlH_4$ (1) in dry ethoxyethane/ether (1)
ii PCl_5 (1)
Steamy/white/misty fumes/valid test for HCl (1)
or Na (1)
Fizzes/gives gas that burns with a squeaky pop (1)

e i $K_a = \dfrac{[H_3O^+]\,[C_9H_{15}COO^-]}{[C_9H_{15}COOH]}$ (1)

ii Amount $= \dfrac{4.62}{168}$

$= 0.0275\,\text{mol}$
Concentration of acid $= 0.0275 \times 4$
$\qquad\qquad\qquad = 0.11(0)\,\text{mol dm}^{-3}$ (1)

iii pH $= 2.91$
thus $[H^+] = 1.23 \times 10^{-3}\,\text{mol dm}^{-3}$ (1)

$K_a = \dfrac{[H^+]^2}{[HA]}$

$= \dfrac{(1.23 \times 10^{-3})^2}{0.11}$ (1)

$= 1.38 \times 10^{-5}\,\text{mol dm}^{-3}$ (1)

Index

Page numbers in *italics* refer to a question.